Environmental and Agricultural Microbiology

Scrivener Publishing
100 Cummings Center, Suite 541J
Beverly, MA 01915-6106

Publishers at Scrivener
Martin Scrivener (martin@scrivenerpublishing.com)
Phillip Carmical (pcarmical@scrivenerpublishing.com)

Environmental and Agricultural Microbiology

Applications for Sustainability

Edited by
Bibhuti Bhusan Mishra
Suraja Kumar Nayak
Swati Mohapatra
Deviprasad Samantaray

Scrivener Publishing

WILEY

This edition first published 2021 by John Wiley & Sons, Inc., 111 River Street, Hoboken, NJ 07030, USA and Scrivener Publishing LLC, 100 Cummings Center, Suite 541J, Beverly, MA 01915, USA

For more information about Scrivener publications please visit www.scrivenerpublishing.com.

Wiley Global Headquarters
111 River Street, Hoboken, NJ 07030, USA

For details of our global editorial offices, customer services, and more information about Wiley products visit us at www.wiley.com.

Library of Congress Cataloging-in-Publication Data

ISBN 978-1-119-52623-0

Cover image: Pixabay.Com
Cover design by Russell Richardson

C9781119526230_160921

Contents

Preface **xvii**

Part 1: Microbial Bioremediation and Biopolymer Technology **1**

1 A Recent Perspective on Bioremediation of Agrochemicals by Microalgae: Aspects and Strategies **3**
Prithu Baruah and Neha Chaurasia
1.1 Introduction 4
1.2 Pollution Due to Pesticides 6
1.2.1 Acute Effects 8
1.2.2 Chronic Effects 9
1.3 Microalgal Species Involved in Bioremediation of Pesticides 9
1.4 Strategies for Phycoremediation of Pesticides 13
1.4.1 Involvement of Enzymes in Phycoremediation of Pesticides 13
1.4.2 Use of Genetically Engineered Microalgae 13
1.5 Molecular Aspects of Pesticide Biodegradation by Microalgae 14
1.6 Factor Affecting Phycoremediation of Pesticides 16
1.6.1 Biological Factor 16
1.6.2 Chemical Factor 16
1.6.3 Environment Factor 17
1.7 Benefit and Shortcomings of Phycoremediation 17
1.7.1 Benefits 17
1.7.2 Shortcomings 17
1.8 Conclusion and Future Prospects 18
References 18

2 Microalgal Bioremediation of Toxic Hexavalent Chromium: A Review **25**
Pritikrishna Majhi, Satyabrata Nayak and Saubhagya Manjari Samantaray
2.1 Introduction 25
2.1.1 Chromium Cycle 27
2.2 Effects of Hexavalent Chromium Toxicity 27
2.2.1 Toxicity to Microorganisms 27
2.2.2 Toxicity to Plant Body 28
2.2.3 Toxicity to Animals 29
2.3 Chromium Bioremediation by Microalgae 30
2.3.1 Cyanobacteria 30
2.3.2 Green Algae 31
2.3.3 Diatoms 31
2.4 Mechanism Involved in Hexavalent Chromium Reduction in Microalgae 32
2.5 Conclusion 33
References 34

3 Biodetoxification of Heavy Metals Using Biofilm Bacteria **39**
Adyasa Barik, Debasish Biswal, A. Arun and Vellaisamy Balasubramanian
3.1 Introduction 40
3.2 Source and Toxicity of Heavy Metal Pollution 41
3.2.1 Non-Essential Heavy Metals 42
3.2.1.1 Arsenic 42
3.2.1.2 Cadmium 43
3.2.1.3 Chromium 43
3.2.1.4 Lead 44
3.2.1.5 Mercury 45
3.2.2 Essential Heavy Metals 45
3.2.2.1 Copper 45
3.2.2.2 Zinc 46
3.2.2.3 Nickel 46
3.3 Biofilm Bacteria 47
3.4 Interaction of Metal and Biofilm Bacteria 47
3.5 Biodetoxification Mechanisms 48
3.5.1 Biosorption 48
3.5.2 Bioleaching 50
3.5.3 Biovolatilization 52
3.5.4 Bioimmobilization 54

3.6 Conclusion 55
References 55

4 Microbial-Derived Polymers and Their Degradability Behavior for Future Prospects 63
Mohammad Asif Ali, Aniruddha Nag and Maninder Singh
4.1 Introduction 63
4.2 Polyamides 65
4.2.1 Bioavailability and Production 66
4.2.2 Biodegradability of Polyamides 66
4.2.3 Degradation of Nylon 4 Under the Soil 67
4.2.4 Fungal Degradation of Nylon 6 and Nylon 66 (Synthetic Polyamide) 67
4.2.5 Itaconic Acid-Based Heterocyclic Polyamide 68
4.2.6 Summary and Future Development 69
4.3 Polylactic Acid 69
4.3.1 Availability and Production 70
4.3.2 Polymerization Method 71
4.3.3 Biodegradability of Polylactic Acid 73
4.3.4 Copolymerization Method 73
4.3.5 Blending Method 73
4.3.6 Nanocomposite Formation 74
4.3.7 Summary 74
4.4 Polyhydroxyalkanoates 74
4.4.1 Biosynthesis of Polyhydroxyalkanoates 75
4.4.2 Application of PHAs 75
4.4.3 Biodegradability of PHAs 76
4.4.4 Degradability Methods 76
4.4.5 Summary 77
4.5 Conclusion and Future Development 77
References 78

5 A Review on PHAs: The Future Biopolymer 83
S. Mohapatra, K. Vishwakarma, N. C. Joshi, S. Maity, R. Kumar, M. Ramchander, S. Pattnaik and D. P. Samantaray
5.1 Introduction 84
5.2 Green Plastic: Biodegradable Polymer Used as Plastic 85
5.3 Difference Between Biopolymer and Bioplastic 88
5.4 Polyhydroxyalkanoates 88
5.5 Polyhydroxyalkanoates and Its Applications 89
5.6 Microorganisms Producing PHAs 90

5.7 Advantages 96
5.8 Conclusion and Future Prospective 96
References 96

6 Polyhydroxybutyrate as an Eco-Friendly Alternative of Synthetic Plastics 101
Shikha Sharma, Priyanka Sharma, Vishal Sharma and Bijender Kumar Bajaj
6.1 Introduction 102
6.2 Bioplastics 104
6.3 Bioplastics vs. Petroleum-Based Plastics 106
6.4 Classification of Biodegradable Polymers 107
6.5 PHB-Producing Bacteria 109
6.6 Methods for Detecting PHB Granules 113
6.7 Biochemical Pathway for Synthesis of PHB 114
6.8 Production of PHB 116
6.8.1 Process Optimization for PHB Production 117
6.8.2 Optimization of PHB Production by One Variable at a Time Approach 118
6.8.3 Statistical Approaches for PHB Optimization 120
6.9 Production of PHB Using Genetically Modified Organisms 123
6.10 Characterization of PHB 125
6.11 Various Biochemical Techniques Used for PHB Characterization 126
6.11.1 Fourier Transform Infrared Spectroscopy 127
6.11.2 Differential Scanning Calorimetry 127
6.11.3 Thermogravimetric Analysis 128
6.11.4 X-Ray Powder Diffraction (XRD) 128
6.11.5 Nuclear Magnetic Resonance Spectroscopy 128
6.11.6 Microscopic Techniques 129
6.11.7 Elemental Analysis 130
6.11.8 Polarimetry 130
6.11.9 Molecular Size Analysis 130
6.12 Biodegradation of PHB 131
6.13 Application Spectrum of PHB 132
6.14 Conclusion 135
6.15 Future Perspectives 135
Acknowledgements 136
References 136

7 Microbial Synthesis of Polyhydroxyalkanoates (PHAs) and Their Applications 151
N.N.N. Anitha and Rajesh K. Srivastava

7.1 Introduction 153
7.2 Conventional Plastics and Its Issues in Utility 156
7.2.1 Synthetic Plastic and Its Accumulation or Degradation Impacts 158
7.3 Bioplastics 159
7.3.1 Polyhydroxyalkanoates 160
7.3.1.1 Microorganisms in the Production of PHAs 164
7.4 Fermentation for PHAs Production 171
7.5 Downstream Process for PHAs 173
7.6 Conclusions 175
References 176

8 Polyhydroxyalkanoates for Sustainable Smart Packaging of Fruits 183
S. Pati, S. Mohapatra, S. Maity, A. Dash and D. P. Samantaray
8.1 Introduction 183
8.2 Physiological Changes of Fresh Fruits During Ripening and Minimal Processing 185
8.3 Smart Packaging 186
8.4 Biodegradable Polymers for Fruit Packaging 188
8.5 Legal Aspects of Smart Packaging 189
8.6 Pros and Cons of Smart Packaging Using PHAs 189
8.7 Conclusion 190
References 191

9 Biosurfactants Production and Their Commercial Importance 197
Saishree Rath and Rajesh K. Srivastava
9.1 Introduction 198
9.2 Chemical Surfactant Compounds 200
9.2.1 Biosurfactant Compounds 202
9.3 Properties of Biosurfactant Compound 205
9.3.1 Activities of Surface and Interface Location 205
9.3.2 Temperature and pH Tolerance 205
9.3.3 Biodegradability 206
9.3.4 Low Toxicity 206
9.3.5 Emulsion Forming and Breaking 206
9.4 Production of Biosurfactant by Microbial Fermentation 206
9.4.1 Factors Influencing the Production of Biosurfactants 209
9.4.1.1 Environmental Conditions 209
9.4.1.2 Carbon Substrates 210
9.4.1.3 Estimation of Biosurfactants Activity 211

9.5 Advantages, Microorganisms Involved, and Applications of Biosurfactants 211
9.5.1 Advantages of Using Biosurfactants 211
9.5.1.1 Easy Raw Materials for Biosurfactant Biosynthesis 211
9.5.1.2 Low Toxic Levels for Environment 211
9.5.1.3 Best Operation With Surface and Interface Activity 212
9.5.1.4 Good Biodegradability 212
9.5.1.5 Physical Variables 212
9.5.2 Microbial Sources 212
9.5.3 Production of Biosurfactants 213
9.5.3.1 Production of Rhamnolipids 213
9.5.3.2 Regulation of Rhamnolipids Synthesis 214
9.5.3.3 Commercial Use of Biosurfactants 214
9.6 Conclusions 215
References 216

Part 2: Microbes in Sustainable Agriculture and Biotechnological Applications 219

10 Functional Soil Microbes: An Approach Toward Sustainable Horticulture 221
C. Sarathambal, R. Dinesh and V. Srinivasan
10.1 Introduction 221
10.2 Rhizosphere Microbial Diversity 222
10.3 Plant Growth–Promoting Rhizobacteria 223
10.3.1 Nitrogen Fixation 224
10.3.2 Production of Phytohormones 225
10.3.3 Production of Enzymes That can Transform Crop Growth 225
10.3.4 Microbial Antagonism 226
10.3.5 Solubilization of Minerals 226
10.3.6 Siderophore and Hydrogen Cyanide (HCN) Production 228
10.3.7 Cyanide (HCN) Production 229
10.3.8 Plant Growth–Promoting Rhizobacteria on Growth of Horticultural Crops 229
10.4 Conclusion and Future Perspectives 235
References 235

11 Rhizosphere Microbiome: The Next-Generation Crop Improvement Strategy **243**
M. Anandaraj, S. Manivannan and P. Umadevi
11.1 Introduction 244
11.2 Rhizosphere Engineering 245
11.3 Omics Tools to Study Rhizosphere Metagenome 246
11.3.1 Metagenomics 246
11.3.2 Metaproteomics 248
11.3.3 Metatranscriptomics 249
11.3.4 Ionomics 250
11.4 As Next-Generation Crop Improvement Strategy 251
11.5 Conclusion 252
References 252

12 Methane Emission and Strategies for Mitigation in Livestock **257**
Nibedita Sahoo, Swati Pattnaik, Matrujyoti Pattnaik and Swati Mohapatra
12.1 Introduction 258
12.2 Contribution of Methane from Livestock 259
12.3 Methanogens 259
12.3.1 Rumen Microbial Community 260
12.3.2 Methanogens Found in Rumen 260
12.3.3 Enrichment of Methanogens from Rumen Liquor 261
12.3.4 Screening for Methane Production 261
12.3.5 Isolation of Methanogens 261
12.3.6 Molecular Characterization 261
12.4 Methanogenesis: Methane Production 262
12.4.1 Pathways of Methanogenesis 262
12.4.2 Pathway of CO_2 Reduction 262
12.4.3 CO_2 Reduction to Formyl-Methanofuran 263
12.4.4 Conversion of the Formyl Group from Formyl-Methanofuran to Formyl-Tetrahydromethanopterin 263
12.4.5 Formation of Methenyl-Tetrahydromethanopterin 263
12.4.6 Reduction of Methenyl-Tetrahydromethanopterin to Methyl-Tetrahydromethanopterin 263
12.4.7 Reduction of Methyl-Tetrahydromethanopterin to Methyl-S-Coenzyme M 264
12.4.8 Reduction of Methyl-S-Coenzyme M to CH_4 264
12.5 Strategies for Mitigation of Methane Emission 264
12.5.1 Dietary Manipulation 264
12.5.1.1 Increasing Dry Matter Intake 264

12.5.1.2 Increasing Ration Concentrate Fraction 265
12.5.1.3 Supplementation of Lipid 265
12.5.1.4 Protozoa Removal 266
12.5.2 Feed Additives 266
12.5.2.1 Ionophore Compounds 266
12.5.2.2 Halogenated Methane Compound 267
12.5.2.3 Organic Acid 267
12.5.3 Microbial Feed Additives 268
12.5.3.1 Vaccination 268
12.5.3.2 Bacteriophages and Bacteriocins 269
12.5.4 Animal Breeding and Selection 270
12.6 Conclusion 270
References 271

13 Liquid Biofertilizers and Their Applications: An Overview 275
Avro Dey
13.1 Introduction 275
13.1.1 Chemical Fertilizer and its Harmful Effect 277
13.2 Biofertilizers "Boon for Mankind" 278
13.3 Carrier-Based Biofertilizers 279
13.3.1 Solid Carrier-Based Biofertilizers 279
13.3.2 Liquid Biofertilizer 279
13.4 Sterilization of the Carrier 282
13.5 Merits of Using Liquid Biofertilizer Over Solid Carrier-Based Biofertilizer 282
13.6 Types of Liquid Biofertilizer 283
13.7 Production of Liquid Biofertilizers 285
13.7.1 Isolation of the Microorganism 285
13.7.2 Preparation of Medium and Growth Condition 285
13.7.3 Culture and Preservation 286
13.7.4 Preparation of Liquid Culture 286
13.7.5 Fermentation and Mass Production 287
13.7.6 Formulation of the Liquid Biofertilizers 287
13.8 Applications of Biofertilizers 288
13.9 Conclusion 290
References 291

14 Extremozymes: Biocatalysts From Extremophilic Microorganisms and Their Relevance in Current Biotechnology 293
Khushbu Kumari Singh and Lopamudra Ray
14.1 Introduction 294
14.2 Extremophiles: The Source of Novel Enzymes 295

14.2.1 Thermophilic Extremozymes 296
14.2.2 Psychrophilic Extremozymes 299
14.2.3 Halophilic Extremozymes 300
14.2.4 Alkaliphilic/Acidiophilic Extremozymes 300
14.2.5 Piezophilic Extremozymes 301
14.3 The Potential Application of Extremozymes in Biotechnology 301
14.4 Conclusion and Future Perspectives 303
References 304

15 Microbial Chitinases and Their Applications: An Overview 313
Suraja Kumar Nayak, Swapnarani Nayak, Swaraj Mohanty, Jitendra Kumar Sundaray and Bibhuti Bhusan Mishra
15.1 Introduction 314
15.2 Chitinases and Its Types 315
15.3 Sources of Microbial Chitinase 317
15.3.1 Bacterial Chitinases 317
15.3.2 Fungal Chitinases 319
15.3.3 Actinobacteria 321
15.3.4 Viruses/Others 322
15.4 Genetics of Microbial Chitinase 322
15.5 Biotechnological Advances in Microbial Chitinase Production 323
15.5.1 Media Components 324
15.5.2 Physical Parameters 325
15.5.3 Modes and Methods of Fermentation 325
15.5.4 Advances Biotechnological Methods 326
15.6 Applications of Microbial Chitinases 327
15.6.1 Agricultural 328
15.6.1.1 Biopesticides 328
15.6.1.2 Biocontrol 328
15.6.2 Biomedical 329
15.6.3 Pharmaceutical 329
15.6.4 Industrial 330
15.6.5 Environmental 330
15.6.5.1 Waste Management 331
15.6.6 Others 331
15.7 Conclusion 332
References 332

16 Lithobiontic Ecology: Stone Encrusting Microbes and their Environment 341
Abhik Mojumdar, Himadri Tanaya Behera and Lopamudra Ray
16.1 Introduction 341

16.2 Diversity of Lithobionts and Its Ecological Niche 342
16.2.1 Epiliths 342
16.2.2 Endoliths 343
16.2.3 Hypoliths 344
16.3 Colonization Strategies of Lithobionts 345
16.3.1 Temperature 346
16.3.2 Water Availability 346
16.3.3 Light Availability 347
16.4 Geography of Lithobbiontic Coatings 348
16.4.1 Bacteria 348
16.4.2 Cyanobacteria 349
16.4.3 Fungi 349
16.4.4 Algae 349
16.4.5 Lichens 350
16.5 Impacts of Lithobiontic Coatings 351
16.5.1 On Organic Remains 351
16.5.2 On Rock Weathering 351
16.5.3 On Rock Coatings 352
16.6 Role of Lithobionts in Harsh Environments 352
16.7 Conclusion 353
Acknowledgement 353
References 353

17 Microbial Intervention in Sustainable Production of Biofuels and Other Bioenergy Products 361
Himadri Tanaya Behera, Abhik Mojumdar, Smruti Ranjan Das, Chiranjib Mohapatra and Lopamudra Ray
17.1 Introduction 362
17.2 Biomass 363
17.3 Biofuel 364
17.3.1 Biodiesel 365
17.3.1.1 Microalgae in Biodiesel Production 365
17.3.1.2 Oleaginous Yeasts in Biodiesel Production 366
17.3.1.3 Oleaginous Fungi in Biodiesel Production 366
17.3.1.4 Bacteria in Biodiesel Production 367
17.3.2 Bioalcohol 367
17.3.2.1 Bioethanol 367
17.3.2.2 Biobutanol 368
17.3.3 Biogas 369

17.3.4 Biohydrogen 369
17.4 Other Bioenergy Products 370
17.4.1 Microbial Fuel Cells 370
17.4.1.1 Microbes Used in MFCs 372
17.4.1.2 Future Aspects of Microbial Fuel Cells 372
17.4.2 Microbial Nanowires in Bioenergy Application 374
17.4.2.1 Pili 375
17.4.2.2 Outer Membranes and Extended Periplasmic Space 375
17.4.2.3 Unknown Type—MNWs Whose Identity to be Confirmed 375
17.4.3 Microbial Nanowires in Bioenergy Production 376
17.5 Conclusion 376
References 376

18 Role of Microbes and Microbial Consortium in Solid Waste Management 383
Rachana Jain, Lopa Pattanaik, Susant Kumar Padhi and Satya Narayan Naik
18.1 Introduction 384
18.2 Types of Solid Waste 384
18.2.1 Domestic Wastes 385
18.2.2 Institutional and Commercial Wastes 385
18.2.3 Wastes From Street Cleansing 385
18.2.4 Industrial Wastes 385
18.2.5 Nuclear Wastes 385
18.2.6 Agricultural Wastes 385
18.3 Waste Management in India 386
18.4 Solid Waste Management 390
18.4.1 Municipal Solid Waste Management 390
18.5 Solid Waste Management Techniques 390
18.5.1 Incineration 392
18.5.2 Pyrolysis and Gasification 392
18.5.3 Landfilling 393
18.5.4 Aerobic Composting 394
18.5.5 Vermicomposting 397
18.5.6 Anaerobic Digestion 401
18.5.6.1 Enzymatic Hydrolysis 402
18.5.6.2 Fermentation 402
18.5.6.3 Acetogenesis 403
18.5.6.4 Methanogenesis 403

18.5.7 Bioethanol From Various Solid Wastes 404
18.6 Conclusion 413
References 413

Index **423**

Preface

The indiscriminate use of agrochemicals, resulting in an increasing concentration of synthetic contaminants in the environment, is a matter of great concern across the globe. These chemicals or their toxic intermediates enter into the food chain and are subjected to biomagnifications. Moreover, contaminants from industrial discharges like effluents, solid waste and exhaust air, radioactive substances, hydrocarbons, etc., worsen the scenario. The sustainability of the ecosystem is strongly influenced by the microbial community from natural resources. The pivotal role played by microbes in effectively degrading these environmental contaminants may even be better than conventional methods. Microbes' innate capacity allows them to degrade, utilize and/or transform a plethora of both organic and inorganic substances, including persistent and recalcitrant agrochemicals, metals and minerals and different xenobiotic polymers which have been successfully used for remediation of pollutants in the biosphere.

In the agricultural milieu, application of microbes is the focus of the day, with an emphasis on plant-microbe interactions. These interactions are complex and may be antagonistic, mutualistic or synergistic, depending on their microbial diversity and association with the host and environment. With the involvement of microbes, complex chemical substances essential for sustaining life are being transformed into available forms that are easily absorbed by plants. In addition, microbes are also involved in biofortification of elements through recycling, resulting in enhanced growth and productivity.

Though books pertaining to soil and agricultural microbiology/environmental biotechnology are available, there is a dearth of comprehensive literature on the behavior of microorganisms in the environmental and agricultural realm covered in this book, which makes it unique. This book is divided into two parts, with chapters that embody the sustenance and life cycles of microorganisms under various environmental conditions, their dispersal, interactions with other inhabited communities, metabolite production and reclamation. Part 1 of the book includes topics on bioremediation of agrochemicals by microalgae, detoxification of chromium and other heavy metals by microbial biofilm, microbial biopolymer technology, including polyhydroxyalkanoates (PHAs) and polyhydroxybutyrates (PHB), their production, degradability behaviors and applications. Biosurfactants production and their commercial importance are also systematically represented in this part. In Part 2 of the book, imperative ideas are presented on approaches for sustainable agriculture through functional soil microbes, next-generation crop improvement strategies via rhizosphere microbiome, production and implementation of liquid biofertilizers, mitigation of methane from livestock, chitinases from microbes, and extremozymes (enzymes from extremophilic microorganisms). Their relevance in current biotechnology, lithobiontic communities and their environmental importance are comprehensively elaborated. Since this is the era of sustainable energy production in which biofuel and other bioenergy products play a key role, also covered is their production from microbial sources—considered the new frontier for researchers. The concluding chapter reveals the importance of microbes and their consortia for management of solid waste in combination with biotechnology.

Because the field of environmental and agricultural biotechnology/microbiology is so large and appeals to those with varied interests, in order to make the topics covered herein more useful, informative and relevant to a vast range of readers, a broad array of current advances is covered in chapters that are supplemented with illustrative diagrams and informative tables, along with the future prospects of microorganisms. Therefore, doctoral and post-doctoral fellows working in the area, and environmental microbiologists and chemical engineers who want to delve into the largely unexplored realm of microorganisms in benign, beneficial agricultural production and environmental remediation will be exposed to the latest findings in these research frontiers in a comprehensive manner. Moreover, readers will be provided with key knowledge on cutting-edge biotechnological methods applied in soil and environmental microbiology.

The editors express their sincere gratitude to all contributors for their excellent cooperation, critical thoughts and contributions that helped to complete this timely edited volume. We also sincerely thank Scrivener Publishing for providing us with a platform to publish this book. Last but not least, we wish that the current and upcoming scientific generations will use the knowledge presented herein for the benefit and development of society. We will definitely appreciate any comments on the book for future consideration.

Bibhuti Bhusan Mishra
Suraja Kumar Nayak
Swati Mohapatra
Deviprasad Samantaray
June 2021

Part 1

MICROBIAL BIOREMEDIATION AND BIOPOLYMER TECHNOLOGY

1

A Recent Perspective on Bioremediation of Agrochemicals by Microalgae: Aspects and Strategies

Prithu Baruah and Neha Chaurasia*

Department of Biotechnology and Bioinformatics, North-Eastern Hill University, Shillong, India

Abstract

With the increasing world's population, enhancement of crop production has become a major target for mankind survival. This leads to extensive use of agrochemicals which has revolutionized the entire pest control system. However, due to their uncontrolled use, the equilibrium between their beneficial effects and harmful consequences has been compromised which lead to severe environmental havoc. To combat their hazardous influences, several remediation methods such as adsorption and ultrasonic irradiation have been developed. But unfortunately, most of them are not cost-effective and environment-friendly. As a result, bioremediation has become a potential alternative to these remediation methods being less expensive and eco-friendly. Microalgae have recently received sufficient attention as a bioremediation candidate due to their cheap nutritional requirements (solar light and CO_2) and versatile metabolic activities. The microalgae-based remediation technologies are ecologically more comprehensive and can be integrated with several other technologies such as biofuel production and carbon mitigation. Regardless of these conveniences, a critical scrutiny of the current status of the technology is required to get an in-depth insight into the applicability of microalgae for remediation of pollutants. The present article is an attempt to provide a crucial look into the microalgae-based removal of agricultural pollutants and an outline of its mechanistic perspectives. Also, molecular aspects of bioremediation by microalgae have been discussed to provide a better understanding of its remediation capabilities.

**Corresponding author*: nchaurasia@nehu.ac.in; cyanoneha@gmail.com

Bibhuti Bhusan Mishra, Suraja Kumar Nayak, Swati Mohapatra, and Deviprasad Samantaray (eds.) Environmental and Agricultural Microbiology: Applications for Sustainability, (3–24) © 2021 Scrivener Publishing LLC

Keywords: Microalgae, phycoremediation, agrochemicals, metabolism, agriculture, fertilizers, pesticides, environmental pollution

1.1 Introduction

The human population is constantly increasing at a fast rate and might reach around 9.7 billion people by 2050 [1]. To satisfy the food requirements of this enormous population, an enhancement in crop production is needed which open doors for the use of various agrochemicals. Agrochemicals (fertilizers and pesticides) are the group of chemicals used in agricultural practices to improve crop yield. Limited availability of macronutrients such as nitrogen, phosphorus, and potassium may result in poor growth of the crop. Thus, commercial fertilizers enriched with these ingredients may be applied to meet the demand of the essential elements. But plants can absorb only a limited amount of these nutrients and the excess fertilizer may be washed down along with the rain into water bodies and thereby causing contamination of the same [2].

Thus, uncontrolled and excessive use of fertilizers (e.g., phosphate fertilizer) may result in eutrophication of canals and reservoirs [3, 4]. In addition to chemical fertilizers, another chemical extensively used in agricultural activities is pesticides. These substances are utilized to control the infestations by crop destroying organisms referred to as pest and thereby enhancing agricultural productivity [5]. Although these agrochemicals are used to benefit humans, they have a hazardous impact on the environment [6]. Thus, with the use of several million tons of agrochemicals every year, the agricultural sector has been considered to be a major source of environmental pollution [7].

Among the various agrochemicals used in modern agricultural practice, pesticide and its residues pose a serious threat to environmental health and stability [8]. As a result, environmental pollution due to pesticide has become a major global concern. Pesticides can be defined as substances (or a mixture of substances) developed to repel or mitigate pests [9]. Pesticides include a wide array of compounds intended to reduce crop destroying agents such as insects, weeds, fungi, and rodents. These pesticides vary in their physical as well as chemical properties, and hence, it is important to classify them which make their study convenient. Although there are various ways of pesticide classification, the one based on their chemical composition are the most used one. This type of classification provides a proper correlation between structural features, activity, toxicity, and degradation mechanisms, among different members [5]. Based on chemical composition, pesticides have been classified into four major classes, namely, organochlorines, organophosphorus, carbamates, and pyrethroids [10]. Table 1.1 shows the chemical composition and general characteristics of important pesticides [11, 12].

Table 1.1 Chemical composition and general characteristics of different pesticide groups [11, 12].

Group	Chemical Composition	General Characteristics	Example
Organochlorines	Composed of C, H, Cl, and sometimes "O" is also present.	Lipid soluble, accumulation in fat rich animal tissue, persistent for a longer period, nonpolar in nature.	Lindane, endosulfan, mirex, DDT
Organophosphate	Phosphorus atom occupies central position within the molecule. They may be heterocyclic, cyclic, and aliphatic.	Shows solubility in water and organic solvents, low persistence compared to organochlorines, the central nervous system gets affected by these compounds.	Diazinon, methyl parathion, malathion
Carbamates	Chemical structure is similar to a plant alkaloid produced by *Physostigma venenosum.*	Derived from carbamate acid; have high vertebrate toxicity; less persistent.	Carbaryl, sevin
Pyrethroids	Chemical structure is based on pyrethrin obtained from *Chrysanthemum cinerariifolium.*	Affect the nervous system; are less persistent compared to other pesticides.	Pyrethrins

Although pesticides benefit human by enhancing agricultural productivity, they adversely affect human beings as well as other non-target organisms (explained in details in Section 1.2). Thus, remediation of these anthropogenic compounds is highly essential. Scientists have developed several physical and chemical remediation methods such as adsorption, oxidation, ozonation, nanofiltration, and membrane filtration ultrasonic irradiation for pesticide elimination from the environmental matrices and thereby minimizing their hazardous influences [13, 14]. But unfortunately, most of these methods are not environment-friendly and the cost associated with them is very high. As such, there is a requirement of an alternative technology devoid of these limitations. Bioremediation being inexpensive and eco-friendly proves itself as a potential replacement to various physical and chemical remediation methods. Earlier researchers have focused mainly on bioremediation using fungal and bacterial strains [15]. But recently, microalgae have received sufficient attention as an efficient bioremediation candidate due to their versatile metabolic activities, low-cost nutritional requirements (solar light and CO_2), and ability to survive in different environmental conditions [13]. The aim of this article is to summarize and evaluate the various aspects of bioremediation of pesticides using microalgae with attention on microalgal species involved, strategies, molecular basis, and factor affecting the process.

1.2 Pollution Due to Pesticides

Pesticides are anthropogenic compounds developed for human welfare by improving agricultural productivity. The estimated loss of agricultural products is 40% worldwide due to the effect of various agents such as plant diseases, pests, and weeds. Accordingly, the utilization of pesticide in agriculture has counteracted increment in this rate [6]. This is a roundabout way lessens the likelihood of price rise due to the decline in food production as a consequence of low agricultural productivity.

In addition to crop protection, pesticides also contribute to human health improvement by killing insect and rodent vectors responsible for spreading diseases. Pesticide application has been found useful in controlling various diseases such as typhus, bubonic plague, encephalitis, typhoid fever, and yellow fever, which are mainly vector-borne [16, 17]. Despite these beneficial effects, pesticides have several harmful consequences which outshadow its beneficial impacts.

Depending on solubility, pesticides can get entry into the ecosystem mainly by two processes: firstly, pesticides which are water soluble directly enter the water bodies such as ponds, rivers, lakes, and streams by getting dissolved in water and thereby adversely affecting the non-target life forms. Secondly, fat soluble pesticides get dissolved in the tissues of animals and move from one trophic level to the next through the food chain. The concentration of the pesticides in each trophic level increases as it passes from one trophic level to the other by the process of bioamplification [18] (Figure 1.1).

Pesticides drifting from land into various water bodies such as rivers and lakes adversely affect the aquatic ecosystem. Aquatic plants are an important component of the aquatic ecosystem and are responsible for providing approximately 80% of the dissolved oxygen [6]. Death of plants due to pesticides (e.g., herbicide) can lower the level of O_2 and aquatic organisms such as fishes can suffer due to oxygen depletion. This may further result in a reduction in fish productivity [19]. In addition to fishes, amphibian species are also affected by pesticide exposure. For instance, Rohr *et al.* [20] demonstrated a toxic impact of herbicide atrazine on some fish and amphibian species. Their mesocosm study revealed a relationship between exposure of herbicide atrazine and abundance alteration of larval trematodes in northern leopard frogs.

In addition to the aquatic ecosystem, terrestrial ecosystems are also adversely affected by the uncontrolled use of pesticide. Both target and non-target plants are affected by pesticide application. For instance, disease susceptibility of plants can be accelerated due to the application of herbicide glyphosate [21]. Further, the yield of non-targeted crops can be adversely affected by herbicides; sulphonamides, sulfonylureas, and imidazolinones [22]. Excessive use of pesticides also has deleterious effects on beneficial microbes present in the soil. Many soil dwelling microbes are

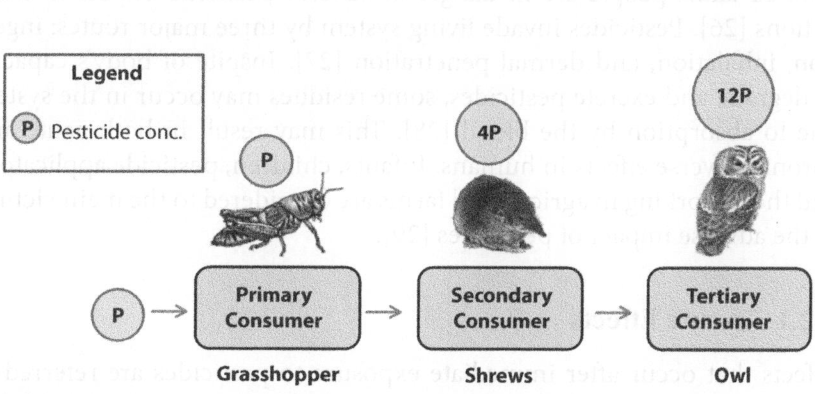

Figure 1.1 A diagrammatic representation of pesticide bioamplification in the environment [45].

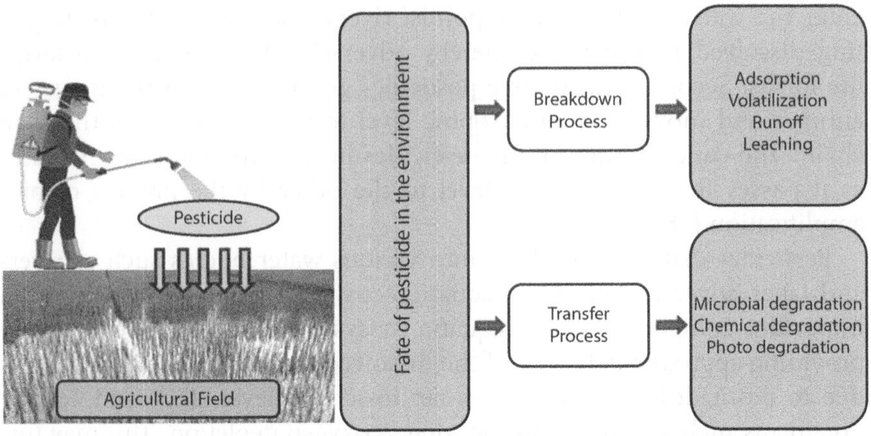

Figure 1.2 A diagrammatic representation of fate of pesticide in the environment [35].

involved in atmospheric nitrogen fixation. Pesticide can have a dangerous impact on these microbial communities. For example, growth and activity of soil dwelling bacteria can be negatively influenced by glyphosate [23]. Furthermore, nitrification and denitrification processes can be drastically altered by chlorothalonil and dinitrophenyl fungicides [24] (Figure 1.2).

Although pesticides contribute to the improvement of human health by controlling disease causing vectors (as mentioned earlier), it has several adverse effects as well. World Health Organization (WHO) states that about 30 lakhs cases of pesticide poisoning and 2 lakhs 20 thousand cases of death is reported annually in developing countries [25, 29]. In addition, 22 lakhs people are in danger of adverse pesticide impact in these nations [26]. Pesticides invade living system by three major routes: ingestion, inhalation, and dermal penetration [27]. Inspite of body's capacity to degrade and excrete pesticides, some residues may occur in the system due to absorption by the blood [28]. This may result in both acute and chronic adverse effects in humans. Infants, children, pesticide applicators, and those working in agricultural farms are considered to the main victims of the adverse impact of pesticides [29].

1.2.1 Acute Effects

Effects that occur after immediate exposure to pesticides are referred to as acute effects. These effects include skin itching, an occurrence of skin blisters and rashes, nose and throat irritation, blurred vision, vomiting and nausea, and diarrhoea. Acute effects are not serious enough to seek medical help and are rarely fatal [29].

1.2.2 Chronic Effects

Chronic effects of pesticides refer to its long term effects which may require even years to appear. Various body organs such as the lungs, liver, and kidney may be adversely affected due to the chronic impact of pesticides [6]. Reduction in motor signalling and visual ability, as well as impaired coordination and memory, can be attributed to the chronic effects of pesticide exposure [25]. Alteration in levels of human reproductive hormones (male and female) due to prolong presence of pesticide in the body may adversely affect reproductive potential and may result in infertility, stillbirth, birth defects, and spontaneous abortion [29]. Prolong exposure to pesticide may negatively affect the immune system and at the same time may cause various ailments such as hypersensitivity, asthma, and allergies [30]. Furthermore, various negative consequences such as nervousness, dizziness, confusion, nausea, vomiting, tremors, and hypersensitivity toward sound, light, and touch may occur due to ingestion of pesticides such as organochlorines [25].

1.3 Microalgal Species Involved in Bioremediation of Pesticides

Agrochemicals find widespread application in modern day agricultural practices to control pests and weeds to accelerate crop productivity. But environmental deterioration created by these chemicals has compelled human beings to look for an eco-friendly technology such as bioremediation. With the establishment of microalgae as an ideal bioremediation candidate, isolation and selection of strains which are resistant as well as have biodegrading potential received sufficient scientific attention. There are number of scientific investigations which reveal the pesticide degradation capabilities of cyanobacteria and algae (Table 1.2). According to Megharaj *et al.* [31] cyanobacteria *Nostoc linckia*, *Phormidium tenue*, and *Synechococcus elongatus* and green algae *Scenedesmus bijugatus* and *Chlorella vulgaris* had the capability to metabolise two organophosphorus insecticide monocrotophos and quinalphos. They concluded that both cyanobacteria and algae had similar biodegradation potential. In another work Megharaj *et al.* [32] also showed the biodegradation of the pesticide methyl parathion (MP)

Table 1.2 Cyanobacterial/microalgal strains involved in biodegradation of pesticide.

Chemical	Microalgae/Cyanobacteria	Reference
Monocrotophos and Quinalphos	*Chlorella vulgaris, Scenedesmus bijugatus, Synechococcus elongatus, Phormidium tenue, Nostoc linckia*	[31]
Methyl parathion	*C. vulgaris, S. bijugatus, N. linckia, N. muscorum, Oscillatoria animalis, P. foveolarum*	[32]
DDT	*Chlorococcum* sp., *Anabaena* sp., *Nostoc* sp.	[77]
α-Endosulfan	*Scenedesmus* sp., *Chlorococcum* sp.,	[76]
Fenamiphos	*Pseudokirchneriella subcapitata, Chlorococcum* sp.	[33]
Dimethomorph and Pyrimethanil	S. quadricauda	[39]
Fluroxypyr	Chlamydomonas reinhardtii	[40]
Chlorpyrifos	*Synechocystis* sp. strain PUPCCC 64	[41]
Prometryne	*C. reinhardtii*	[43]
Anilofos	*Synechocystis* sp. strain PUPCCC 64	[42]
Acephate, Imidaclorpid	*C. mexicana*	[44]
Diazinon	*C. vulgaris*	[13]
Methyl parathion	*Fischerella* sp.	[45]

by cyanobacteria *P. foveolarum*, *N. muscorum*, *N. linckia*, and *Oscillatoria animalis* and green algae *S. bijugatus* and *C. vulgaris*. The study showed that they were capable of hydrolyzing the insecticide in 20 days while *C. vulgaris*, *N. linckia*, and *S. bijugatus* could hydrolyze the same in 30 days. Thus, it concluded that the biodegradation capabilities of selected microalgal and cyanobacterial strain followed the following order: *C. vulgaris* < *S. bijugatus* < *N. linckia* < *N. muscorum* < *O. animalis* < *P. foveolarum*.

In addition to this, five green algae (*Chlorella* sp., *Scenedesmus* sp. MM1, *Stichococcus* sp., *Scenedesmus* sp. MM2, and *Chlamydomonas* sp.) and five cyanobacteria (*Anabaena* sp., *Nostoc* sp. MM1, *N. muscorum*, *Nostoc* sp. MM3, and *Nostoc* sp. MM2) have been reported to degrade fenamiphos which is an organophosphorus pesticide [33].

2,4-dichlorophenol (2,4-DCP) is often used as an intermediate in synthesis of insecticides and herbicides such as 2,4-D. Thus, the release of chlorophenols as industrial waste or by degradation of chlorinated pesticides have cause serious environmental threat [34]. Yang *et al.* [35] reported biotransformation and enzymatic responses of 2,4-dichlorophenol in *Skeletonema costatum* (diatom). They demonstrated that Cytochrome P-450, a key enzyme in biotransformation and metabolization, did not play an important role in 2,4-DCP detoxification.

Popular pest control agents such as chlorinated agrochemicals cause serious environmental problems such as accumulation in non-target organisms as well as in water and soil. Considering the high persistence and toxicity of chlorinated pesticide like lindane, many countries have prohibited its direct application [36]. Thus, there is a requirement of potential microalgal strain for eco-friendly remediation of chlorinated pesticides. Kuritz and Wolk [37] evaluated the lindane degrading potential of cyanobacteria *N. ellipsosporum* and *Anabaena* sp. genetically manipulated to biodegrade another contaminant 4-chlorobenzoate. Biodegradation of the pesticide lindane by the cyanobacterial strains *Synechococcus* sp., *Oscillatoria* sp., *Cyanothece* sp., *Nodularia* sp., *Synechococcus* sp., *Nostoc* sp., *Microcystis aeruginosa*, *A. cylindrical*, *M. aeruginosa*, *A. spiroides*, and *A. flos-aquae* has been reported [38].

Dosnon-Olette [39] demonstrated the removal of fungicides dimethomorph and pyrimethanil and herbicide isoproturon by the microalgae *S. quadricauda* and *S. obliquus*. The study showed that *S. quadricauda* removed dimethomorph and pyrimethanil more effectively than *S. obliquus*. Fluroxypyr (pesticide) accumulation and degradation by green alga *C. reinhardtii* was reported by Zhang [40]. They noted that *C. reinhardtii* had the potential to degrade more than 57% of bioaccumulated fluroxypyr within 5 days.

Singh *et al.* [41] demonstrated the potential of the cyanobacterium *Synechocystis* sp. to biodegrade the organophosphorus pesticide chlorpyrifos. The study showed that the organism could tolerate chlorpyrifos up to 15 mg L^{-1}. Maximum removal of chlorpyrifos was achieved at a temperature of 30°C, pH 7.0, and 100 mg $protein^{-1}$ biomass. Metabolization of the pesticide by the cyanobacteria resulted in production of 3,5,6-trichloro-2-pyridinol as degradation product. The same cyanobacterial strain was later reported to degrade anilofos by Singh *et al.* [42]. In the study, the organism was found to tolerate high concentration of anilofos (25 mg L^{-1}).

The influenced of the pesticide on photosynthetic pigment content was dose-dependent. The herbicide was uptaken rapidly by the organism during the first 6 hours after which there was slow uptake until 5 days. The cyanobacterium utilized anilfos as a source of phosphate with maximum removal of anilofos at temperature of 30°C, pH 8.0, and 100 mg protein L^{-1}. In addition to cyanobacteria, microalgae are reported to degrade herbicides. For instance, the green alga *C. reinhardtii* was found to accumulate and biodegrade the pesticide prometryne. The study demonstrated that *C. reinhardtii* had the capacity to degrade prometryne at a moderate concentration of 5 g L^{-1}. This uptake and degradation of herbicide by *C. reinhardtii* reflect the internal tolerance mechanism of the green algae and establish it as a potential strain for remediation of prometryne from contaminated water [43].

In a recent study, Kurade *et al.* [13] found that *C. vulgaris* has the capacity of bioremediation of diazinon (Figure 1.3). In the study, the rate constant of degradation (k) of diazinon (0.5–100 mg L^{-1}) ranged between 0.2304 to 0.049 d^{-1} and the half-life ($T_{1/2}$) ranged between 3.01 and 14.06 d^{-1}. According to gas chromatography mass spectroscopic (GC-MS) study, metabolism of diazinon by microalgal strain resulted in the formation of 2-isopropyl-6-methyl-4-pyrimidinol (IMP) which is a by-product with low toxcity. In another work, Kumar *et al.* [44] studied the degradation of pesticide acephate and imidacloprid by the microalgae *C. mexicana*. They concluded that *C. mexicana* was able to remove 25% and 21% of acephate and imidacloprid, respectively. In another recent work, Tiwari *et al.* [45] demonstrated that cyanobacterium *Fischerella* sp. isolated from paddy

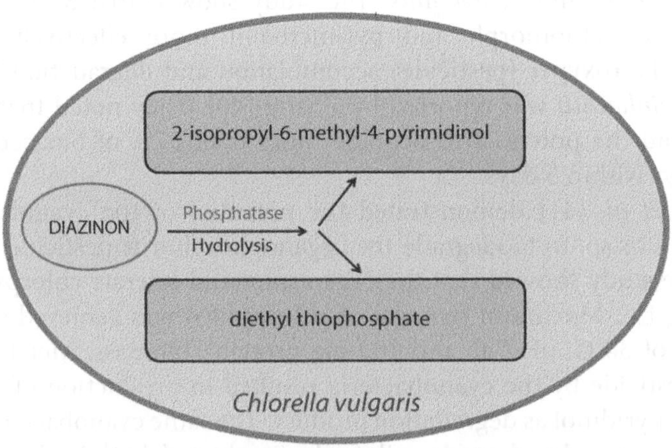

Figure 1.3 Schematic representation diazinon degradation by *Chlorella vulgaris* [38].

fields has the capacity to degrade organophosphorus pesticide MP. Based on their study, they recommend the organism as a potential candidate for pesticide bioremediation.

1.4 Strategies for Phycoremediation of Pesticides

1.4.1 Involvement of Enzymes in Phycoremediation of Pesticides

Biodegradation involves the breakdown of organic compounds into its inorganic constituents. Enzymes are one of the important biomolecules involved in the degradation of pesticides. The three main enzymes involved in pesticide degradation are hydrolases, esterases (also hydrolases), and the mixed function oxidases (MFOs). These enzyme systems are involved in the first metabolism stage of the pesticide and the glutathione S-transferase (GST) system, in the second phase [46]. In general pesticide, metabolism involves three main phases. During the Phase I of pesticide metabolism, the parent compound is converted into a more water-soluble and less toxic form by various processes such as oxidation, reduction, or hydrolysis. In the second phase, the water solubility and toxicity of the pesticide is further reduced by conjugation of the pesticide or pesticide metabolite to an amino acid or sugar. In the third phase, Phase II metabolites are converted into non-toxic secondary conjugates [46, 47]. Microalgae are photosynthetic organisms equipped with efficient enzyme system to metabolize and degrade various organic pollutants such as pesticides. Pertaining to their potential to degrade pesticides, microalgal species are recommended for remediation of the site contaminated with highly toxic pesticide like lindane [8]. Degradation of organophosphorus pesticide in presence of microbial enzymes has attracted the attention of scientist across the world. For instance, the enzyme alkaline phosphatase secreted by *Spirulina platensis* can hydrolyze chlorpyrifos, an organophosphorus pesticide, into 3,5,6-trichloro-2-pyridinol (TCP) [48]. Thus, immobilization of these pesticide degrading enzymes secreted form microalgae on solid matrix can be employed for remediation of pesticide contaminated sites [8].

1.4.2 Use of Genetically Engineered Microalgae

Development of genetically manipulated microalgae is a modern technology. This involves overexpression of contains proteins and enzymes which can combat the toxic effect of the contaminant. Extensive sequence information and good background knowledge about molecular, biochemical, physiological, and ecological characteristics of the microalgal species are required for the development of transgenic species to be used for bioremediation [49]. According to studies, *Anabaena* sp. strain PCC7120 and *N. ellipsossorum* are capable of degrading γ-Hexachlorocyclohexane (HCH) [37]. These two strains showed enhanced degradation of lindane when they are genetically modified using Lin A gene [37]. Thus, microorganisms can be genetically modified to develop highly efficient pesticide degradation strains which can be employed for eco-friendly remediation of pesticides.

1.5 Molecular Aspects of Pesticide Biodegradation by Microalgae

Several scientific studies are available in which algae and cyanobacteria have been reported to be highly efficient in detoxification of xenobiotics such as pesticides. Singh *et al.* [41] reported the degradation of the organophosphorus insecticide chlorpyrifos by the cyanobacterium *Synechocystis* PUPCCC. According to the author, the degradation mechanism of chlorpyrifos by cyanobacteria might be similar to bacteria. In bacteria, phosphotriesterases are the major group of enzymes involved in degradation of organophosphate pesticides [50]. These enzymes are encoded by a gene called *opd* (organophosphate-degrading). Mulbry and Karns [51] cloned and sequenced the gene. Phosphotriesterases are responsible for hydrolysis of phosphoester bonds, such as P–O, P–F, P–NC, and P–S [52]. The *opd* gene encoding organophosphorus hydrolase (the enzyme responsible for degradation of organophosphate pesticide) has 996 nucleotides, a typical promoter sequence of the promoter TTGCAA N17 TATACT from *E. coli* [53]. Chungjatupornchai and Fa-Aroonsawat [54] expressed opd gene from *Flavobacterium* sp. both on the surface and intracellularly in the cyanobacterium *Synechococcus* PCC7942 and used it for biodegradation of organophosphate pesticide. This reflects the importance of *opd* gene in biodegradation of organophosphate pesticides.

Exposure of plants to toxic organic substances provokes production of intracellular reactive oxygen species (ROS) which may adversely affect various cellular functions such as peroxidation of lipids and oxidation of proteins [55]. In order to minimize the adverse effects of ROS, plants possess an elaborate defense system consisting of antioxidant enzymes. Scavenging of ROS depends on the coordinated function of antioxidant enzymes such as Superoxide dismutase (SOD), Catalase (CAT), and Ascorbate peroxidise (APX) [56]. SOD is involved in the dismutation of superoxide anion O_2^- to

H_2O_2 and O_2. H_2O_2 is further scavenged by the catalytic activity of CAT and APX. APX is involved in the ascorbate-glutathione cycle to reduce stress [55, 57]. Jin *et al.* [43] reported an upregulation of the genes encoding Mn-SOD, CAT, and APX in green alga *C. reinhardtii* exposed to the herbicide prometryne. An efficient scavenging/detoxification system is responsible for quick accumulation and degradation of pesticide by microalgae [40]. Jin *et al.* [43] also noted an upregulation of the inducible gene HO-1 (Heme Oxygenase-1) in *C. reinhardtii* exposed to the herbicide prometryne suggesting its involvement in the tolerance of the microalgae toward the herbicide. Kumari *et al.* [58] evaluated butachlor toxicity in *Aulosira fertilissima* using a proteomic approach. They concluded that out of eight proteins altered during butachlor exposure, downregulation of GroES (associated with protein folding), and overexpression of NusB (associated with transcription termination) are curtail for cell death. Molecular docking studies confirm that interaction of butachlor with GroES and NusB is responsible for its toxicity [59].

Agrawal *et al.* [60] demonstrated the molecular basis of butachlor toxicity/tolerance in three *Anabaena* species using comparative proteomics. The study showed that 75 proteins involved in photosynthesis, C, N and protein metabolism, redox homeostasis, and signal transduction were differentially expressed in each *Anabaena* sp. Agrawal *et al.* [61] reported that a novel aldo-keto reductase (AKR17A1) from *Anabaena* sp.7120 has the capacity to degrade chloroacetanilide herbicide butachlor. The study demonstrated that, in addition to combating multiple stresses, aldo-keto reductase encoding open reading frame all 2,316 plays a significant role in butachlor degradation. The gene can be used to develop transgenics with butachlor degradation and stress tolerance capabilities [61].

For evaluation of biodegradation and biotransformation of pesticide by microalgae, time-dependent environmental risk assessment is very essential [62]. Esperanza *et al.* [63] evaluated the toxicity of the widespread herbicide atrazine to the green alga *C. reinhardtii* by the transcriptomic and proteomic approach. They found that exposure of the microalgae to sublethal concentration of atrazine (0.25 μM) for 3 h resulted in differential expression of 185 genes, of this 124 showed upregulation and 61 genes showed downregulation. These genes belonged to 13 different categories of function such as photosynthesis, metabolism, gene expression, energy, amino acids, cell cycle, redox, lipid, regulation, ROS and stress, proteases, other and unknown [64]. They also noted that nine genes related to photosynthesis were differentially expressed, of which three genes (HLA3, LCIA, and ELI3) showed significant upregulation and six genes (LHCBM8, LHCSR3, LI818R-1, PTOX2, CAH4, and CAH5) showed significant downregulation. In a recent study Tiwari *et al.* [65] demonstrated the tolerance strategy of cyanobacteria *Fischerella* sp. exposed to organophosphorus insecticide MP by analyses of proteome and transcriptome. Proteome analysis revealed a differential expression of proteins connected to various metabolic activities such as photosynthesis, energy and protein metabolism, redox homeostasis, signal transduction, and cellular defense. Transcript analyses showed differential expression of genes such as phycocyanin α subunit (*cpcA*), ribulose bisphosphate carboxylase (*rbcl*), F0F1 ATP synthase subunit α, F0F1 ATP synthase subunit β, SOD (*sod*), NifH (nif H), DnaK (*dnaK*), and Peptidase S8 in *Fischerella* sp. exposed to MP. In addition, some hypothetical proteins related to signaling and carbohydrate metabolism were also found to be upregulated in the cyanobacterium exposed to MP stress. One hypothetical protein was found to be homologous to lectin with an MP binding pocket. The author suggests that this carbohydrate binding protein might have been involved in metabolism and degradation of the pesticide.

1.6 Factor Affecting Phycoremediation of Pesticides

Microalgae have the capacity to degrade a wide range of pesticides owing to their robust metabolic machinery. However, several factors influence pesticide degradation by microalgae. Some of the key factors are discussed below.

1.6.1 Biological Factor

Phycoremediation of pollutants such as pesticides by a selected microalgal strain depends on its physiology, survival and growth behaviors, species density, tolerance, and previous exposure to the specific pollutant. Moreover, a good synergy and compatibility of the organism with the existing microbiota play a key role is phycoremediation [66–68]. According to previous reports, a consortium of algae and bacteria performs better as a bioremediating candidate than individual algal or bacterial strain [67, 69, 70].

1.6.2 Chemical Factor

The characteristic features of the xenobiotic compounds such as physical and chemical properties (properties, i.e., hydrophobicity, solubility, and volatility) and concentration play a key role in phycoremediation [70–72].

For instance, light aromatic and saturated compounds are more easily degraded than polar and high molecular weight compounds [73].

1.6.3 Environment Factor

Environmental factors such as temperature, pH, light duration and intensity, and oxidation-reduction potential, salinity, and dissolved oxygen of the medium are key players in the process of phycoremediation of pollutants such as pesticides. These factors may limit the growth and survivability of the microalgae and may influence the media geochemistry and consequently affecting the efficacy of the process [71, 70, 74].

1.7 Benefit and Shortcomings of Phycoremediation

The major benefits [49] and shortcomings of phycoremediation are discussed below.

1.7.1 Benefits

1. Phycoremediation technology is a cost-effective technology. There is no requirement of sophisticated instruments and expensive chemicals. Microalgae can efficiently remediate environmental contamination without any extra cost.
2. The biomass generated during the process of remediation can act as a potential feedstock for the production of various products such as bio-chemicals (e.g., pharmaceuticals), bio-fertilizer, and bio-fuel.
3. Microalgae are photosynthetic creatures; thus, they consume the CO_2 generated during the phycoremediation process and help in maintaining CO_2 balance.
4. Conventional remedial methods generate a large amount of sludge which may be hazardous for the environment. But the sludge generated after phycoremediation contains algal biomass which can be used for energy generation and production of other value-added products.

1.7.2 Shortcomings

1. Bioremediation has several shortcomings. For instance, bioremediation depends a lot on the nature of the organism.

Biodegradation of xenobiotics such as pesticide is not a benign response of the microorganism; on the contrary, it is a survival strategy. Most microorganisms carry out biodegradation under conditions which fulfils its necessities. Thus, certain modification of environment might be required to enable the organism to degrade pollutant in an efficient manner [75].

2. Low compatibility of the microalgal strain with the existing microflora and fauna can significantly affect the phycoremediation process.
3. Environmental factors such as pH, temperature, and salinity may influence the feasibility and success of the phycoremediation process.
4. Phycoremediation of pesticide is a slow process which makes its practical feasibility questionable.

1.8 Conclusion and Future Prospects

Bioremediation has proved to be an excellent tool for environmental remediation of pesticides originating from agricultural activities. There are a number of conventional techniques which are employed for pesticide remediation. But the cost associated with these methods is huge which made humans look for alternative remediation methods such as bioremediation. Traditionally, bacteria and fungi have been exploited for bioremediation but recently scientists and researchers have given sufficient attention to microalgae as a bioremediation candidate pertaining to its low nutritional requirements and versatile metabolic activity. Further, microalgae-based remediation may be integrated with other technology such as biofuel production, making them superior to its fungal and bacterial counterparts. However, there is an urgent need of more advance studies using proteomics and genomic tools to identify key genes involved in pesticide degradation. These genes can be used for development of transgenic microalgae for an efficient bioremediation of pesticides.

References

1. Halstead, N.T., Hoover, C.M., Arakawa, A., Civitello, D.J., De Leo, G.A., Gambhir, M., Johnson, S.A., Jouanard, N., Lorenz, K.A., McMahon, T.A., Ndione, R.A., Nguyen, K., Raffel, T.R., Remais, J.V., Riveau, G., Sokolow,

S.H., Rohr, J.R., Agrochemicals increase risk of human schistosomiasis by supporting higher densities of intermediate hosts. *Nat. Commun.*, 9, 837, 2018.

2. Wimalawansa, S.A. and Wimalawansa, S.J., Protection of Watersheds, and Control and Responsible use of Fertiliser to Prevent Phosphate Eutrophication of Reservoirs. *IJRES.*, 1, 1, 2015.
3. Carpenter, S.R., Eutrophication of aquatic ecosystems: biostability and soil phosphorus. *Proc. Natl. Acad. Sci.*, 102, 10002, 2005.
4. Lewis, W.M., Wurtsbaugh, W.A., Paerl, H.W., Rationale for control of anthropogenic nitrogen and phosphorus to reduce eutrophication of inland waters. *Environ. Sci. Technol.*, 45, 10300, 2011.
5. Ortiz-Hernández, M.L., Sánchez-Salinas, E., Dantán-González, E., María Luisa Castrejón-Godínez, M.L., Pesticide Biodegradation: Mechanisms, Genetics and Strategies to Enhance the Process, in: *Biodegradation - Life of Science*, R. Chamy and F. Rosenkranz (Eds.), pp. 251–287, IntechOpen, United Kingdom, 2013.
6. Mahmood, I., Imadi, S.R., Shazadi, K., Gul, A., Hakeem, K.R., Effects of Pesticides on Environment, in: *Plant, Soil and Microbes*, K.R. Hakeem (Eds.), pp. 253–269, Springer, Switzerland, 2016.
7. Schwarzenbach, R.P., Escher, B.I., Fenner, K., Hofstetter, T.B., Johnson, C.A., von Gunten, U., Wehrli, B., The Challenge of Micropollutants in Aquatic Systems. *Science*, 313, 1072, 2006.
8. Kumar, P.S., Carolin, C.F., Varjani, S.J., Pesticides Bioremediation, in: *Bioremediation: Applications for Environmental Protection and Management, Energy, Environment, and Sustainability*, S.J. Varjani (Eds.), pp. 197–222, Springer Nature, Singapore, 2018.
9. United States Environmental Protection Agency(n.d) What are Pesticides? from https://www.epa.gov/pesticides.
10. Buchel, K.H. (Ed.), (translated by Holmwood G), *Chemistry of Pesticides*, Wiley Interscience Publications, New York, 1983.
11. Badii, M. and Landeros, J., *Plaguicidas que afectan la saludhumana y la sustentabilidad*, pp. 4–19, CULCYT/ Toxicología De Plaguicidas, Mexico, 2007.
12. Ortiz-Hernández, M.L., *Biodegradación de plaguicidasorganofosforados por nuevasbacteriasaisladas del suelo*, Thesis, Biotechnology PhD, Universidad Autónoma del Estado de Morelos, México, 2002.
13. Kurade, M.B., Kim, J.R., Govindwar, S.P., Jeon, B.H., Insights into microalgae mediated biodegradation of diazinon by *Chlorella vulgaris*: Microalgal tolerance to xenobiotic pollutants and metabolism. *Algal. Res.*, 20, 126, 2016.
14. Rani, M., Shanker, U., Jassal, V., Recent strategies for removal and degradation of persistent & toxic organochlorine pesticides using nanoparticles: A review. *J. Environ. Manage.*, 190, 208, 2017.
15. Nayak, S.K., Dash, B., Baliyarsingh, B., Microbial Remediation of Persistent Agro-chemicals by Soil Bacteria: An Overview, in: *Microbial Biotechnology*, J.K. Patra (Eds.), pp. 275–301, Springer, Singapore, 2018.

16. Benefits of pesticides and crop protection chemicals, in: *Crop life America*, http://www.croplifeamerica.org/crop-protection/benefits.
17. Cunningham, M., Use of pesticides: benefits and problems associated with pesticides, in: *Education portal*, n.d, http://education-portal.com/academy/lesson/use-ofpesticides-benefits-and-problems-associated-with-pesticides.html.
18. Warsi, F., How do pesticides affect ecosystems, in: *Pesticides*, http://farhanwarsi.tripod.com/id9.html.
19. Helfrich, L.A., Weigmann, D.L., Hipkins, P., Stinson, E.R., *Pesticides and aquatic animals: a guide to reducing impacts on aquatic systems*, Virginia Polytechnic Institute and State University, United States, 2009, https://pubs.ext.vt.edu/420/420-013/420-013.html.
20. Rohr, J.R., Schotthoefer, A.M., Raffel, T.R., Carrick, H.J., Halstead, N., Hoverman, J.T., Johnson, C.M., Johnson, L.B., Lieske, C., Piwoni, M.D., Schoff, P.K., Beasley, V.R., Agrochemicals increase trematode infections in a declining amphibian species. *Nature*, 455, 1235, 2008.
21. Brammall, R.A. and Higgins, V.J., The effect of glyphosate on resistance of tomato to Fusarium crown and root rot disease and on the formation of host structural defensive barriers. *Can. J. Bot.*, 66, 1547, 1988.
22. Fletcher, J.S., Pfleeger, T.G., Ratsch, H.C., Potential environmental risks associated with the new sulfonylurea herbicides. *Environ. Sci. Technol.*, 27, 2250, 1993.
23. Santos, A. and Flores, M., Effects of glyphosate on nitrogen fixation of free-living heterotrophic bacteria. *Lett. Appl. Microbiol.*, 20, 349, 1995.
24. Lang, M. and Cai, Z., Effects of chlorothalonil and carbendazim on nitrification and denitrification in soils. *J. Environ. Sci.*, 21, 458, 2009.
25. Lah, K., Effects of pesticides on human health, in: *Toxipedia*, http://www.toxipedia.org/display/toxipedia/Effects+of+Pesticides+on+Human+Health. 2011.
26. Hicks, B., *Agricultural pesticides and human health*, National Association of GeoscienceTeachers, http://serc.carleton.edu/NAGTWorkshops/health/case_studies/pesticides.html., United States, 2013.
27. Spear, R., Recognised and possible exposure to pesticides, in: *Handbook of pesticide toxicology*, W.J. Hayes and E.R. Laws (Eds.), pp. 245–274, Academic, San Diego, CA, 1991.
28. Jabbar, A. and Mallick, S., *Pesticides and environment situation in Pakistan (Working Paper Series No. 19)*, Sustainable Development Policy Institute, Pakistan, 1994.
29. Boudh, S. and Singh, J.S., Pesticide Contamination: Environmental Problems and Remediation Strategies, in: *Emerging and Eco-Friendly Approaches for Waste Management*, R.N. Bhargava and P. Chowdhary (Eds.), pp. 245–269, Springer Nature, Singapore, 2019.
30. Culliney, T.W., Pimentel, D., Pimentel, M.H., Pesticides and natural toxicants in foods. *Agric. Ecosyst. Environ.*, 41, 297–320, 1992.

31. Megharaj, M., Venkateswarlu, K., Rao, S., Metabolism of Monocrotophos and Quinalphos by Algae Isolated from Soil. *Bull. Environ. Contam. Toxicol.*, 39, 251, 1987.
32. Megharaj, M., Madhavi, D.R., Sreenivasulu, C., Umamaheswari, A., Venkateswarlu, K., Biodegradation of Methyl Parathion by Soil Isolates of Microalgae and Cyanobacteria. *Bull. Environ. Contam. Toxicol.*, 53, 292, 1994.
33. Cáceres, T.P., Megharaj, M., Naidu, R., Biodegradation of the Pesticide Fenamiphos by Ten Different Species of Green Algae and Cyanobacteria. *Curr. Microbiol.*, 57, 643, 2008.
34. Petroutsos, D., Wang, J., Katapodis, P., Kekos, D., Sommerfeld, M., Hu, Q., Toxicity and metabolism of p-chlorophenol in the marine microalga *Tetraselmis marina. Aquat. Toxicol.*, 85, 192, 2007.
35. Yang, S., Wu, R.S.S., Kong, R.Y.C., Biodegradation and enzymatic responses in the marine diatom *Skeletonemacostatum* upon exposure to 2,4-dichlorophenol. *Aquat. Toxicol.*, 59, 191, 2002.
36. Ghasemi, Y., Rasoul-Amini, S., Fotooh-Abadi, E., Review: The Biotransformation, Biodegradation and Bioremediation of Organic Compounds by Microalgae. *J. Phycol.*, 47, 969, 2011.
37. Kuritz, T. and Wolk, C.P., Use of Filamentous Cyanobacteria for Biodegradation of Organic Pollutants. *Appl. Environ. Microbiol.*, 61, 234, 1995.
38. El-Bestawy, E., El-Salam, Z., Mansy, E.R.H., Potential Use of Environmental Cyanobacterial Species in Bioremediation of Lindane-Contaminated Effluents. *Int. Biodeterioration. Biodegrad.*, 59, 180, 2007.
39. Dosnon-Olette, R., Trotel-Aziz, P., Couderchet, M., Eullaffroy, P., Fungicides and herbicide removal in *Scenedesmus* cell suspensions. *Chemosphere.*, 79, 117, 2010.
40. Zhang, S., Qiu, C.B., Zhou, Y., Jin, Z.P., Yang, H., Bioaccumulation and degradation of pesticide fluroxypyr are associated with toxic tolerance in green alga. *Chlamydomonas reinhardtii. Ecotoxicology.*, 20, 337, 2012.
41. Singh, D.P., Khattar, J.I.S., Nadda, J., Singh, Y., Garg, A., Kaur, N., Gulati, A., Chlorpyrifos degradation by the cyanobacterium *Synechocystis* sp. strain PUPCCC 64. *Environ. Sci. Pollut. Res.*, 18, 1351, 2011.
42. Singh, D.P., Khattar, J.I.S., Kaur, M., Kaur, G., Gupta, M., Singh, Y., Anilofos Tolerance and Its Mineralization by the Cyanobacterium *Synechocystis* sp. Strain PUPCCC 64. *PLoS ONE.*, 8, e53445, 2013.
43. Jin, Z.P., Luo, K., Zhang, S., Zheng, Q., Yang, H., Bioaccumulation and catabolism of prometryne in green algae. *Chemosphere.*, 87, 278, 2012.
44. Kumar, M.S., Aabra, A.N., Min, B., Ei-Dalatony, M.M., Xiong, J., Thajuddin, N., Lee, D.S., Jeon, B.H., Insecticide induced biochemical changes in freshwater microalga. *Chlamydomonas mexicana. Environ. Sci. Pollut. Res.*, 23, 1091, 2016.
45. Tiwari, B., Chakrabortya, S., Srivastava, A.K., Mishra, A.K., Biodegradation and rapid removal of methyl parathion by the paddy field cyanobacterium. *Fischerella sp. Algal. Res.*, 25, 285, 2017.

46. Li, X., Schuler, M.A., Berenbaum, M.R., Molecular mechanisms of metabolic resistance to synthetic and natural xenobiotics. *Ann. Rev. Entomol.*, 52, 231, 2007.
47. Ortiz-Hernández, M.L., Sánchez-Salinas, E., Olvera-Velona, A., Folch-Mallol, J.L., Pesticides in the Environment: Impacts and its Biodegradation as a Strategy for Residues Treatment, in: *Pesticides - Formulations, Effects, Fate*, M. Stoytcheva (Ed.), pp. 551–574, IntechOpen, United Kingdom, 2011.
48. Thengodkar, R.R. and Sivakami, S., Degradation of Chlorpyrifos by an alkaline phosphatase from the cyanobacterium *Spirulina platensis*. *Biodegradation*, 21, 637, 2010.
49. Paliwal, R., Gusain, P., Kaphaliya, B., Uniyal, S., Algal Potential for Inorganic and Organic Pollutants Decontamination, in: *Water Biology*, D.R. Khanna and R. Bhutiani (Eds.), pp. 309–317, Discovery Publishing House Pvt. Ltd., New Delhi, 2018.
50. Chino-Flores, C., Dantán-González, E., Vázquez-Ramos, A., Tinoco-Valencia, R., Díaz-Méndez, R., Sánchez-Salinas, E., Castrejón-Godínez, M.L., Ramos-Quintana, F., Ortiz-Hernández, M.L., Isolation of the opdE gene that encodes for a new hydrolase of *Enterobacter* sp. capable of degrading organophosphorus pesticides. *Biodegradation*, 23, 387, 2012.
51. Mulbry, W.W. and Karns, J.S., Parathion hydrolase specified by the *Flavobacterium* opd gene: Relationship between the gene and protein. *J. Bacteriol.*, 171, 6740, 1989.
52. Ortiz-Hernández, M.L., Quintero-Ramírez, R., Nava-Ocampo, A.A., Bello-Ramírez, A.M., Study of the mechanism of *Flavobacterium* sp. for hydrolyzing organophosphate pesticides. *Fundam. Clin. Pharmacol.*, 17, 717, 2003.
53. Zhang, R., Cui, Z., Jiang, J., Gu, X., Li, S., Diversity of organophosphorus pesticides degrading bacteria in a polluted soil and conservation of their organophosphorus hydrolase genes. *Can. J. Microbiol.*, 5, 337, 2005.
54. Chungjatupornchai, W. and Fa-Aroonsawat, S., Biodegradation of Organophosphate Pesticide Using Recombinant Cyanobacteria with Surface- and Intracellular-Expressed Organophosphorus Hydrolase. *J. Microbiol. Biotechnol.*, 18, 946, 2008.
55. Mittler, R., Oxidative stress, antioxidants and stress to tolerance. *Trends. Plant. Sci.*, 7, 405, 2002.
56. Maurino, V.G. and Flugge, U., Experimental systems to assess the effects of reactive oxygen species in plant tissues. *Plant. Signal. Behav.*, 3, 923, 2008.
57. Alscher, R.G., Erturk, N., Heath, L.S., Role of superoxide dismutases (SODs) in controlling oxidative stress in plants. *J. Exp. Bot.*, 53, 1331, 2002.
58. Kumari, N., Narayan, O.P., Rai, L.C., Understanding butachlor toxicity in *Aulosirafertilissima* using physiological, biochemical and proteomic approaches. *Chemosphere.*, 77, 1501, 2009.
59. Kumari, N., Singh, V.K., Narayan, O.P., Rai, L.C., Toxicity of butachlor assessed by molecular docking to NusB and GroES protein. *Onl. J. Bioinform.*, 12, 289, 2011.

60. Agrawal, C., Sen, S., Singh, S., Rai, S., Singh, P.K., Singh, V.K., Rai, L.C., Comparative proteomics reveals association of early accumulated proteins in conferring butachlor tolerance in three N (2)-fixing *Anabaena* spp. *J. Proteomics.*, 96, 271, 2014.
61. Agrawal, C., Sen, S., Yadav, S., Rai, S., Rai, L.C., A Novel Aldo-Keto Reductase (AKR17A1) of *Anabaena* sp. PCC 7120 Degrades the Rice Field Herbicide Butachlor and Confers Tolerance to Abiotic Stresses in *E. coli. PLoS ONE.*, 10, e0137744, 2015.
62. Gupta, S.K., Shriwastav, A., Kumari, S., Ansari, F.A., Malik, A., Bux, F., Phycoremediation of Emerging Contaminants, in: *Algae and Environmental Sustainability, Developments in Applied Phycology*, vol. 7, B. Singh (Eds.), pp. 129–146, Springer, India, 2015.
63. Esperanza, M., Seoane, M., Rioboo, C., Herrero, C., Cid, A., Early alterations on photosynthesis-related parameters in *Chlamydomonas reinhardtii* cells exposed to atrazine: A multiple approach study. *Sci. Total. Environ.*, 554, 237, 2016.
64. Hemschemeier, A., Casero, D., Liu, B., Benning, C., Pellegrini, M., Happe, T., Merchant, S.S., Copper response regulator1-dependent and -independent responses of the *Chlamydomonas reinhardtii* transcriptome to dark anoxia. *Plant Cell.*, 25, 3186, 2013.
65. Tiwari, B., Verma, E., Chakrabortya, S., Srivastava, A.K., Mishra, A.K., Tolerance strategies in cyanobacterium *Fischerella* sp. under pesticide stress and possible role of a carbohydrate-binding protein in the metabolism of methyl parathion (MP). *Int. Biodeterioration. Biodegrad.*, 127, 217, 2018.
66. Corner, T.R., Synergism in the inhibition of *Bacillus subtilis* by combinations of lipophilic weak acids and fatty alcohols. *Antimicrob. Agents. Chemother.*, 19, 1082, 1981.
67. Horvath, R.S., Microbial co-metabolism and the degradation of organic compounds in nature. *Bacteriol. Rev.*, 36, 146, 1972.
68. Rosenzweig, W.D. and Stotzky, G., Influence of environmental factors on antagonism of fungi by bacteria in soil: nutrient levels. *Appl. Environ. Microbiol.*, 39, 354, 1980.
69. Chekroun, K.B., Sánchez, E., Baghour, M., The role of algae in bioremediation of organic pollutants. *Int. Res J. Public. Environ. Health.*, 1, 19, 2014.
70. Kobayashi, H. and Rittmann, B.E., Microbial removal of hazardous organic compounds. *Environ. Sci. Technol.*, 16, 170A, 1982.
71. Gibson, D.T., Microbial transformation of aromatic pollutants, in: *Aquatic pollutants*, O. Hutzinger, L.H. Van Lelyveld, B.C.J., Zoeteman (Eds.), Pergamon Press, New York, 1978.
72. Steen, W.C., Paris, D.F., Baughman, G.L., Effects of sediment sorption on microbial degradation of toxic substances. *Contam. Sediment.*, 1, 477, 1980.
73. Leahy, J.G. and Colwell, R.R., Microbial degradation of hydrocarbons in the environment. *Microbiol. Rev.*, 54, 305, 1990.

74. Colwell, R. and Sayler, G., Microbial degradation of industrial chemicals. *Water. Pollut. Microbiol.*, 2, 111, 1978.
75. Singh, D.K., Biodegradation and bioremediation of pesticide in soil: concept, method and recent developments. *Indian J. Microbiol.*, 48, 35, 2008.
76. Sethunathan, N., Megharaj, M., Chen, Z.L., Williams, B.D., Lewis, G., Naidu, R., Algal degradation of a known endocrine disrupting insecticide, α-endosulfan, and its metabolite, endosulfansulfate, in liquid medium and soil. *J. Agric. Food. Chem.*, 52, 3030, 2004.
77. Megharaj, M., Kantachote, D., Singleton, I., Naidu, R., Effects of long-term contamination of DDT on soil microflora with special reference to soil algae and algal transformation of DDT. *Environ. Pollut.*, 109, 35, 2000.

2

Microalgal Bioremediation of Toxic Hexavalent Chromium: A Review

Pritikrishna Majhi[1], Satyabrata Nayak[2] and Saubhagya Manjari Samantaray[1]*

[1]*Department of Microbiology, College of Basic Science & Humanities, Odisha University of Agriculture and Technology, Bhubaneswar, Odisha, India*

[2]*Silviculture Division, Rayagada, Odisha, India*

Abstract

Chromium is the seventh most abundant metal in earth crust which is used in leather tanning, electroplating, pigment manufacturing, dying and production of stainless steel, refractory, ceramics, chemicals, electrode, alloy production, and wood preservation. Increased soil run off from the mining area and dumping of industrial waste increases the chromium concentration of the soil. Among the different oxidative states, Cr(III) and Cr(VI) are very stable and commonly found in nature. Consequently, hexavalent chromium at a high concentration is toxic for the plant, animal, human, as well as microbes. The microalgae would be an option for the removal and detoxification of Cr from chromium-rich soil. Chemical methods used for Cr removal from soil are quite costly with severe side effects for which this review emphasizes on the methods of biological reduction of Cr(VI) to Cr(III) using microalgae.

Keywords: Microalgae, Cr(VI), bioremediation, chromium toxicity, mechanism

2.1 Introduction

Chromium is a silver-gray colored, lustrous, hard, and brittle metal. N. L. Vauquelin discovered Chromium in 1798 from the Siberian red ore (crocoites) [1]. South Africa and Zimbabwe account for 85% and 10% of total earth's chrome ore reserve (7,500 million tons). In addition, 2.5%

**Corresponding author*: saubhagyasamantaray@yahoo.com

Bibhuti Bhusan Mishra, Suraja Kumar Nayak, Swati Mohapatra, and Deviprasad Samantaray (eds.) Environmental and Agricultural Microbiology: Applications for Sustainability, (25–38) © 2021 Scrivener Publishing LLC

the world's chromium resources are found in India (i.e., about 186 million tons). India accounts for around 97% of world chromium reserve and it is mostly stuffed in Sukinda ultramafic belt of Odisha [2]. Chromium is found in variable valence forms ranging from −2 to +6, and the most stable forms are Cr(VI) and Cr(III) which are amalgamated with oxygen. Chromium oxide (CrO_3) and chromium hydroxide [$Cr(OH)_3$] are two different forms of Cr(III) while Cr(VI) is available in the form of dichromate ($Cr_2O_7^{2-}$), chromate (CrO_4^{2-}) or hydro-chromate ($HCrO_4^-$). Along with various industrial applications, it is also attributed with some strategic importance in the field of defence, aero-space, and aviation [2]. Mertz (1969) [3] reported that chromium (III) is responsible for the glucose tolerance factor. Cr(III) is necessary for the metabolism of carbohydrate in human and animal nutrition [3, 4]. Intake of a large amount of Cr(III) beyond permissible limit may also responsible for many diseases like lung's cancer [5, 6]. Cr(III) is quite different from Cr(VI) on the basis of their mobility, bioavailability, and toxicity. It has been observed that the hexavalent form of chromium is more soluble and toxic than the trivalent form. Daily consumption/intake of Cr(VI) beyond the permissible limit present in the contaminated water and food products results in the entry of chromate into the human body [7, 8]. Serious environmental complicacies have been observed due to the rampant usage of toxic forms of chromium which cause vitiation in soil and water. Open cast mining of chromium cause the assemblage of chromite ores and waste rock materials which are further discarded in to the open ground without considering its effect on the environment [9]. This toxic metal percolates to the groundwater systems through rain and enters into the surface water bodies through soil run off. Hence, it is the dire need of the time to ameliorate the toxicity of the metal using some eco-friendly and economic sources. In this regard, the present scenario is highly demanding a suitable biological alternative which can also overcome the dangerous side effects of expensive chemical treatments.

Chromium toxicity can be lessened using biosorbents prepared from various microbes like bacteria, fungi, yeasts, moulds, and algae [10]. However, involvement of microalgae for the reduction of toxicity is highly recommendable due to the availability of some exclusive properties in them. Presence of different binding groups, polysaccharides, proteins, and vacuoles collectively provide a higher binding affinity with the metal and, hence, facilitate the process of bioremediation [11]. Moreover, these microbes also possess numerous advantages like high efficiency in eliminating heavy metals even from very low concentration, cheaper cost, high adsorbing capacity, larger surface area, greater mucilage area, and high binding affinity with simple nutrient requirement. Besides these, they are

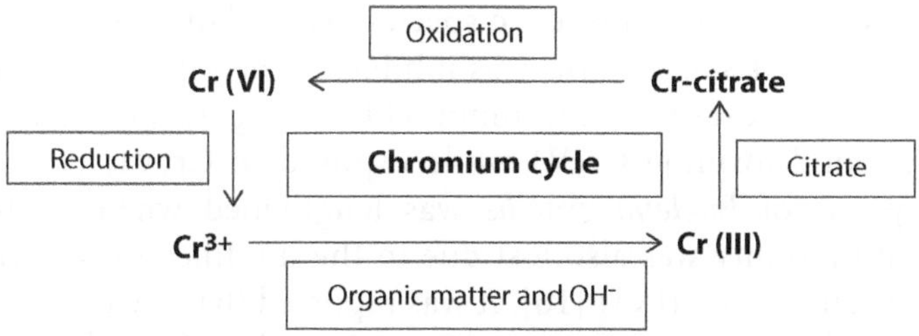

Figure 2.1 The chromium cycle.

capable of growing in both aquatic and terrestrial area. In diverse ecosystems, algae play significant roles for which they are regarded as cosmopolitan microorganisms. They can synthesize low molecular weight thiol-peptides and reduced glutathione and phytochelatin when grown in a heavy metal polluted environment [12].

2.1.1 Chromium Cycle

The chromium cycle mostly comprises oxidation and reduction of Cr(III) and Cr(VI), respectively (Figure 2.1). Although both the forms of chromium are found to have opposite characteristics (e.g., toxicity, mobility, and reactivity) but eventually both are highly dangerous when the concentration becomes high. Moreover, this cycle depicts the simultaneous reduction of Cr(VI) through some carbon compounds and the oxidation of Cr(III) in the presence of manganese oxide available in soil and sediments represented in the below equations [13].

$$C_6H_6O_2 + CrO_4^{2-} + 2H_2O \rightarrow 0.5Cr_2O_3 + 1.5C_6H_4O_2 + 2.5H_2O + 2OH^- \quad (2.1)$$

$$Cr^{3+} + 1.5MnO_2 + H_2O \rightarrow HCrO_4 + 1.5Mn^{2+} + H^+ \quad (2.2)$$

2.2 Effects of Hexavalent Chromium Toxicity

2.2.1 Toxicity to Microorganisms

Chronic exposure to hexavalent chromium has many deleterious effects on the structure and function of the microbial cells, and in some cases, it also causes dormancy. It leads to species loss and disturbs the diversity. Growth of *Scenedesmus acutus* was inhibited when it treated with more than 15 ppm of hexavalent chromium [14]. *Spirogyra* sp. and *Mougeotia* sp. were found forming Cr(V) while exposed to Cr(VI) [15]. The lag growth phase of *Euglena gracilis* was lengthened when treated with Cr(VI) and motility was also lost due to the modifications in the cytoskeleton induced by Cr(VI) [16]. It was reported that the photosynthesis was inhibited due to the presence of Cr in the cells of *Scenedesmus* sp. and *Chlorella* sp. [17, 18]. The sulfate transport system mediated transport of chromate ions has diverse toxic effects in the cytoplasm of *Salmonella typhimurium*, *Alkaligenes eutrophus*, *Escherichia coli*, and *Pseudomonas fluorescens*. According to Viamajala *et al.*, (2002) [19], the minimum concentration of Cr (VI) (0.015 mM) has slowed down the growth of *Shewanella oneidensis*. The reduction of growth was observed in mycelium of fungi due to the toxic effect of hexavalent chromium. Interference of chromium causes gene mutation and conversion which further lead to growth inhibition in fungal cell [16].

2.2.2 Toxicity to Plant Body

Hexavalent chromium diffuses across the cell membrane due to the structural resemblance of chromate ions to phosphate or sulphate. It can easily enter inside the cell and where the reduction takes place producing Cr(V) and then Cr(III) reactive oxygen species and free radicals [20]. Cr(III) is impermeable, so unable to cross the cellular membrane and prefers to bind the protein molecules available on the membrane surface with greater affinity causing DNA damage, inhibition of DNA replication, and RNA transcription [21]. Plant growth, development, and plant physiology (mineral nutrition, water relations, and photosynthesis) are greatly affected by hexavalent chromium [22]. The amount of chlorophyll (Chl) content, nitrate reductase activity, and δ-aminolevulinic acid contents were also reduced in plants growing in chromium contaminated soil [23]. Hexavalent chromium induces the inhibition of photosynthesis rate in terms of CO_2 fixation, electron transport processes, enzyme activities, and photophosphorylation in plants [24, 25]. Bishnoi *et al.*, (1993) [26] has observed that Cr(VI) was influencing the PS I and PS II by isolating the chloroplasts from peas. The direct effect of Cr exposure has also been found on enzymes or other metabolites that may cause increased oxidative stress and lipid peroxidation [27–29]. Consequently, herein, we can conclude three key roles of Cr on plants as follows:

(i) Production of a new metabolites to change the metabolic pool which would providetolerance of Cr stress (e.g., phytochelatins and histidine) [30].
(ii) Variation of the production in several pigments (like chlorophyll and anthocyanin) for the sustenance of plants [31].
(iii) Cr stress induces the production of metabolites like glutathione and ascorbic acid which may cause damage to the plants [32, 33].

2.2.3 Toxicity to Animals

People those are directly exposed to chromium show nasal irritation, perforation of the nasal septum, nasal ulcers, "chrome holes" [34], and hypersensitivity reactions in the skin. But some other cases reported that the normal people who are not practically exposed to chromium but ingested chromium through food and water show deposition of chromium in different organelles like kidney, adrenals, lungs, liver, spleen, plasma, bone marrow, and red blood cells in due to low pH of the stomach. Ingestion of Cr(VI) poses a significant carcinogenic risk because of the solubility of particulate chromate at low pH which is weakly carcinogenic to the lungs [34]. Enduring exposure of low level of Cr(VI) between 4 and 25 ppm to skin can cause a long lasting sensitisation that leads allergic contact dermatitis (ACD) while 20 to 25ppm of Cr(VI) can cause inflammation, eczema, and open sores (ulcers) [35]. Similarly, there are some significant observations of Cr(VI) dusts exposure [36, 37]. According to these reports, inhalation of even only 2 μg of Cr(VI) dust leads irritation of nose, throat, and lungs along with respiratory inflammation, nosebleeds, ulceration, and perforation (holes) in the septum when come in contact with 0.09μg of Cr(VI). Some noteworthy observations were also documented in a group of women who were exposed to industrial chromium contamination showed irregularity in menstruation cycle, birth complications, and increases in post-birth haemorrhage [38, 39]. A remarkable study revealed that symptoms like mouth sores, diarrhoea, stomach pains, indigestion, vomiting, and higher levels of white blood cells were found when a group of individuals were exposed to approximately in drinking water that contaminated by a ferrochrome plant [40]. According to the survey of US EPA (Environment Protection Agency) in 1998, it was observed that the contamination of drinking water with 20,000 μg L^{-1} of Cr(VI) caused many diseases like mouth sores, vomiting, indigestion and diarrhoea [41]. Men exposed to chromium released from welding fumes exhibited toxicity in testes and blood, increased semen abnormalities, and reduced sperm concentrations [42]. It has explained when adult female rats take Cr(VI) contaminated drinking water; it is found to be toxic to the ovaries. It damages the ovarian tissues, reduces the number of follicles and ovum which ultimately, increases the chances of infertility. In mice, it has been observed that Cr(VI) is toxic to foetus, embryos (250, 500, and 750 mg L^{-1}) and also increases skeletal abnormalities (250 and 500 mg L^{-1}) [43]. Cr(VI) concentrations at 100, 200, and 400 mg L^{-1} was found to be toxic to reproductive organs, changed endocrine organ weight, testis enzymes levels and sperms when given to male monkeys through drinking water [44, 45].

The summary of hexavalent chromium effects optimistically made us to find out a significant bio-remediating agent to convert it to non-toxic form which would be cost-effective, easily available, and without any side effects. Herein, we can deliberate the microbes as an alternative of chemical agents. Numbers of reports are proposed basing upon the chromium removal strategy with strains of bacteria, fungi, virus, microalgae, and seaweeds. But in this present piece of work, emphasis has been given on microalgae as a potent source of bioremediation.

2.3 Chromium Bioremediation by Microalgae

Microalgae play an important role in the chromium bioremediation. Biosorption is a method of bioremediation where sorption is taking place either by using dead or living biomass, and it has various significant advantages as follows:

(i) High efficiency in eliminating heavy metals even from very low concentrations
(ii) Cost effective
(iii) High metal adsorbing capacity
(iv) The ability of recovering the important metals adsorbed

Algal cells are considered as natural ion-exchange matter as they contain various anionic groups on their surface and this allows them to eliminate heavy metal ions efficiently [46, 47]. It has been observed that various strains of algae like blue-green algae, green algae, red algae, and diatoms are able to remove hexavalent chromium from soil and water.

2.3.1 Cyanobacteria

According to Elhaddad and Mahmoud (2015) [48], a blue-green alga *Spirulina platensis* acts as a good biosorbent that help in reducing hexavalent chromium. When its chromium reduction efficiency was studied at different pH (1.0 to 7.0), at different time period (5, 10, 20, 30, 60, 80, 90, and 120 minutes), pH 6.0 and 40 minutes of time period are suitable conditions for chromium reduction. Another *S. platensis* was also studied by Khuntia *et al.*, (2011) [49] for the reduction of hexavalent chromium. In addition, 99.7% of chromium was reduced at an optimum pH of 0.5 and biomass dose of 5 g L^{-1}. According to Katircioğlu *et al.*, (2012) [50], a cyanobacterial strain *Oscillatoria* sp. H1 obtained from the Mogan Lake in Ankara, Turkey, was also observed for hexavalent chromium (VI) removal from aqueous solutions. Maximum biosorption of Cr(VI) was found at pH 6.0 which remain unaffected to temperature between 20°C and 40°C. Increase in biosorption was observed by increase the dried free and immobilized live and heat inactivated biomass (0.04 g) and 30 beads, respectively.

2.3.2 Green Algae

The smaller freshwater green algae *Pseudokirchneriella subcapitata*, formerly known as *Selenastrum capricornutum* Pintz, amplifies metal binding sites, leading to an increase in bioaccumulation and consequential increase the capacity to accumulate chromium [51]. *Spirogyra* sp. was found to be a cost effective and eco-friendly biosorbent while studied using different concentrations of chromium (1.0, 5.0, 15.0, and 25.0 mg L^{-1}), different dosages of dead algal biomass (0.1, 0.2, and 0.3g) with variation time, pH, and temperature [52]. Chatterjee and Abraham, 2015 [53] observed maximum biosorption in the dried biomass of the *Spirogyra* sp. (2.5 g L^{-1}) at pH 6.0 when it was treated with 10 mg L^{-1} chromium concentration for one hour. *Sphaeroplea* sp. was treated with different chromium concentration with variation in time period, in its natural and acid treated form to study the biosorption capacity. Maximum result was observed at pH 5.0 in the acid treated alga (158.9 mg g^{-1}) than its natural form (29.85 mg g^{-1}) [54].

2.3.3 Diatoms

Sbihi *et al.*, (2012) [56] have observed maximum Cr(VI) biosorption capacity 93.45 mg g^{-1} of Cr(VI) in *Planothidium lanceolatum* at a concentration of 0.4-g dried diatoms per liter with a Cr(VI) concentration of 20 mg L^{-1}. Hence, it was proved to be a potent microalga for biosorption of hexavalent chromium.

2.4 Mechanism Involved in Hexavalent Chromium Reduction in Microalgae

Polysaccharide is the basic building block present in the cell walls of prokaryotic and eukaryotic microalgae following proteins and lipids [57]. They contain functional groups like phosphate, amino, sulfhydryl, thiol, and carboxylic groups which are mostly capable of binding to the heavy metals as per their specificity and affinity as seen in Figure 2.2.

Cyanobacteria are able to synthesize metallothioneins (intracellular metal binding proteins) [58]. These are low molecular weight proteins (6,000 to 8,000 amu) rich in cystein residue and bind to metal ion in metal thiolate cluster. It has been reported that cyanobacterial species like *Oscillatoria* sp., *Gleocapsa* sp., and *Spirulina* sp. have the ability of synthesizing siderophores possessing metal chelating properties [50]. It has also been noticed that the heavy metals get deposited in polyphosphate bodies (intracellular storage compartments). Besides this they have also some other advantages like larger surface area, high binding affinity, simple nutrient requirement, and greater mucilage volume which help them

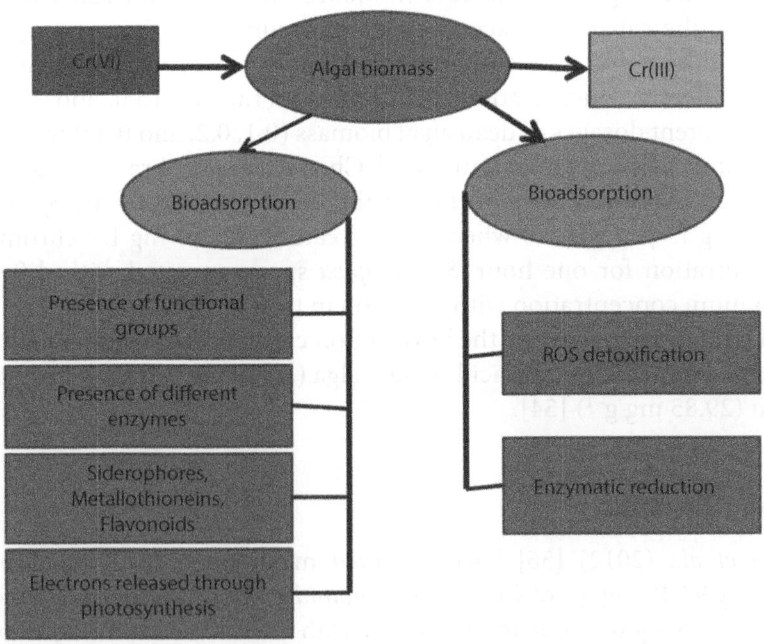

Figure 2.2 Schematic diagram: Mechanism of Cr(VI) reduction through micro-algal biomass.

to act as biosorbents [55]. Besides this, another mechanism involves the partitioning of metal ion between cell wall and exopolymer sheath [58].

Except extracellular chromium reduction, intracellular reduction can also be taken as a major mechanism. Algal cells are found to be better source of Cr(VI) reduction, so there must be presence of Cr(VI) reducing enzymes in their cells like bacteria and fungi. The protoplasm of these cells contains some components such as NADH, proteins, low molecular weight carbohydrates, fatty acids, amino acids, and flavoproteins which can completely reduce Cr(VI) to Cr(III). Generally, in chromium-rich region an oxidative stress condition is created inside the cell leading to the generation of several harmful reactive oxygen species (ROS). In order to avoid this situation, the cell starts to produce special kind of protein, enzyme, or any substance which can able to reduce, remove, or transform the Cr(VI). Besides this, microalgae also release electron through photosynthesis and they have a very unique metabolic process compensating the electron for the reduction of Cr(VI) [16, 59].

According to the findings of Nacorda *et al.*, (2010), there is an initial rapid phase of passive extracellular biosorption process [60]. It was carried out following a slower active intracellular bio-absorption. This method is quite similar to the biphasic uptake take place in bacteria, fungi and other microbes. It is also reported that the longer is the incubation time the higher is the amount of Cr(VI) absorbed by *Chlorella vulgaris*. Another reason behind this bio-absorption may be due to the high storing capacity of the protoplasm.

2.5 Conclusion

Although, chromium is pervasive metal in the environment and Cr(VI) is reported as toxic with several carcinogenic, mutagenic, and a few more hazards, which are affected to behavioral, physiological, biochemical, and immunological aspects. Although bacteria, fungi, and other algal forms are able to convert the hexavalent chromium to trivalent chromium (non-toxic form) but in addition to the common mechanism found in bacteria and all other microbes, the microalgae uses some special mechanism like the residues of flavonoids and the electrons release during photosynthesis for the conversion of hexavalent chromium to trivalent form. Microalgae are potential candidate for the detoxification of Cr(VI), which would be used for the treatment of chromium contaminated water and soil in an eco-friendly manner.

References

1. Hayat, S., Khalique, G., Irfan, M., Wani, A.S., Tripathi, B.N., Ahmad, A., Physiological changes induced by chromium stress in plants: an overview. *Protoplasma*, 249, 599, 2012.
2. Choudhury, S., Role of Chromite Mineralisation in Orissa. *Orissa Rev.*, 57–60, 2006.
3. Mertz, W., Chromium occurance and its function in biological systems. *Physiol. Rev.*, 49, 165, 1969.
4. Anderson, A.J., Mayer, D.R., Mayer, F.K., Heavy metal toxicities: levels of nickel, cobalt and chromium in the soil and plants associated with visual symptoms and variation in growth of an oat crop. *Aust J Agric Pest.*, 89, 47, 1972.
5. Zhitkovich, A., Voitkun, V., Costa, M., Formation of the amino acid-DNA complexes by hexavalent and trivalent chromium *in vitro:* Importance of trivalent chromium and the phosphate group. *Biochem.*, 35, 7275, 1996.
6. Costa, M., Toxicity and carcinogenicity of Cr(VI) in animal models and humans. *Crit. Rev. Toxicol.*, 27, 431, 1997.
7. Jordão, C.P., Pereira, J.L., Jham, G.N., Chromium contamination in sediment, vegetation and fish caused by tanneries in the State of Minas Gerais, Brazil. *Sci. Total Environ.*, 207, 1, 1997.
8. Khasim, D.I., Kumar, N.V., Hussain, R.C., Environmental contamination of chromium in agricultural and animal products near a chromate industry. *Bull. Environ. Contam. Toxicol.*, 43, 742, 1989.
9. Dhakate, R. and Singh, V.S., Heavy metal contamination in groundwater due to mining activities in Sukinda valley, Orissa - A case study. *J. Geogr. Reg. Plann.*, 1, 58, 2008.
10. Nigam, H., Das, M., Chauhan, S., Pandey, P., Swati, P., Yadav, M., Tiwari, A., Effect of chromium generated by solid waste of tannery and microbial degradation of chromium to reduce its toxicity: A review. *Adv. Appl. Sci. Res.*, 6, 129, 2015.
11. Zeraatkar, A.K., Ahmadzadeh, H., Talebi, A.F., Moheimani, N.R., McHenry, M.P., Potential use of algae for heavy metal bioremediation, a critical review. *J. Environ. Manage.*, 181, 817, 2016.
12. Majhi, P. and Samantaray, S.M., Thermotolerant microalgal diversity in the chromium metal polluted sites of Sukinda mining area. *Int. J. Curr. Microbiol. Appl. Sci.*, 9, 1109, 2020.
13. Jobby, R., Jha, P., Yadav, A.K., Desai, N., Biosorption and biotransformation of hexavalent chromium [Cr(VI)]: A comprehensive review. *Chemosphere*, 207, 255, 2018.
14. Travieso, L., Canizarez, R.O., Borja, R., Benitez, F., Dominguez, A.R., Dupeyron, R., Valiente, V., Heavy metal removal by microalgae. *Bull. Environ. Contam. Toxicol.*, 62, 144, 1999.
15. Liu, K.J., Jiang, J., Shi, X., Gabrys, H., Walczak, T., Swartz, M., Low-frequency EPR study of chromium (V) formation from chromium(VI) in living plants. *Biochem. Biophys. Res. Commun.*, 206, 829, 1995.

16. Cervantes, C., Garc, J.C., Devars, S., Corona, F.G., Tavera, H.L., Torres-Guzma, J.C., Moreno-Sainchez, R., Interactions of chromium with microorganisms and plants. *FEMS Microbiol. Rev.*, 25, 335, 2001.
17. Corradi, M.G., Gorbi, G., Ricci, A., Torelli, A., Bassi, A.M., Chromium-induced sexual reproduction gives rise to a Cr-tolerant progeny in *Scenedesmus acutus. Ecotoxicol. Environ. Safety*, 32, 12, 1995.
18. Wong, P.T. and Trevors, J.T., Chromium toxicity to algae and bacteria, in: *Chromium in the Natural and Human Environments*, NriaguJO, Nieboer E (eds), vol. 305, John Wiley and Sons, Wiley, New York, 1988.
19. Viamajala, S., Peyton, B.M., Apel, W.A., Petersen, J.N., Chromate reduction in Shewanellaoneidensis MR-1 is an inducible process associated with anaerobic growth. *Biotechnol. Prog.*, 18, 290, 2002.
20. Ahemad, M., Enhancing phytoremediation of chromium stressed soils through plant growth promoting bacteria. *J. Genet. Eng. Biotechnol.*, 13, 51, 2015.
21. Cervantes, C., Garcia, J.C., Nies, D., Silver, S., *Molecular Microbiology of heavy metals*, vol. 407, Springer-Verlag, Berlin, 2007.
22. Subrahmanyam, D., Effects of chromium toxicity on leaf photosynthetic characteristics and oxidative changes in wheat (*Triticum aestivum* L.). *Photosynthetica*, 46, 339, 2008.
23. Vajpayee, P., Tripathi, R.D., Rai, U.N., Ali, M.B., Singh, S.N., Chromium (VI) accumulation reduces chlorophyll biosynthesis, nitrate reductase activity and protein content in Nymphaea alba. *Chemosphere*, 41, 1075, 2000.
24. Hörcsik, Z., Kovács, L., Láposi, R., Mészáros, I., Lakatos, G., Garab, G., Effect of chromium on photosystem 2 in the unicellular green alga, *Chlorella pyrenoidosa. Photosynthetica*, 45, 65, 2007.
25. Clijsters, H. and Van Assche, F., Inhibition of photosynthesis by heavy metals. *Photosynth. Res.*, 7, 31, 1985.
26. Bishnoi, N.R., Dua, A., Gupta, V.K., Sawhney, S.K., Effect of chromium on seed germination, seedling growth and yield of peas. *Agri. Eco. Environ.*, 47, 47, 1993.
27. Shankar, A.K., Cervantes, C., Loza-Tavera, H., Avudainayagam, S., Chromium toxicity in plants. *Environ. int.*, 1, 739, 2005.
28. Montes-Holguin, M.O., Peralta Videa, J.R., Meitzner, G., Martinez-Martinez, A., de la Rosa, G., Castillo-Michel, H.A., Gardea-Torresdey, J.L., Biochemical and spectroscopic studies of the response of Convolvulus arvensis L. @ to chromium (III) and chromium (VI) stress. *Environ. Toxicol. Chem.*, 25, 220, 2006.
29. Sinha, S., Saxena, R., Singh, S., Chromium induced lipid peroxidation in the plants of *Pistia stratiotes* L. role of antioxidants and antioxidant enzymes. *Chemosphere*, 58, 595, 2005.
30. Schmfger, M.E.V., *Phytochelatins: complexation of metals and metalloids, studies on the phytochelatin synthase*, PhD Thesis, Munich University of Technology (TUM), Munich, 2001.

31. Boonyapookana, B., Upatham, E.S., Kruatrachue, M., Pokethitiyook, P., Singhakaew, S., Phytoaccumulation and phytotoxicity of cadmium and chromium in duckweed *Wolffia globosa. Int. J. Phytorem.*, 4, 87, 2002.
32. Shanker, A.K., Djanaguiraman, M., Pathmanabhan, G., Sudhagar, R., Avudainayagam, S., Uptake and phytoaccumulation of chromium by selected tree species. *Proceedings of the international conference on water and environment held in Bhopal, India*, 2003.
33. Nagajyoti, P.C., Lee, K.D., Sreekanth, T.V.M., Heavy metals, occurrence and toxicity for plants: a review. *Environ Chem Lett.*, 8, 199, 2010.
34. Holmes, A.L., Wise, S.S., Wise, J.P., Carcinogenicity of hexavalent chromium. *Indian J. Med. Res.*, 128, 353, 2008.
35. Potential Toxic Effects of Chromium, Chromite Mining and Ferrochrome Production: A Literature Review, Mining Watch Canada, 2012.
36. Lindberg, E. and Hedenstierna, G., Chrome plating: symptoms, findings in the upper airways and effects on lung function. *Arch. Environ. Health*, 38, 367, 1983.
37. Barceloux, D.G., Chromium. *Clin. Toxicol.*, 37, 173, 1999.
38. Shmitova, L., The course of pregnancy in women engaged in the production of chromium and its compounds. *Sverdlovsk*, 19, 108, 1978.
39. Shmitova, L., Content of hexavalent chromium in the biological substrates of pregnant women and women in the immediate postnatal period engaged in the manufacture of chromium compounds. *Gig. Trud. Prof. Zabol.*, 2, 32, 1980.
40. Zhang, L., Zhao, L., Yu, Y., Chen, C., Removal of Pb^{2+} from aqueous solution by non-living *Rhizopus nigricans. J. Water Res.*, 32, 1437, 1998.
41. EPA US, *Integrated Risk Information System (IRIS) [online electronic data file]*, U.S. Environment Protection Agency, Office of research and development, National Center for Environment Assessment. U.S., Last updated December 1, 1996; Available at:http://toxnet.nim.nih.gov 1996, accessed in September, 2009.
42. Danadevi, K., Rozati, R., Reddy, P.P., Grover, P., Semen quality of Indian welders occupationally exposed to nickel and chromium. *Reprod. Toxicol.*, 17, 451, 2003.
43. Junaid, M., Murthy, R.C., Saxena, D.K., Embryo- and fetotoxicity of chromium in pregestationally exposed rats. *Bull. Environ. Contam. Toxicol.*, 57, 327, 1996.
44. Subramanian, S., Rajendiran, G., Sekhar, P., Gowri, C., Govindarajulu, P., Aruldhas, M.M., Reproductive toxicity of chromium in adult bonnet monkeys (*Macaca radiata* Geoffrey). Reversible oxidative stress in the semen. *Toxicol. Appl. Pharmacol.*, 215, 237, 2006.
45. Aruldhas, M.M., Subramanian, S., Sekar, P., Hasan, G.C., Govindarajulu, P., Akbarsha, M.A., Microcanalization in the epididymis to overcome ductal obstruction caused by chronic exposure to chromium – a study in the mature bonnet monkey (*Macaca radiata* Geoffroy). *Reproduction*, 128, 127, 2004.

46. Gupta, V.K. and Rastogi, A., Equilibrium and kinetic modelling of cadmium [II] biosorption by nonliving algal biomass *Oedogonium* sp. from aqueous phase. *J. Hazard. Mater.*, 153, 759, 396–402, 2008a.
47. Gupta, V.K. and Rastogi, A., Biosorption of lead from aqueous solutions by green algae *Spirogyra* sp.: kinetics and equilibrium studies. *J. Hazard. Mater.*, 152, 407, 2008b.
48. Elhaddad, E. and Mahmoud, A.M.A., Isotherm studies of removal of Cr (III) and Ni (II) by *Spirulina* algae. *Int. J. Adv. Pharm. Biol. Chem.*, 4, 247–251, 2015.
49. Khuntia, S., Gorgai, M.K., Das, C., Golder, A., Effect of background ions on reduction of Cr(VI) to Cr(III) using saline water algae. *Res. J. Chem. Environ.*, 15, 450, 2011.
50. Katircioğlu, H., Aslim, B., Tunçeli, A., Chromium(VI) Biosorption from Aqueous Solutions by Free and Immobilized Biomass of *Oscillatoria* sp. H1 Isolated from Freshwater. *ISIJ Int.*, 52, 1173, 2012.
51. Giloni-lima, D., Delello, M.L.M., Cremonez, M.N., Éler, V.A.L., Assessment of chromium bioaccumulation in Pseudokirchneriellasubcapitata (Korshikov) Hindak by the Central Composite Design (CCD) and Response Surface Methodology (RSM). *J. Braz. Soc. Ecotoxicol.*, 7, 43, 2012.
52. Yaqub, A., Mughal, M.S., Adnan, A., Khan, W.A., Anjum, K.M., Biosorption of hexavalent chromium by *Spirogyra* sp.: equilibrium, kinetics and thermodynamics. *J. Anim. Plant Sci.*, 22, 408, 2012.
53. Chatterjee, A. and Abraham, J., Biosorption Capacity of Dried *Spirogyra* on Heavy Metals. *Int. J. Chemtech. Res.*, 8, 382, 2015.
54. Popuri, S.R., Kalyani, S., Kachireddy, R., Krishnaiah, A., Biosorption of hexavalent chromium from aqueous solution by using prawn pond algae (*Sphaeroplea*). *Indian J. Chem. Technol.*, 46, 284, 2006.
55. Anjana, K., Kaushik, A., Kiran, B., Nisha, R., Biosorption of Cr(VI) by immobilized biomass of two indigenous strains of cyanobacteria isolated from metal contaminated soil. *J. Hazard. Mater.*, 148, 383, 2007.
56. Sbihi, K., Lange-Bertalot, C.O., Bertrand, M., Toxicity and biosorption of chromium from aqueous solutions by the diatom *Planothidium lanceolatum* (Brébisson). *Am. J. Sci. Ind. Res.*, 3, 27, 2012.
57. Dönmez, G.C., Aksu, Z., Ozturk, A., Kutsal, T., A comparative study on heavy metal biosorption characteristics of some algae. *Process Biochem.*, 34, 885, 1999.
58. Prasanna, R., Jaiswal, P., Kaushik, B.D., Cyanobacteria as potential options for environmental sustainability – promises and challenges. *Indian J. Microbiol.*, 48, 89, 2008.
59. Malik, A., Metal bioremediation through growing cells. *Environ. Int.*, 30, 261, 2004.
60. Nacorda, J.O., Martinez- Goss, M.R., Torreta, N.K., Bioremoval and Bioreduction of Chromium (VI) by the Green Microalgae, *Chlorella vulgaris* Beij., isolated from Laguna de Bay, Philippines. *Philipp. J. Sci.*, 139, 181, 2010.

3

Biodetoxification of Heavy Metals Using Biofilm Bacteria

Adyasa Barik[1], Debasish Biswal[2], A. Arun[1] and Vellaisamy Balasubramanian[1]*

[1]Department of Microbiology, Alagappa University, Karaikudi, Tamil Nadu, India
[2]Department of Biotechnology, Ravenshaw University, Cuttack, Odisha, India

Abstract

Heavy metal wastes are produced from various sources including anthropogenic and industrial activities. These metals create severe problem to our environment and cause different diseases in human such as cancer, skin lesions, birth defect, cerebral and bodily retention, disability to gain knowledge, and malfunction of liver and kidney. Therefore, heavy metals detoxification is a big challenge for researchers. Strategies have been employed to exploit the biofilm bacteria for detoxification of heavy metal. The drastic growth of biofilm bacteria occurs in polluted water environment through accumulating heavy metals. It is resistance to heavy metal through extra polymeric substances (EPSs) play a major role in detoxification of heavy metal. Polysaccharides, uronic acid, and sugar have functional group such as carboxylic acid and amino acid groups. These are the chemical composition of EPS. These functional groups could be acidic and retain the ability to bind or detoxify the heavy metal ions. The proteinaceous part of EPS plays an important role in complexation of metal ions. Several studies demonstrated that, the metal resistance genes (MRGs) and antibiotic resistance genes (ARGs) co-occur in bacteria isolated from water bodies polluted with heavy metal wastes. These kinds of studies give a little clue about the heavy metal resistance potential of antibiotic resistance strains. The stability and structure of biofilm together with diverse range of arrays will have more number of unexplored metabolic characteristics features of biofilm bacterial community's toward the biofilm-mediated detoxification of heavy metal.

Keywords: Heavy metals, detoxification, biofilm bacteria, EPS, polysaccharides, functional groups, acidic, metal ions, MRGs, ARGs

**Corresponding author*: microbaalu8211@gmail.com

Bibhuti Bhusan Mishra, Suraja Kumar Nayak, Swati Mohapatra, and Deviprasad Samantaray (eds.) Environmental and Agricultural Microbiology: Applications for Sustainability, (39–62) © 2021 Scrivener Publishing LLC

3.1 Introduction

In current days, industrial, domestic, agriculture, and anthropogenic activity increase due to increasing population and demand of society. Therefore, new pollutants are present in different sources. Some of them are not degradable or take several years for degradation like xenobiotic and heavy metals. They are present in many forms in soil and water, enter in food chain of animal, plant, and human, and create diseases and several physical and physiological disorders. In recent days, heavy metal waste production is more from anthropogenic and industrial sources and these are creating serious problem to our surrounding and causing several diseases like cancer, skin lesions, birth defect, cerebral and bodily retention, gain knowledge disability, and malfunction of liver and kidney [1]. Therefore, heavy metal pollution has become a headache to our society. The metals and metalloids are heavy metals and biologically classified into two categories: essential [manganese (Mn), zinc (Zn), cobalt (Co), copper (Cu), and chromium (Cr)] and nonessential [cadmium (Cd), lead (Pb), and mercury (Hg)] elements [2]. The low concentration of essential heavy metal is required for animal, plant, and human nutrition, and non-essential element is generally known as toxic element for living beings [3, 4]. The treatment of toxic elements by biological process is better than physical and chemical process because of cost effectiveness and environmental compatibility. The potential of biofilm communities for biodetoxification methods has currently been realized [5]. The biodetoxification of heavy metals can be possible by immobilization, volatilization, concentration, and separating to an environmental part, thereby reducing estimated vulnerabilities [6, 7]. Development of biofilm of microbes and formation of their Extracellular Polymeric Substance (EPS) are commonly linked with resistance, ability to tolerate, and bioremediation of metal [8]. EPS of microbes has vital importance on development of biofilm and cell mass that gives safeguard to cells against antagonistic atmosphere and can tie substantial amount of heavy metals [8]. Biofilm EPS has high resistance capacity to entrapment of metal precipitate like cupper reducing bacteria, sulfur reducing bacteria, and another some bacteria. In polluted water, growth of biofilm is easy and accumulates heavy metals and resistance to heavy metal [9]. Polysaccharides, uronic acid, sugar, and proteins have functional groups such as carboxylic acid and amino acid groups, which are the composition of EPS, and these functional groups could be acidic and have ability to bind metal ions [1, 9]. In some studies, the polysaccharide part of EPS is essential fraction for metal removal. Some other authors stated that the proteinaceous part of EPS plays an important role in complexation of metal ions [10]. Several studies demonstrated that the metal resistance genes (MRGs) and antibiotic resistance genes (ARGs) co-occur in bacteria isolated from water bodies polluted with heavy metal wastes. These kinds of studies give a little clue about the heavy metal resistance potential of antibiotic resistance strains [11]. Metal transformation by microorganisms serves various biological functions. Anaerobic respiration of microbes reduces metals, causing in detoxification, and the reduced forms are less toxic and little soluble as well [12]. Some other methods like biosorption, bioleaching, and precipitation are found to be very efficient for detoxification [12–14]. The application of mono- or multi-species of biofilm gives the microorganisms a best existence slot and their metabolic abilities also increases in presence of high amount of lethal compounds [1]. Indeed, microorganisms can simply remediate polluted water in several water bodies and waste streams by removing metals, separating metals in soil and sediments by different processes including enzymatic actions [15]. Moreover, using bacteria over other microorganisms helps in reducing other contaminants present in waste materials. Therefore, biodetoxification of heavy metal by using biofilm bacteria is a more efficient process, eco-friendly, cost effective, and possess no side effect to living beings. Generally, heavy metals present everywhere such as air, water, soil, and sediment. In this chapter, we discuss about some biofilm bacteria and their role in detoxification of heavy metals.

3.2 Source and Toxicity of Heavy Metal Pollution

The multiple applications of heavy metals in industrial, domestic, agricultural, medical, and technology sectors is the main reason for their wide spreading in environment [16]. Generally, the heavy metals exist all over the earth surface. The social contact to environment results anthropogenic activities like mining and smelting operation, industrial manufacture and application, and metal and metal containing compound application in domestic and agriculture field [17, 18]. Sometimes, natural incidence like volcanic eruptions on land as well as on the ocean beds are reported to be responsible for heavy metal pollution in soil and water bodies [16]. Industrial sources have a large contribution toward heavy metal pollution from activities including metal melting out in processing plants, coal flaming in power plants, incineration of petroleum products, nuclear power stations and high-tension lines, textiles, plastics, wood conservation, microelectronics, and paper processing plant [19, 20]. The wastage from livestock systems can disturb the micro- and macro-environment such as water, soil, and food chain [21]. The metals' presence in water reduces their quality and causes human disease, even the essential metals at high concentration gives negative effect and toxicity [21]. The metals and metalloids are common pollutant in waste water [22]. Soil accumulate heavy metals and metalloids by production from quickly growing industrial areas, mine tailings, high metal waste disposal, leaded gasoline and paint, fertilizers applied in land, animal manures, sewage sludge, pesticides in agriculture, coal incineration deposits, petrochemical spillage, and atmospheric deposition [23]. Heavy metals enter to ecosystem and hence human through direct contact with contaminated soil, food chain, and drinking of contaminated ground water. It causes significant reduction in food quality by phytotoxicity, decrease the quality, and hence the fertility of land used for cultivation purpose affecting food safety and land occupation difficulties [23]. Metal ions combine with biological factors such as DNA and nuclear protein result in deterioration of DNA and conformational change which may indicates to variation of cell cycle, carcinogenesis, or apoptosis [16]. The nonessential heavy metals have direct or indirect negative effect on human from tissue level to organ system and from nucleic acid to physiology level.

3.2.1 Non-Essential Heavy Metals

The toxicity and carcinogenicity potential of some frequently present nonessential heavy metals like mercury, chromium, lead, arsenic, and cadmium are described in this section.

3.2.1.1 Arsenic

Arsenic present in periodic table of period 4 and group VA in metalloid state. The inorganic form includes trivalent arsenite (AsIII) and pentavalent arsenate (AsV) and methylated metabolites are organic form of arsenic, e.g., monomethylarsonic acid (MMA), dimethylarsinic acid (DMA), and trimethylarsine (TMA) oxide. Atmospheric pollutions occur through arsenic due to volcanic eruption, soil erosion, and anthropogenic activities [24]. In ores, arsenic generally exists in powdery amorphous and crystalline forms. It enters in to the environment through withstand of rocks, mining and smelting methods, pesticide practice in agriculture, and coal ignition. It causes ground water as well as surface water contamination and exists as arsenate (AsV) and arsenite (AsIII) in maximum groundwater. Its high concentrations in drinking water create toxic effect to animal and human [22].

Industrial product in agricultural application such as herbicides, insecticides, sheep dip, dye-stuffs, preservatives of wood and algicides contains arsenic compounds or components. Arsenic is also used in veterinary medicines and medical treatments in drugs to treat syphilis, yaws, amoebic dysentery, and *Trypanosomiasis* like diseases [25, 26]. Arsenic compound also create genotoxicity by inhibiting DNA repair, promote chromosomal aberration, exchanges of sister chromatid, and micronuclei development in both rodent and human cells [27, 28].

3.2.1.2 Cadmium

Cadmium is a highly toxic and nonessential heavy metal for environment. Moderate concentration of cadmium (around 0.1/kg) is commonly found in the soil crust. The maximum amount of cadmium compounds are accumulated in sedimentary rock and phosphates of marine (contain nearly 15 mg/kg) [29] and naturally released to environment by abrasion of rocks and soil, forest fires, and volcanic eruption. The anthropogenic activities are also responsible for cadmium pollution such as metal plating, metallurgical alloying, ceramics, mining, and other industrial operations. It is used as a protecting guard on alloys and steel, in paints and plastic solder and braces color and nickel-cadmium rechargeable batteries in stabilizer. It is also present in fungicides, super phosphate fertilizer, cigarette, and ash. Anthropogenically, their concentrations arise mainly by minerals used in agriculture and industries [30]. The humans are exposed to cadmium via the rout of inhalation or smoking of cigarette, ingestion contaminated food, working in cadmium contaminated place, but skin absorption is rare and smoking is the main influencer [31, 32]. Cadmium is moreover exist in trace amount in particular foods like green and leafy vegetables, potatoes, seeds, grains, mushrooms, and in some sea foods such as kidney and liver of mollusks and crustaceans, shellfish, mussels, cocoa powder, and dried seaweeds [16]. Cadmium causes severe health problems such as erosion in gastrointestinal tract and internal damage in pulmonary, hepatic, or renal systems, depending on the rout of contamination [33, 34]. Cadmium is highly carcinogenic but mainly it causes pulmonary cancer and other parts such as adrenals, testes, and the hematopoietic system [35].

3.2.1.3 Chromium

Chromium occupies a position in the first row of d-block in the periodic table and is a transition metal of group VIB. It does not exist in elemental form, so it forms compound and is less commonly available element [23].

Naturally, it is present in the earth with oxidation state ranging from chromium (II) to chromium (VI) [36]. The trivalent form of chromium compound [Cr(III)] is stable and accumulated in ores such as ferro-chromite. Hexavalent [Cr(VI)] is another form of chromium compound and is second most stable state form [37]. Chromium entered in different environments (air, water, and soil) through the release of waste from industrial and other anthropogenic activities. The different chromium industries such as metal melting out, tannery services, chromate manufacture, stainless steel repairing, and ferrochrome and chrome pigment manufacture industries are mainly responsible for chromium contamination in the environment [16]. Chromium is also used in paper, pulp, and rubber manufacturing applications [22]. The hexavalent form of chromium [Cr(IV)] is the toxic compound from industrial pollutant which is classified as human carcinogen by various regulatory and non-regulatory agencies [16]. For drinking water, World Health Organization (WHO) restricted 50 μg of Cr(VI) per liter, but currently, naturally Cr(VI) have been found above the WHO's limit in ground and surface water [38]. The toxicity of chromium causes liver and kidney damage and skin ulceration and affects the central nervous system, and it is also connected with the effects on hematological problem and immune response in fresh water fishes. Its toxicity also associated with plant species, because it decreases the rate of photosynthesis [22].

3.2.1.4 Lead

Lead occupied sixth period and group IV in periodic table [23]. It is a metal generally occurring gray-bluish in color and found as a mineral form binding with other element like sulfur (PbS) or oxygen ($PbCO_3$) and present in small amount in earth crust [16, 23]. It has occupied fifth rank in the production of heavy metals from industry and commonly used in lead storage batteries, solders, bearings, cable covers, ammunition, plumbing, pigments, and caulking [23]. It is also used in many different industrial, agricultural, and domestic applications [16]. Lead exposure to humans and animals occurs through lead contaminated dust particle and consumption of lead contaminated food stuffs and water [39, 40]. Adult peoples are taking 35% to 50% lead particles by drinking water but children are taking more than 50% [41]. In children, it causes toxicity of blood, deficient brainpower, poorer intelligence quotient-IQ, late or diminished growth development, neurobehavioral deformities, reduced auditory perception, speaking and wording disabilities, and unsocial and inattentive activities. In adults, it affects reproduction, such as, in men, reduces sperm count, and in women, continuous miscarriage have been reported due to long time exposure to lead contamination [16]. Sever expose to lead causes brain injury, malfunction of kidney, and different diseases in gastrointestinal track, while the prolonged expose may cause damage to blood, blood pressure, antagonistic effect on central nervous system, damage of kidneys, and trouble in metabolism of vitamin D [16, 39, 40, 42].

3.2.1.5 Mercury

In periodic table, mercury belongs to the transition element series. It occurs in three forms in the nature as elemental, inorganic, and organic form with individual toxicity character [43]. Generally, it exists in liquid form [23]. It can also exist as a cation through oxidation state of +1 (mercurous) or +2 (mercuric) [44]. It is used in electrical industry, dental amalgams, and in various industrial methods containing the manufacture of caustic soda, antifungal agent, manufacture of nuclear reactors, usage as a solvent for reactive and expensive metal, as a preservative of pharmaceutical products, etc. [45]. Mercury is released and contaminates the environment from combustion of coal, manometers at gas pressure measuring stations, and gas/oil pipelines as well. It exists in mercuric (Hg^{2+}), mercurous (Hg_2^{+2}), elemental (HgO), or alkylated (methyl/ethyl mercury) form [23]. Human and animals expose to mercury and other chemicals by calamities, atmospheric pollution, contamination of food, dental repair, precautionary medical applications, farming, and industrial processes [17, 30]. Mercury entered to water by natural procedure of gassing from ground of earth and by industrial pollutants and accumulates in fish and tiny organisms inhabiting in the water bodies. Due to its lipid soluble nature, it can easily cross placenta and blood brain barrier. By eating methyl mercury affected fish, it enters to gastrointestinal tract and also affects kidney, neurological tissue and liver of human and it causes gastrointestinal toxicity, neurotoxicity, and nephrotoxicity [16].

3.2.2 Essential Heavy Metals

The essential heavy metals such as nickel, zinc, and copper are require for biological metabolism but in high concentration they show toxic and harmful effects and on living organisms.

3.2.2.1 Copper

Copper (Cu) is a transition element, occupied a place in group IB of period 4 in periodic table and ranked as third highest used metal in the world [23, 46].

It is a fundamental micronutrient for animal and plant development. In human, it helps for production of blood hemoglobin and in plant, Cu is essential for seed germination, resistant to disease and water regulation [23]. Spontaneously high amount of exposure of cupper dust causes eye, nose, and mouth irritation and often causes nausea and diarrhea [22]. Continuously exposure to high doses of Cu causes anemia, malfunction of liver, kidney, and impatience in stomach and intestine [23]. Generally, copper (Cu) is present in drinking water because flow pipes are made up of copper to control algal growth [23]. Mining, metallurgy, and industrial applications are the important causes of copper contamination in the environment [22].

3.2.2.2 Zinc

Zinc occupied a place in group IIB of periodic table. It is a transition element normally present in soil approximately 70 mg kg^{-1} in ground rocks [47]. It is an essential micro nutrient in our diet but higher concentration is toxic and may cause anemia and cholesterol complications in human beings and nausea and vomiting in children [22]. Currently, Zn concentration is found to be rising because of increasing anthropogenic activities. Mostly, industrial sectors and other human activities like mining, incineration of coal and waste, and steel dispensation are major causes for increase in Zn concentration in the environment. Other sources of Zn contamination are crops or drinking water stored in metal chambers [23].

3.2.2.3 Nickel

Nickel is the transition element that is present in environment only at very less amount and small doses are require for biological systems, but it can be hazardous when the concentration exceeds the permissible limit [23]. The higher concentration of Ni causes different types of cancer in various parts of animal body, mostly of those staying nearby industries [48]. It also causes damage to cells, reduces body weight, and damages the liver and heart [22]. Nickel is utilized in the steel industries, nickel-cadmium battery industries on a large scale, and in other metallic products [22, 48]. It is also found in paint formulation and cigarettes. The industrial wastes containing nickel enter into the water bodies contaminate the water and affect aquatic living organisms [49]. Earlier studies reported that microorganisms have problems for their growth and development in the presence of nickel. But, some studies in current days reported that some microorganisms have developed resistance to nickel [23].

3.3 Biofilm Bacteria

Biofilms are communities of one or more species of microorganisms living within the protection of an extracellular matrix composed of polysaccharides, proteins, DNA, and other molecules, collectively termed as the extracellular polymeric substances (EPS) [12, 19, 22].

Microbial EPS is crucial for the formation of biofilm and cell aggregates, which contribute to protect cells from hostile environments and can bind significant amounts of heavy metals [53–56]. Biofilm and planktonic cells have distinct heavy metal and metalloid susceptibility [57–59]. It is suggested that the complexation or sequestration of heavy metals and retarding their diffusion in to the biofilm may be responsible for protecting cells from heavy metal toxicity [58]. Microbial EPS are also of particular interest and relevance to the bioremediation process due to their involvement in flocculation and binding of heavy metals from solutions [53, 60–62].

3.4 Interaction of Metal and Biofilm Bacteria

The availability of heavy metal ions is the hazardous factor for environment. Availability of metal ions in different components of environment like water, soil, microorganisms, aquatic lives, and other forms of lives are dependent on several factors like industrial activity, natural sources like volcanic eruption, and unlimited anthropogenic activities. Again, its presence, concentration and effects are influenced by several environmental factors and circumstances like pH, alkalinity, redox potential, and action of microorganisms. Earlier, it was thought that metals have toxic effect only on microbial metabolic process or mechanism. But later, it is discovered that they are not only lethal but their existence can persuade different mechanism of metal resistance in microorganisms. Microbes and metal ion communication can take place through different mechanisms. These mechanisms are classified depending on the pathway of communication of metal ions with microbes such as active and passive uptake of metal ions [63]. The biological interaction of microbes and metal ions transform the ions from toxic to less toxic or few accessible forms or arrest metal ions to inhibit their opening into bioprocess (Figure 3.1). The various interactive mechanisms are available such as biosorption, bioleaching, biovolatilization, bioimmobilization, and bioaccumulation. The EPS of biofilm bacterial cell also interact with metal ions, due to communication among positive charge metal ions and negative charge EPS of cell surface [64].

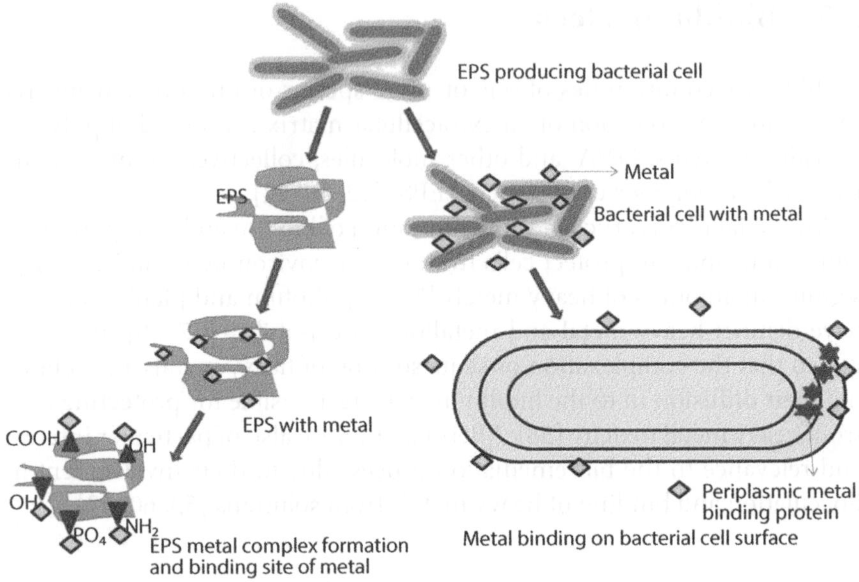

Figure 3.1 Interaction of metal with EPs and binding on bacterial cell surface.

3.5 Biodetoxification Mechanisms

The accumulation of heavy metals in food chain and their toxicity affects to biological system creates various problems. These can also enter to water bodies and contaminate soil through agricultural extract, industrial wastes, domestic runoff, and other commercial activities. We can eliminate or reduce heavy metal from contaminated sources. Therefore, there are different types of detoxification technology that have been utilized to eliminate heavy metals from contaminated sources. These detoxification technologies are briefly described as follows:

- Biosorption
- Bioleaching
- Biovolatilization
- Bioimmobilization

3.5.1 Biosorption

The capability of biological materials to accumulate or bind heavy metals present in the wastewater of polluted water bodies through metabolically facilitated or physico-chemical pathways is called biosorption.

All microorganisms (algae, bacteria, fungi, and yeast) are proved to be potential metal biosorbents. This method of treatment is having some advantages such as cost effectiveness, highly efficient and effective, reduction chemical and biological slurry, no extra nutrient necessity, revival of biosorbent, and probability of metal reclamation [65]. The sorption of metal can take place by microorganisms following two different processes: active process and passive process [66].

Active process: This process is metabolism dependent and also called as bioaccumulation process. In this process, transport of metal through the membrane of cell with a subsequent accumulation of intracellular metal facilitated through metabolism of cell. Only viable cells can perform bioaccumulation, which are also often linked with a mechanism of resistance initiated through microorganism in the existence of a toxic metal [66].

Passive process: This process is a metabolism-independent process, otherwise known as biosorption. This is a physic-chemical process, normally includes four mechanism (adsorption, ion-exchange, complexation, and precipitation) and this mechanism helps to transport metal inside the cell [66].

Adsorption: The adsorption occurs with the help of van der Waals' force [52, 64]. The selective materials for adsorption of Cr(VI) and Ni(II) are "crushed initiated carbon > bagasse > fly ash" and "crushed initiated carbon >fly ash >bagasse", respectively. The lower pH of 6.0 is suitable for removal of Cr(VI) and pH 8.0 is appropriate condition for removal of Ni(II) ions. The limitation of adsorption is that the ability is very low and their use for industrial runoffs treatment cannot be defensible [22].

Ion-exchange: In biosorption process, the ion-exchange method was first introduced by Volesky and Holan (1995) and is backing through numerous current studies [66]. In passive absorption, the ion exchange method has essential role. In this mechanism, the biomass is displayed toward metal because the first metal aliquots are continuously discharged in to the solution while the second metal is combined, and a portion of second metal is combined to the bio-sorbent. This assay is suitable for Cu^{2+} and Pb^{2+} removal [67].

Complexion: In this method, the complex formation on the cell surface after communication between metals and functional groups of microorganisms occurs for metal removal from the solution. The magnesium, copper, calcium, mercury, zinc, and cadmium accumulation via *Pseudomonas syringae* takes place and is removed by simply complexion mechanism. The organic acid may produce by microorganisms may chelate toxic metals, resulting in makeup of metallo-organic molecules [65].

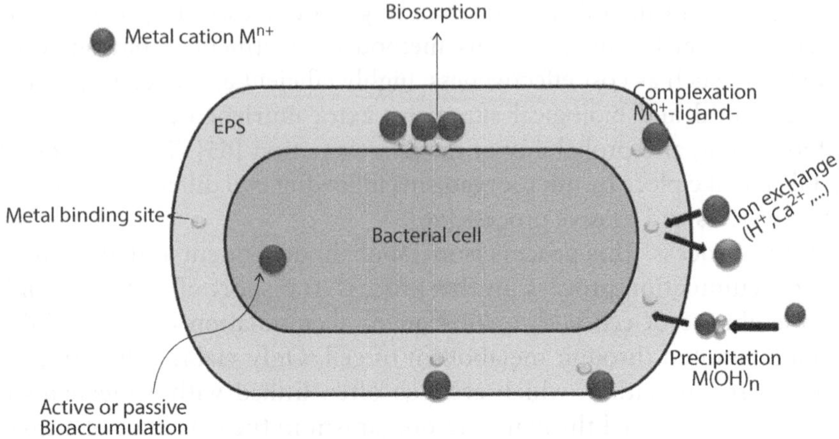

Figure 3.2 Mechanism of biosorption.

Precipitation: The precipitation may be dependent or autonomous to metabolism process. The metal elimination from solution is frequently linked by the functional defense system of microorganisms. The microorganisms behave in the occurrence of noxious metal generating compounds such as metallothionein and phytochelatins and induce the precipitation method. This method is not dependent upon the cellular metabolism; rather, it may be a chemical reaction significantly occurs among the metal and the cell surface, with an indigenous aggregation of ions of metal and their subsequent precipitation. The mechanism of biosorption is described overhead may occur concurrently (Figure 3.2) [65, 66].

3.5.2 Bioleaching

In bioleaching, metal cations are mobilized from almost insoluble ores by complexation and biological oxidation method. The application of microorganisms for recovery of heavy metals is currently a universal enhanced biotechnological method [68]. This technique is mainly accepted by mining industries in order to extract metals incorporate in low-grade sulfide ores. In this method, solid remains are discarded as waste material, while metals are transformed to solution phase. Presently, bioleaching method is used in several metal removal applications from ground water, sludge, soil, and sediments [69]. The biomining is a universal word for two techniques such as bioleaching and bio-oxidation techniques. There are two types of leaching method such as contact and non-contact. In non-contact leaching method, metals are mainly removed via planktonic bacteria that oxidize the surface mineral ions in solution. The Fe^{3+} ions arise from the bacteria and interact with the surface of mineral, where they decreased and oxidized the moiety of sulfide and release the Fe^{2+}. So, again Fe^{2+} ions pass into the cycle to continue the reaction again. In contact leaching method, the maximum cells adhere to the surface of sulfide minerals. This is an electrochemical method in which the suspension of sulfide minerals occur between the borders of bacterial cell and the sulfide mineral surface and this area is occupied by EPS [68].

The EPS plays an important role in this bioleaching process through an interfacial method, which takes place in natural environment. During the process of up-taking of heavy metal (Fe^{3+}) ions, uronic acids facilitate to produce exopolymers among cell wall and surface of metal sulfide. These complexes involved in electrostatic interaction to form primary attachment, oxidizing attacks to deterioration of metal sulfide and act as nutrient [70] (Figure 3.3). The chemo-lithotrophic bacteria such as *Acidothiobacillus* sp., *Leptosperillium* sp., and archaea are involved in metal recovery from sulfide minerals. These bacteria are gram negative, aerobic, and able to survive below pH 3.0 and temperature 25°C–35°C. Leaching bacteria are mostly belongs to proteobacteria (*Acidithiobacillus* sp., *Acidiphilium* sp., *Acidiferrobacter* sp., and *Ferrovum* sp.), *Nitrospirae* (*Leptospirillum* sp.), *Firmicutes* (*Alicyclobacillus* sp. and *Sulfobacillus* sp.), and Actinobacteria (*Ferrimicrobium* sp., *Acidimicrobium* sp., and *Ferrithrix* sp.) [68, 69]. But among all bioleaching bacteria, *Thiobacillus* sp. involved in solubilization of metal sulfide because it takes carbon

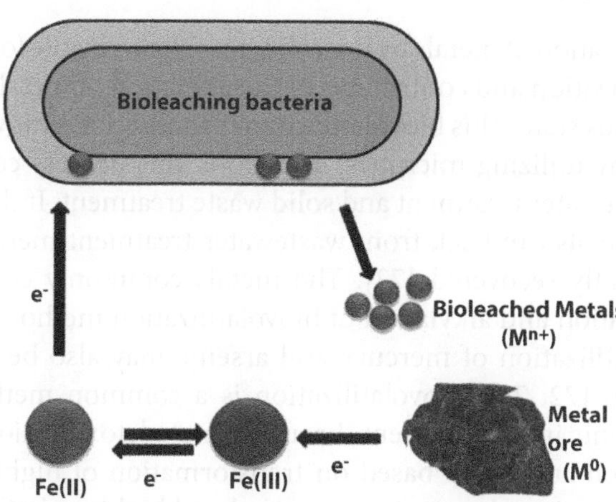

Figure 3.3 Mechanism of bioleaching.

dioxide from atmosphere for cellular synthesis, pulls their energy from oxidation of elemental sulfur, reduces sulfur compounds, and results in production of ferric ions and sulfuric acids, which are entangled in heavy metal extraction. The frequently used bacteria for bioleaching process are *Acidothiobacillus thiooxidans*, *Acidothiobacillus ferrooxidans*, and *Leptospirillum ferrooxidans*, and these are able to grow in high acidic condition (pH 1.5–3.0) [69].

The oxidation of metal sulfide by Fe/S oxidizing bacteria is defined through two distinct pathways such as polysulfide and thiosulfate pathway [68, 69]. These mechanisms depend on metal sulfide reactivity with protons (acid solubility) [69]. In case of thiosulfate pathway, metals are acid-insoluble such as pyrite (FeS_2), molybdenite (MoS_2), and tungstenite (WS_2), and Fe^{3+} ions occur through metal sulfide extraction. This reaction results the production of metal cations (M^+) and thiosulfate that oxidizes to sulfuric acid. The production of sulfuric acid creates acidic condition so *T. ferrooxidans* and *L. ferroxidans* catalyze Fe^{3+} ions for recycling. In case of polysulfide pathway, metals are acid soluble such as sphalerite (ZnS), galena (PbS), arsenopyrite (FeAsS), chalcopyrite ($CuFeS_2$), and hauerite (MnS_2) through electron extraction by iron(III) ions and proton attack. In this mechanism, polysulfide is the main intermediate form and can be oxidized to sulfuric acid by using bacteria *A. ferroxidans* and *A. thiooxidans* [71]. In bioleaching process, maintenance of acidic condition is essential because the optimum action of Fe/S oxidizing bacteria and to retain metals constant in solution phase.

3.5.3 Biovolatilization

The transformation of metals by microbes into their volatile forms is known as bio-volatilization and contributes in the alteration of metal from soluble state to gaseous state. This biovolatilization process can remove metal from solid phase by utilizing microbes. Therefore, this process can be applied for both wastewater treatment and solid waste treatment. If the gas form of volatilized metals can trick from wastewater treatment method, they can be consequently recovered [72]. The metals commonly connected with their methylation and alkylation of biovolatilization method by microbes, whereas volatilization of mercury and arsenic may also be facilitated by their removal [72, 73]. Biovolatilization is a common method for mercury and arsenic in environment through which detoxification approaches applied on soil and water based on transformation of highly toxic compound to nontoxic or less toxic compound and highly volatile for removal of metals (Figure 3.4) [73].

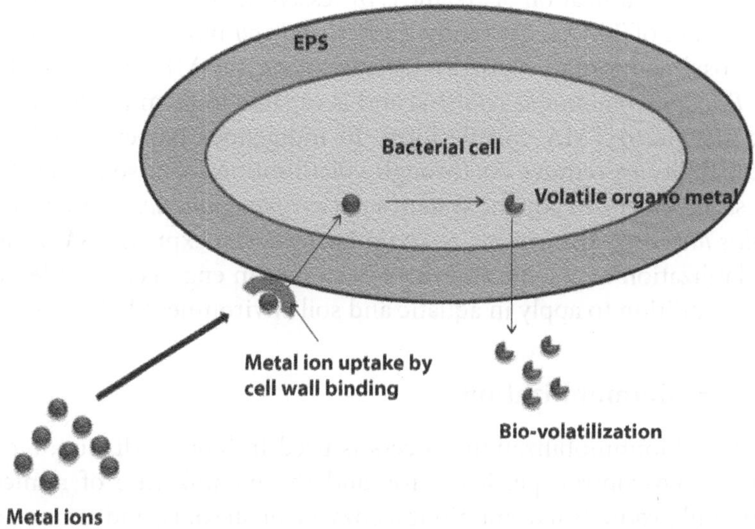

Figure 3.4 Mechanism of biovolatilization.

In contaminated environment, bacteria developed resistance resulting due to the aforesaid mechanism which further leads to mercury detoxification. The reductase enzyme (Mercury(II)reductase) of the bacteria causes a reduction of Hg^{2+} to nontoxic Hg^0, and hence, a diffusional loss of Hg^0 from bacterial cell takes place. The mercuric reductase coded by *merA* gene is important for reduction of inorganic Hg while cytosolic mercuric lyase enzymes coded by *merB* gene breaks the C-Hg bond of most organomercury [69]. Earlier studies reported that bacteria involved in this mechanism and resistance to Hg such as *Bacillus* sp., *Pseudomonas* sp., *Psychrobacter* sp., *Halmonas* sp., *Luteimonas* sp., and *Micrococcus* sp. are isolated from highly polluted area [74]. The elemental mercury is highly volatile and the gas phase needs some special treatment to immobilize it. The Hg^0 produced by volatilization and it is removed into gas phase by fast oxidative absorption process and recovered. This technique can be applied on soil, wastewater, and sediment [69].

The biovolatilization process also involve in arsenic removal from contaminated soil and water. In soil, arsenic could be converted into volatile byproducts and removed. Both aerobic and anaerobic microorganisms are involved in the evolution of volatile arsenicals. The volatilization of arsenic by microorganisms depends upon several factors like arsenic compound, concentration, and moisture of soil, organic materials, temperature, other similar components, growth of microbes, and ability of volatilization of arsenic. Biovolatilization of arsenic is by lessening of As(V) to As(III) with end product of TMAs. Currently, *Escherichia coli* have expressed arsenite S-adenosylmethionine methyltransferase gene (arsM), which is cloned from *Rhodopseudomonas palustris* and is capable to form methylate inorganic arsenic to TMA volatile form. In indigenous bacteria, *arsM* gene has capability to remove As through volatilization from soil. The strains express *arsM* gene in aquatic system such as *Sphingomonas desiccabilis* and *Bacillus idriensis*. The arsenic resistant bacteria can express *arsM* gene for biovolatilization of arsenic and these bacteria can engineered under laboratory condition to apply in aquatic and soil environment [73].

3.5.4 Bioimmobilization

Currently, bioimmobilization process is used in bioremediation, biodegradation, bio-control, pesticide use, and the manufacture of numerous compound products like antibiotics, enzyme or steroids, and amino acids. In this technique, metal can immobilized using microbial biomass by biosorption to cell walls or by extracellular substances and some common procedures are using for immobilization such as adsorption on exteriors, flocculation, cross connecting of cells, nanocoating, entrapment, covalent bonding to carriers, and encapsulation. The bacteria persuade immobilization mechanism to reduce the heavy metal concentration [69, 75]. The metabolism and intrinsic property of some bacteria associated with cell wall structure and the presence of extracellular polymeric substances are able to tolerate metal ions. Some other bacteria resist to metal by using resistance mechanisms such as active transport, efflux pump, intra- and extracellular sequestration, methylation, toxic chemical transfer to less toxic chemical through enzymatic transformation of redox reaction, and sensitivity reduction of cellular targets to metal (Figure 3.5) [76].

The heavy metals are reduced by using immobilization process. The Cr(VI) is reduced to Cr(III) by using both anaerobic and aerobic microorganisms. The presence of oxygen in aerobic condition and the reduction of Cr(VI) by microbes are generally catalyzed though soluble enzyme and lessening of Cr(VI) to Cr(III) by microbes as an eco-friendly method [77]. The bacterial strain such as *E. coli*, *Pseudomonas putida*, *Desulfovibrio* sp., *Bacillus* sp., *Shewanella* sp., *Arthobacter* sp., *Microbacterium* sp., and *Cellulomonas* sp., which reduce Cr(VI) isolated from contaminated area [78]. Arsenic compound used as an electron donor or accepter by microorganisms and possess the detoxification of arsenic, with pushes up to the membrane level of cells to eradicate As(III) from cells and metabolites of cell, finally As(V) removal arise [69]. Anaerobic bacteria are capable

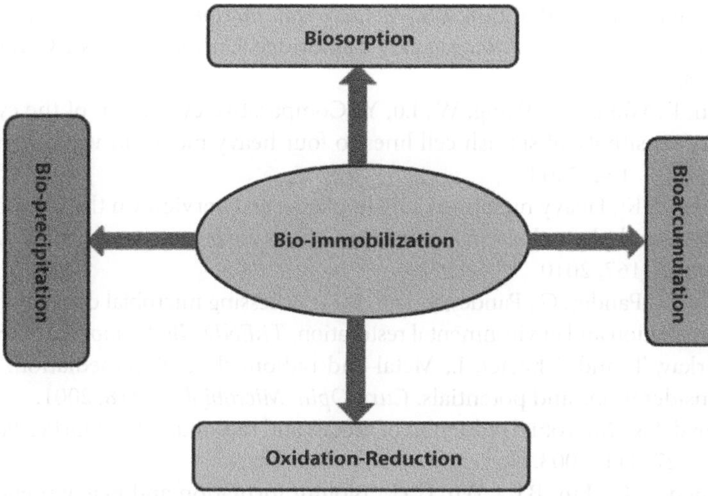

Figure 3.5 Schematic diagram of bioimmobilization.

to reduce contaminated As(V) to As(III) and sulfate to elemental sulfur and precipitates in the form of arsenite sulfide [79]. Therefore sulfide precipitation is a useful mechanism for reduction of arsenic. The EPS of *Chryseomonas luteola* immobilized the metal ions such as cadmium, cobalt, nickel, and copper through adsorption [64].

3.6 Conclusion

Biodetoxification is mainly treated by biosorption, bioleaching, biovolatilization, bioimmobilization, and bioaccumulation mechanism of bacterial cell. These processes are economically significant. EPSs present in bacterial cell are involved in bacteria and metal ion interaction and established the process of biosorption. The metals transform toxic to less toxic or less available or removed from environment by using these mechanisms. Among all mechanisms, biosorption mechanism is more effective and beneficial, and it includes ion-exchange and precipitation mechanisms. These detoxification mechanisms are eco-friendly and cost effective.

References

1. Singh, P. and Cameotra, S.S., Enhancement of metal bioremediation by use of microbial surfactants. *Biochem. Biophys. Res. Commun.*, 319, 291, 2004.

2. Luoma, S.N. and Rainbow, P.S., *Metal contamination in aquatic environments. Science and lateral management*, Cambridge University press, Cambridge, 2008.
3. Tan, F., Wang, M., Wang, W., Lu, Y., Comparative evaluation of the cytotoxicity sensitivity of six fish cell lines to four heavy metals *in vitro*. *Toxicol. In Vitro*, 22, 164, 2008.
4. Yadav, S.K., Heavy metals toxicity in plants: an overview on the role of glutathione and phytochelatins in heavy metal stress tolerance of plants. *S. Afr. J. Bot.*, 76, 167, 2010.
5. Paul, D., Pandey, G., Pandey, J., Jain, R.K., Accessing microbial diversity for bioremediation and environmental restoration. *TRENDS Biotechnol.*, 23, 135, 2005.
6. Barkay, T. and Schaefer, J., Metal and radionuclide bioremediation: issues, considerations and potentials. *Curr. Opin. Microbiol.*, 4, 318, 2001.
7. Lloyd, J.R., Microbial reduction of metals and radionuclides. *FEMS Microbiol. Rev.*, 27, 411, 2003.
8. Chien, C.C., Lin, B.C., Wu, C.H., Biofilm formation and heavy metal resistance by an environmental Pseudomonas sp. *Biochem. Eng. J.*, 78, 132, 2013.
9. Ancion, P.Y., Lear, G., Lewis, G.D., Three common metal contaminants of urban runoff (Zn, Cu & Pb) accumulate in freshwater biofilm and modify embedded bacterial communities. *Environ. Pollut.*, 158, 2738, 2010.
10. Beech, I.B. and Cheung, C.S., Interactions of exopolymers produced by sulphate-reducing bacteria with metal ions. *Int. Biodeter. Biodegr.*, 35, 59, 1995.
11. Baker-Austin, C., Wright, M.S., Stepanauskas, R., McArthur, J.V., Co-selection of antibiotic and metal resistance. *Trends Microbiol.*, 14, 176, 2006.
12. Costley, S.C. and Wallis, F.M., Treatment of heavy metal-polluted wastewaters using the biofilms of a multistage rotating biological contactor. *World J. Microbiol. Biotechnol.*, 17, 71, 2001.
13. Nayak, S.K., Nayak, S., Patra, J.K., Rhizobacteria and its biofilm for sustainable agriculture: A concise review, in: *New and Future Developments in Microbial Biotechnology and Bioengineering: Microbial Biofilms*, M.K. Yadav and B.P. Singh (Eds.), pp. 165–175, Elsevier, Amsterdam, Netherlands, 2020.
14. Lovley, D.R. and Coates, J.D., Bioremediation of metal contamination. *Curr. Opin. Biotechnol.*, 8, 285, 1997.
15. Tchounwou, P.B., Yedjou, C.G., Patlolla, A.K., Sutton, D.J., Heavy metal toxicity and the environment, in: *Molecular, clinical and environmental toxicology*, A. Luch (Ed.), pp. 133–164, Springer, Basel, 2012.
16. He, Z.L., Yang, X.E., Stoffella, P.J., Trace elements in agroecosystems and impacts on the environment. *J. Trace Elem. Med. Bio.*, 19, 125, 2005.
17. Shallari, S., Schwartz, C., Hasko, A., Morel, J.L., Heavy metals in soils and plants of serpentine and industrial sites of Albania. *Sci. Total Environ.*, 209, 133, 1998.
18. Arruti, A., Fernández-Olmo, I., Irabien, Á., Evaluation of the contribution of local sources to trace metals levels in urban PM2. 5 and PM10 in the Cantabria region (Northern Spain). *J. Environ Monit.*, 12, 1451, 2010.

19. Pacyna, J.M., Monitoring and assessment of metal contaminants in the air, in: *Toxicology of Metals*, L.W. Chang, L. Magos, T. Suzuli (Eds.), pp. 9–28, CRC Press, Boca Raton, FL, 1996.
20. Rajaganapathy, V., Xavier, F., Sreekumar, D., Mandal, P.K., Heavy metal contamination in soil, water and fodder and their presence in livestock and products: a review. *J. Environ. Sci. Technol.*, 4, 234, 2011.
21. Gautam, R.K., Sharma, S.K., Mahiya, S., Chattopadhyaya, M.C., Contamination of heavy metals in aquatic media: transport, toxicity and technologies for remediation, in: *Heavy Metals In Water: Presence, Removal and Safety*, S. Sharma (Ed.), pp. 1–24, Royal Society of Chemistry, London, United Kingdom, 2014.
22. Wuana, R.A. and Okieimen, F.E., Heavy metals in contaminated soils: a review of sources, chemistry, risks and best available strategies for remediation. *Isrn Ecology*, 2011, 402647, 2011.
23. Agency for Toxic Substances and Disease Registry (ATSDR), *Toxicological Profile for Arsenic TP-92/09*, Georgia: Center for Disease Control, Atlanta, 2000.
24. Tchounwou, P., Development of Public Health Advisories for Arsenic in Drinking Water. *Environ. Health Rev.*, 14, 211, 1999.
25. Centeno, J.A., Tchounwou, P.B., Patlolla, A.K., Mullick, F.G., Murakata, L., Meza, E., Yedjou, C.G., Environmental pathology and health effects of arsenic poisoning, in: *Managing Arsenic in the Environment: From Soil to Human Health*, R. Naidu, E. Smith, G. Owens, P. Bhattacharya, P. Nadebaum (Eds.), pp. 311–327, CSIRO Publishing, Collingwood, Australia, 2006.
26. Li, J.H. and Rossman, T.G., Inhibition of DNA ligase activity by arsenite: a possible mechanism of its comutagenesis. *Mol Toxicol.*, 2, 1, 1989.
27. Hartmann, A. and Speit, G., Comparative investigations of the genotoxic effects of metals in the single cell gel (SCG) assay and the sister chromatid exchange (SCE) test. *Environ. Mol. Mutagen.*, 23, 299, 1994.
28. IMO/FAO/UNESCO/WMO/WHO/IAEA/UN/UNEP, *Joint Group of Experts on the Scientific Aspects of Marine Pollution: Report of the seventeenth session, Gesamp*, World Health Organization, Geneva, Switzerland, (Reports and Studies No. 31), 1987.
29. Rajaram, R., Mathivanan, K., Rakesh, O.R., Bioremediation of cadmium pollution –A bacterial approach, in: *Bacterial Biochemistry and Biotechnology*, R.K. Gaur, H.K. Gautam, V.K. Gupta (Eds.), pp. 130–143, Lambert Academic Publishing, Saarbrücken, Germany, 2012.
30. Nayak, S.K., Dash, B., Baliyarsingh, B., Microbial Remediation of Persistent Agro-chemicals by Soil Bacteria: An Overview, in: *Microbial Biotechnology*, vol. II, J.K. Patra, G. Das, H.S.,. Shin (Eds.), pp. 275–301, Springer, Singapore, 2018.
31. Paschal, D.C., Burt, V., Caudill, S.P., Gunter, E.W., Pirkle, J.L., Sampson, E.J., Miller, D.T., Jackson, R.J., Exposure of the US population aged 6 years and older to cadmium: 1988–1994. *Arc. Environ. Cont. Toxicol.*, 38, 377, 2000.

32. Baselt, R.C. and Cravey, R.H., *Disposition of Toxic Drugs and Chemicals in Man*, 4th Edn., pp. 105–107, Year Book Medical Publishers, Chicago, IL, 1995.
33. Baselt, R.C., Verapamil, in: *Disposition of Toxic Drugs and Chemicals in Man*, 5th ed., pp. 876–878, Chemical Toxicology Institute, Foster City, CA, 2000.
34. IARC, Working Group on the Evaluation of Carcinogenic Risks to Humans, in: *Some Industrial Chemicals*, International Agency for Research on Cancer, Lyon (FR), (IARC Monographs on the Evaluation of Carcinogenic Risks to Humans, No. 60.), 1994.
35. Jacobs, J.A. and Testa, S.M., Overview of chromium (VI) in the environment: background and history, in: *Chromium (VI) handbook*, J. Guertin, J.A. Jacobs, C.P. Avakian (Eds.), pp. 1–21, CRC Press, Boca Raton, USA, 2005.
36. Patlolla, A.K., Barnes, C., Yedjou, C., Velma, V.R., Tchounwou, P.B., Oxidative stress, DNA damage, and antioxidant enzyme activity induced by hexavalent chromium in Sprague-Dawley rats. *Environ. Toxicol.*, 24, 66, 2009.
37. Velma, V., Vutukuru, S.S., Tchounwou, P.B., Ecotoxicology of hexavalent chromium in freshwater fish: a critical review. *Environ. Health Rev.*, 24, 129, 2009.
38. Agency for Toxic Substances and Disease Registry (ATSDR), *Public Health Service*, Department of Health and Human Services, Toxicological Profile for Lead, Atlanta: U.S, 1999.
39. Agency for Toxic Substances and Disease Registry (ATSDR), *Case Studies in Environmental Medicine - Lead Toxicity*, Public Health Service, U.S. Department of Health and Human Services, Atlanta, 1992, 2012.
40. Flora, S.J., Flora, G., Saxena, G., Environmental occurrence, health effects and management of lead poisoning, in: *Lead*, vol. 158, 2006.
41. Apostoli, P., Kiss, P., Porru, S., Bonde, J.P., Vanhoorne, M., Male reproductive toxicity of lead in animals and humans. ASCLEPIOS Study Group. *Occup. Environ. Med.*, 55, 364, 1998.
42. Clarkson, T.W., Magos, L., Myers, G.J., The toxicology of mercury—current exposures and clinical manifestations. *New Engl. J. Med.*, 349, 1731–1737, 2003.
43. Guzzi, G. and La Porta, C.A., Molecular mechanisms triggered by mercury. *Toxicology*, 244, 1, 2008.
44. Tchounwou, P.B., Ayensu, W.K., Ninashvili, N., Sutton, D., Environmental exposure to mercury and its toxicopathologic implications for public health. *Environ. Toxicol.*, 18, 149, 2003.
45. VCI, *Copper history/Future*, Van Commodities Inc, http://trademetalfutures.com/copperhistory.html, USA, 2011.
46. Davies, B.E. and Jones, L.H.P., Micronutrients and toxic elements, in: *Russell's Soil Conditions and Plant Growth*, S.E.J. Russell and A. Wild (Eds.), p. 780, Longman Scientific & Technical, London, 1988.
47. Khodadoust, A.P., Reddy, K.R., Maturi, K., Removal of nickel and phenanthrene from kaolin soil using different extractants. *Environ. Eng. Sci.*, 21, 691, 2004.

48. Wong, J.P.K., Wong, Y.S., Tam, N.F.Y., Nickel biosorption by two chlorella species, *C. Vulgaris* (a commercial species) and *C. Miniata* (a local isolate). *Bioresour. Technol.*, 73, 133, 2000.
49. Nies, D.H., Efflux-mediated heavy metal resistance in prokaryotes. *FEMS Microbiol. Rev.*, 27, 313, 2003.
50. Harrison, J.J., Ceri, H., Stremick, C., Turner, R.J., Differences in biofilm and planktonic cell mediated reduction of metalloid oxyanions. *FEMS Microbiol. Lett.*, 235, 357, 2004.
51. Harrison, J.J., Ceri, H., Stremick, C., Turner, R.J., Biofilm susceptibility to metal toxicity. *Environ. Microbiol.*, 6, 1220, 2004.
52. Gutnick, D.L. and Bach., H., Engineering bacterial biopolymers for the biosorption of heavy metals; new products and novel formulations. *Appl. Microbiol. Biotechnol.*, 54, 451, 2000.
53. Fang, H.H.P., Xu, L.-C., Chan, K.-Y., Effects of toxic metals and chemicals on biofilm and biocorrosion. *Water Res.*, 36, 4709, 2002.
54. Gadd, G.M., Biosorption: critical review of scientific rationale, environmental importance and significance for pollution treatment. *J. Chem. Technol. Biotechnol.*, 84, 13, 2009.
55. De Philippis, R., Colica, G., Micheletti, E., Exopolysaccharide-producing cyanobacteria in heavy metal removal from water: molecular basis and practical applicability of the biosorption process. *Appl. Microbiol. Biotechnol.*, 92, 697, 2011.
56. Pal, A. and Paul, A.K., Microbial extracellular polymeric substances: central elements in heavy metal bioremediation. *Indian J. Microbiol.*, 48, 49, 2008.
57. Salehizadeh, H. and Shojaosadati, S.A., Removal of metal ions from aqueous solution by polysaccharide produced from *Bacillus firmus*. *Water Res.*, 37, 4231, 2003.
58. Gavrilescu, M., Removal of heavy metals from the environment by biosorption. *Eng. Life Sci.*, 4, 219, 2004.
59. Sundar, K., Mukherjee, A., Sadiq, M., Chandrasekaran, N., Cr (III) bioremoval capacities of indigenous and adapted bacterial strains from Palar river basin. *J. Hazard. Mater.*, 187, 553, 2011.
60. O'Toole, G.A., Pratt, L.A., Watnick, P.I., Newman, D.K., Weaver, V.B., Kolter, R., Genetic approaches to study of biofilms. *Methods Enzymol.*, 310, 91, 1999.
61. Gupta, P. and Diwan, B., Bacterial exopolysaccharide mediated heavy metal removal: a review on biosynthesis, mechanism and remediation strategies. *Biotechnol. Rep.*, 13, 58, 2017.
62. Gupta, P. and Diwan, B., Bacterial Exopolysaccharide mediated heavy metal removal: a review on biosynthesis, mechanism and remediation strategies. *Biotechnol Rep.*, 13, 58, 2017.
63. Alluri, H.K., Ronda, S.R., Settalluri, V.S., Bondili, J.S., Suryanarayana, V., Venkateshwar, P., Biosorption: An eco-friendly alternative for heavy metal removal. *Afr. J. Biotechnol.*, 6, 25, 2007.

64. Ahalya, N., Ramachandra, T.V., Kanamadi, R.D., Biosorption of heavy metals. *Res. J. Chem. Environ.*, 7, 71, 2003.
65. De Philippis, R., Colica, G., Micheletti, E., Exopolysaccharide-producing cyanobacteria in heavy metal removal from water: molecular basis and practical applicability of the biosorption process. *Appl Microbiol Biot.*, 92, 697, 2011.
66. Pereira, S., Micheletti, E., Zille, A., Santos, A., Moradas-Ferreira, P., Tamagnini, P., De Philippis, R., Using extracellular polymeric substances (EPS)-producing cyanobacteria for the bioremediation of heavy metals: do cations compete for the EPS functional groups and also accumulate inside the cell? *Microbiology*, 157, 451, 2011.
67. Vera, M., Schippers, A., Sand, W., Progress in bioleaching: fundamentals and mechanisms of bacterial metal sulfide oxidation—part A. *Appl Microbiol Biotechnol.*, 97, 7529, 2013.
68. Christine, L. and Mamindy, Y., Bioremediation of heavy metals in sediment, in: *Bioremediation: Processes, Challenges and Future Prospects*, J.B. Velázquez-Fernández and S. Muñiz-Hernández (Eds.), pp. 3–18, Nova science publisher, Hauppauge, New York, US, 2014.
69. Sand, W. and Gehrke, T., Extracellular polymeric substances mediate bioleaching/ biocorrosion via interfacial processes involving iron (III) ions and acidophilic bacteria. *Res. Microbiol.*, 157, 49, 2006.
70. Schippers, A. and Sand, W., Bacterial leaching of metal sulfides proceeds by two indirect mechanisms via thiosulfate or via polysulfides and sulfur. *Appl. Environ. Microbiol.*, 65, 319, 1999.
71. Ike, M., Yamashita, M., Kuroda, M., Microbial Removal and Recovery of Metals from Wastewater, in: *Applied Bioengineering: Innovations and Future Directions*, T. Yoshida (Ed.), pp. 573–595, Wiley Publisher, Weinheim, Germany, 2017.
72. Liu, S., Zhang, F., Chen, J., Sun, G., Arsenic removal from contaminated soil via biovolatilization by genetically engineered bacteria under laboratory conditions. *J. Environ. Sci.*, 23, 1544, 2011.
73. Pepi, M., Gaggi, C., Bernardini, E., Focardi, S., Lobianco, A., Ruta, M., Nicolardi, V., Volterrani, M., Gasperini, S., Trinchera, G., Renzi, P., Gabellini, M., Focardi, S.E., Mercury-resistant bacterial strains *Pseudomonas* and *Psychrobacter* spp. isolated from sediments of Orbetello Lagoon (Italy) and their possible use in bioremediation processes. *Int. Biodeter. Biodegr.*, 65, 85, 2011.
74. Żur, J., Wojcieszyńska, D., Guzik, U., Metabolic Responses of Bacterial Cells to Immobilization. *Molecules*, 21, 958, 2016.
75. Gadd, G.M., Bioremedial potential of microbial mechanisms of metal mobilization and immobilization. *Curr. Opin. Biotechnol.*, 11, 271, 2000.
76. Raspor, P., Batič, M., Jamnik, P., Josić, D., Milačič, R., Paš, M., Recek, M., Režić-Dereani, V., Skrt, M., The influence of chromium compounds on yeast physiology: (A review). *Acta Microbiol. Immunol. Hung.*, 47, 143, 2000.

77. Focardi, S., Pepi, M., Focardi, S.E., Microbial reduction of hexavalent chromium as a mechanism of detoxification and possible bioremediation applications, in: *Biodegradation-Life of Science*, R. Chamy (Ed.), pp. 321–347, InTech Open, London, UK, 2013.
78. Lee, J.H., Kim, M.G., Yoo, B., Myung, N.V., Maeng, J., Lee, T., Dohnalkova, A.C., Fredrickson, J.K., Sadowsky, M.J., Hur, H.G., Biogenic formation of photoactive arsenic-sulfide nanotubes by *Shewanella* sp. strain HN-41. *PNAS*, 104, 20410, 2007.

4

Microbial-Derived Polymers and Their Degradability Behavior for Future Prospects

Mohammad Asif Ali[1,2*†], **Aniruddha Nag**[1,3†] **and Maninder Singh**[1†]

[1]*Graduate School of Advanced Science and Technology, Energy and Environment Area, Japan Advanced Institute of Science and Technology, Nomi, Ishikawa, Japan*
[2]*Soft Matter Sciences and Engineering Laboratory, ESPCI Paris, PSL University, CNRS, Paris, France*
[3]*School of Energy Science and Engineering, Vidyasirimedhi Institute of Science and Technology (VISTEC), Payupnai, Wangchan, Rayong, Thailand*

Abstract
This chapter will focus on the development of bio-based polymers such as polyamides (PA), polylactide (PLA), and polyhydroxyalkanoates (PHAs) produced from renewable resources. Nylon™ plastic is a kind of PA is a long chain fiber-forming recalcitrant and biodegradable and degradable polymer with diverse applications. Polylactic acid (PLA) is biodegradable aliphatic polyester derived from a naturally occurring organic acid (lactic acid). On the other hand, PHAs are the high molecular weight biodegradable polyesters synthesized by a wide array of microbes. However, they have an undesirable influence on the environment and substantially impact waste deposition and utilization. This chapter will emphasize the application and microbial degradability of these three kinds (PLA, PHA, and PA) of plastics.

Keywords: Degradation, polyamides microbes, polylactide, polyhydroxyalkanoates

4.1 Introduction

Research and development in the field of materials science over the past two decades have dramatically increased the production of synthetic or

**Corresponding author*: asif.alidu@gmail.com
†*Equal contribution as first author*

Bibhuti Bhusan Mishra, Suraja Kumar Nayak, Swati Mohapatra, and Deviprasad Samantaray (eds.) Environmental and Agricultural Microbiology: Applications for Sustainability, (63–82) © 2021 Scrivener Publishing LLC

bio-based polymers [1, 2]. Plastics are usually hard to decompose but marginally affected by exposure to natural conditions such as sunlight, heat, oxygen, or biological and hydrolytic processes. Long-term undesirable accumulation of plastics in soil and water imposes a significant threat to the environment, and it can take from a year to decades for them to decompose [1–3]. Rising demands of plastics in both commercial and domestic sectors make it challenging for its proper disposal [4]. Although, several methods such as modification of molecular structure, side chain, reinforcement, and addition of some side-chain in plastics were developed, aiming to plastic waste management [5]. However, these modifications affect its molecular chain, morphology, chemistry, degree of polymerization, thermal properties, and hydrolytic behaviors [5]. There are widely used synthetic or bio-based polymers like polyamide (PA), polylactic acid (PLA), and polyhydroxyalkanoates (PHAs) with low melting temperature and degraded faster than other high-performance polymers such as PBO (polybenzoxazole), PBI (polybenzimidazole), and PI (polyimide) [1–3]. PAs were considered to be exciting materials since 1930 due to its molecular structure (-CONH-). However, its fibers are resistant to biodegradation because of chain symmetry [6]. But nylon degradation can occur after side-chain modification; scission of the chain, as well as a low oligomer, can be degraded [6, 7]. Hashimoto and Naoka *et al.* reported about especially nylon 4 degrade under the soil in the presence of marine bacteria [7, 8]. Recently, Kaneko *et al.* (2014) reported itaconic acid-based heterocyclic PAs, which are environmentally degradable under soil and water in the presence of UV light [8]. Naturally occurring polymers, bio-derived plastics, and synthetic bio-based plastics from renewable resources are the existing bases for establishing a sustainable society [9, 10]. Replacing bio-sourced materials over the existing fossil-fuel-based plastics are the prime focus of recent research [9]. However, different microbial-derived well-known bio-polyester like PHAs and PLA has low melting temperature tensile strength, and its biodegradable nature can replace the petrochemical-based plastics. There are various methods for the degradation of the polymer.

1. Thermal degradation
2. Photo-oxidative degradation
3. Hydrolytic degradation
4. Mechanochemical degradation
5. Soil degradation
6. Biodegradation

Some of the factors that affect the degradation process such as

- Color changes
- Scission of the backbone
- Modification of one or more end-groups
- Disruption of a side chain
- Mechanical
- Photo/thermal
- Chemical
- Cracking and charring (weight loss)
- The effect of light, heat, air, and moisture reflects the polymer structure.

Biodegradation can be defined as the breaking down material when exposed to microbes such as bacteria, fungi, actinobacteria, or other biological means anaerobically or aerobically [2]. Moreover, polymers biodegradation is possible by different enzymatic and non-enzymatic hydrolysis without thermal oxidation, radiolysis, or photolysis [3]. Alternatives to the existing fossil fuel and other non-renewable sources are the prime focus of today now. This chapter describes the broad spectrum of bioavailability, biosynthesis and biodegradability of PA, PLA, and PHAs.

4.2 Polyamides

Definition

Nylon™ is a necessary term that represents an important class of PAs. PA with amide linkage exhibits high thermomechanical properties and higher softening temperature because of hydrogen bonding, which provides chain symmetry [9]. Presence of amide linkages in PA used as engineering thermoplastics as a film or fibers form. Nylon 6-6 is the most commercialized polymer widely used because of its high thermomechanical properties [11].

The Nylon-P,Q (Figure 4.1) refers to the number of carbon atom used in the monomeric chain, which was commercialized as Nylon™ is petroleum-derived nylon-6,6 and nylon-6 [6, 9]. PAs are generally a non-biodegradable polymer, although amide linkages were degraded by disrupting the hydrogen bonding [6]. Petroleum-based polymers are posing a significant threat to the environment and its sustainability. Bio-based materials can be alternatives to these petroleum-based polymers. Bio-derived PAs are very much sustainable. The establishment of these

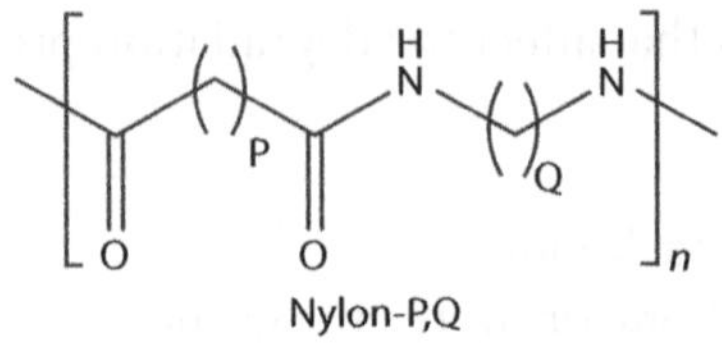

Figure 4.1 Structures of polyamide with trade name NylonTM (Nylon-P, Q).

sustainable bio-based PAs reduces the use of petroleum-based polymers and reduces ecological problems. PAs, on the other hand, are one of the most consumed polymers as it consumed globally around 7.4 million tons/annum for the year 2016. Total consumption of PA in 2016 is divided into two parts: PA fibers and film shares 55% globally but 45% share textiles industries used as seats, carpet, sportswear, and different clothes. Some of the bio based, especially fatty acids in vegetable oils, are the source of various monomers to achieve PA.

4.2.1 Bioavailability and Production

Nondegradable polymers are one of the big issues and create lots of stress over the environment because of PA dumping. A sustainable polymer helps in reducing this stress over the environment. Variety of bio-PAs are derived from renewable raw materials such as PA 4,6; PA 4,4; PA 4,10; PA 4; PA 6,10; PA, 10,10; PA 10,12; and PA 11 [6, 9]. Five decades back, European company Arkema first developed Rilsan, which is 100% castor oil-based PA (PA-11) [6]. Many bio-based polymers were synthesized from castor oil-based with 60% sebacic acid, which exhibits superior performances than petroleum-derived PA 6 and PA 6,6. Nylon 4 is synthesized after ring-opening polymerization of 2-pyrrolidone [2, 5a, 7]. Recently, itaconic acid-based heterocyclic PA has been introduced environmentally degradable, which can reduce the burden of polymer waste [8].

4.2.2 Biodegradability of Polyamides

PAs containing amide linkage have strong hydrogen bonding that is less susceptible to the degradation, but some bacteria can attack their low molecular chain [6, 11, 12]. It should be kept in mind that according to IUPAC terminology, the biodegradable polymer is able to undergo chain scissions, resulting in a decrease in molar mass due to enzymatic process from the action of cells; however, *in vitro* activity of isolated enzymes cannot be considered as biological activity [12, 13]. Even though bio-based material is composed or derived in whole or natural products issued from the biomass, it does not mean that the material is biodegradable. Certain aliphatic PAs are susceptible to biodegradation by microorganisms (fungi or bacterium) [2]. Some of the thermophilic bacteria isolated from the soil favor the degradability of PA 12 and PA 66 in the culture medium [2]. Some of the white-rot fungal strains have these three kinds of enzyme which are able to degrade the nylon [5b, 12, 13]. Some of the marine bacteria degraded the PA 6 and PA 66, such as *Bacillus sphericus*, *Vibrio furnisii* and *Brevundimonas vesicularis*. The PA can be degraded due to the endogenous enzymatic hydrolysis of an amide linkage [3]. The ^{14}C-labeled nylon-6,6 exposed to various enzyme solutions *in vitro*, but it was unaffected by some of the enzyme-like esterases, but it degraded after exposure of chymotrypsin, trypsin, and papain.

4.2.3 Degradation of Nylon 4 Under the Soil

Nylon 4 is synthesized from 2-pyrrolidone, which means it is lactam of γ-aminobutyric acid (GABA). It has been reported that nylon 4 is different from other nylon because it degrades under the soil in the activated sludge [7, 12, 14]. Further, nylon 4 was blended with nylon 6, and its degradability was investigated, only nylon 4 part was degraded. Further, Yamano *et al.* found to degrade the nylon 4 inside the activated sludge further isolated *Pseudomonas* sp. with the strain ND-10 and ND-11 and GABA as a byproduct [14].

4.2.4 Fungal Degradation of Nylon 6 and Nylon 66 (Synthetic Polyamide)

Nylon 6 is synthesized via ring-opening polymerization of the ε-caprolactam (Figure 4.2) and comes with various commercial names such as perlon, nylon, and steelon. A semi-crystalline linear PA is obtained from ring-opening polymerization of ε-caprolactam in the presence of a tin octoate catalyst [12–14]. In the presence of carbon and nitrogen loving fungi, ε-caprolactam is also degraded. Due to the strong interaction of hydrogen bonding, the degradation rate is slow, but some microorganisms can be degraded PAs, including the bacterial genera *Pseudomonas*, *Achromobactor*, and *Corynebacterium* and *Bjerkandera adusta* [12–15]. However, some of the authors reported the two kinds of fungal genera, which are lignolytic fungus *Phanerochaete chrysosporium* NCIM 1073 and *Tarmetes versicolor* NCIM 1086 in submerged cultivation using nitrogenous nutrient as a stimulator.

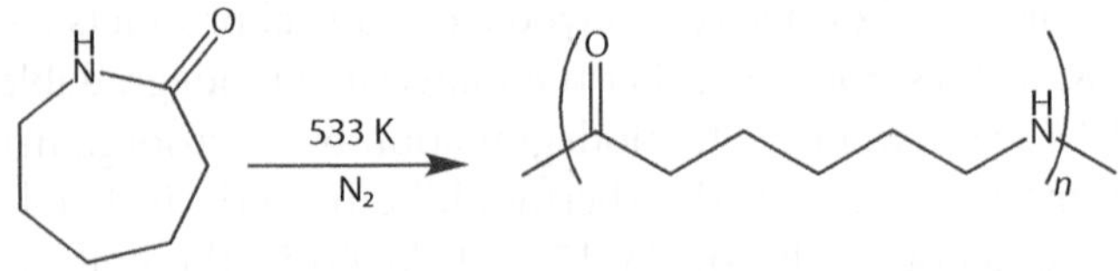

Figure 4.2 Ring-opening polymerization of caprolactam.

Both white-rot fungi are well known for lignolytic activity, which attacks the lignin. PA sheets were exposed through submerges cultivation process for microbial degradation, and nylon sheets decreased its thickness and molecular weight. Jozefa Friedrich *et al.* (2007) had tested 58 fungi for their degradation ability; out of these fungal strain, two fungi were more labile toward the degradation. The white-rot fungi, *B. adusta* and *P. chrysosporium* can degrade the polymer, but especially the *Bjerkandera adusta* disintegrated the fibers. U. Klun *et al.* (2003) were also tested the same kind of fungus *P. chrysosporium* that is well known for its lignolytic activity [12–15]. Abiotic (PA-6 placed without fungus) showed a partially weight loss, which means lesser than biotic (PA-6/fungus). Degradation of nylon-6 was observed in the culture of the basidiomycete *B. adusta* inside the submerged medium, initially break the surface part of the PA. Nylon 6 and Nylon 66 are also degraded in the presence of bacteria *Pseudomonas aeruginosa* NCIM 2242, which only targets the chain which contains an amide linkage.

4.2.5 Itaconic Acid-Based Heterocyclic Polyamide

Itaconic acid-based PA (Figure 4.3) kept inside the soil for 1 year results in decreased shape, size, and color of polymer resins and photo-solubilization behavior under UV light, which favors the ring-opening phenomenon which reduces the threats of waste disposal [8].

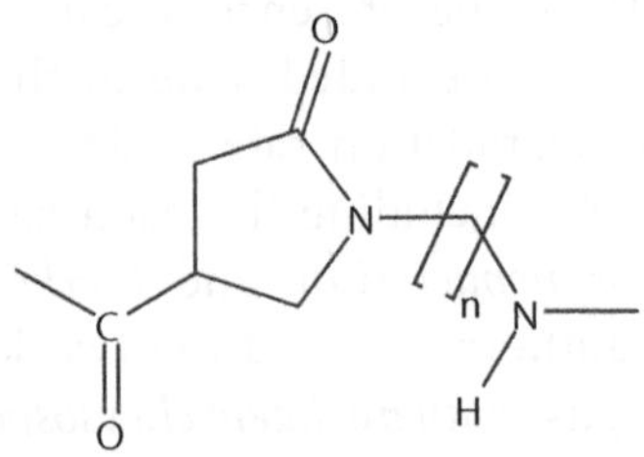

Figure 4.3 Itaconic acid-based heterocyclic polyamide.

4.2.6 Summary and Future Development

The demand for PA degradation has been increased, but the PA industry is still not accepted as a biodegradable polymer, but some of the PAs can be degraded in the soil, which can fulfill future development for degradation.

4.3 Polylactic Acid

Definition
PLA is the most commonly used bio-plastic, and it is a kind of thermoplastic aliphatic polyester. Lactic acid, the precursor of PLA, can be obtained very easily from various raw materials (like corn and starch), which is then polymerized to PLA. Several kinds of PLA are available to include PDLA (Poly-D-lactic Acid), regular PLLA (Poly-L-lactic Acid), PDLA (Poly-D-lactic acid), and Racemic PDLLA (Poly-DL-lactic Acid). They have slightly different characteristics properties but are produced from the same renewable resource (lactic acid).

Naturally occurring polymers, bio-derived plastics, and synthetic bio-based plastics from renewable resources are the existing bases for establishing a sustainable society. Replacing bio-sourced materials over the existing fossil-fuel-based plastics are the prime focus of recent research. Cheap raw materials such as maize, potato, starchy materials, and lignocellulose biomass are feasible for economic lactic acid production. Poly-lactide or polylactic acid (PLA) is the front-runner in the emerging bioplastics market with the best availability and the most attractive cost structure. Theophile-Jules Pelouze first synthesized PLA in 1845 by the lactic acid polycondensation method [16–18]. Later on, Wallace Hume Carothers introduced another method in 1932, which was patented by DuPont in 1954. PLA polymers change from amorphous glassy state to highly crystalline with high glass transition temperature and mechanical property. PLA can be processed into several materials like fused filament fabrication in 3D printers, medical implants (like anchors, screw), and packaging materials. Biodegradability of PLA is a natural phenomenon which is even faster as compared with other bioplastics. Mechanical property and biodegradability can be improved by several methods like-annealing, blending, the composite formation, side-chain modification, etc.

4.3.1 Availability and Production

Lactic acid also has a long invention history with the first reported discovery by Scheele [16, 17] on 1780 as a milk component, later on, Lavoisier named this milk component "acid lactic" in 1789 and Pasteur in 1857 confirmed it as a fermented metabolite rather than milk component [18]. Lactic acid produced by microorganism fermentation or via a synthetic chemical pathway. The demand for lactic acid-based products increasing globally and estimated to be raised around 2,000 kilotons by 2020 [19]; the largest consumer markets in the world are the United States, followed by China and Western Europe [17]. Lactic acid consists of two optical isomers: L (+)-lactic acid and D(−)-lactic acid, which can be prepared as optically pure isomers, i.e., L(+)- or D(−)-lactic acid by microbial fermentation (Figure 4.4) of renewable resources with the correct choice of the microorganisms. Each isomer is advantageous over the other depending upon the application. Optically pure lactic acids are the best choice to make high molecular weight commercial grade bio-plastic rather than the plastics derived from the racemic mixture in the chemical synthesis method (Figure 4.4). Apart from those, other numerous bioresources are available for lactic acid production, like glycerol (a by-product of bio-diesel) and

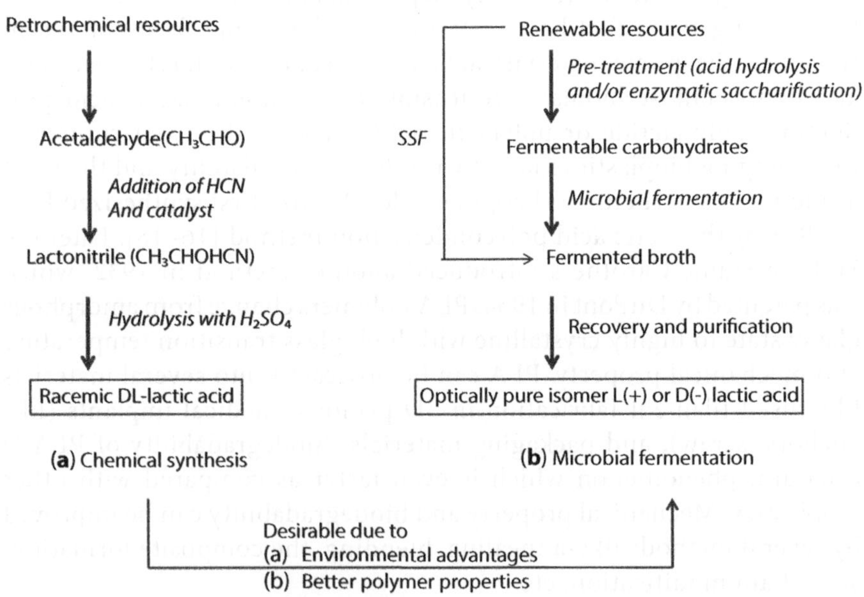

Figure 4.4 Overview of different manufacturing methods of lactic acid (a) chemical synthesis and (b) microbial fermentation [20].

Table 4.1 Reports in the literature about recent investigations on the biotechnological production of lactic acid from cheap raw materials.

Substrate	Microorganism	Fermentation method	Lactic acid Process productivity g/(L.h)	Yield (g/g)	References
Sugarcane bagasse hemicellulose hydrolysate	*Bacillus sp.* 75 strain 17C5	Batch	0.8	0.93	Patel *et al.*, 2004 [49]
Corn fiber hydrolysate	*Bacillus coagulans* MXL-9	Fed-batch	0.21	0.46	Bischoff *et al.*, 2010 [50]
Biomass derived xylose	*Bacillus coagulans* NL01	Batch	1.04	0.75	Ouyang *et al.*, 2012 [51]
Various carbohydrates	*Enterococcus faecalis* RKY1	Batch	5.1	0.96	Yun *et al.*, 2003 [52]

microalgae (harvesting can be possible anywhere with a concise harvesting cycle). Microorganisms producing lactic acid are classified into two groups: bacteria and fungi, and their use depend on the substrates to be fermented. However, lactic acid bacteria (LAB) is the most popular method over the fungal production in terms of production rate caused due to mass transfer limitation and by-products formation. LAB can be classified into two categories depending upon the end fermentation product, homo-fermentative. It converts glucose into lactic acid as the sole product, whereas in the case of hetero-fermentative predominating side products like CO_2 and ethanol are also formed along with the desired lactic acid. Several efforts have been given to optimizing lactic acid production through microorganism engineering, and in Table 4.1, a few of them are listed.

4.3.2 Polymerization Method

PLA can be synthesized using three ways: (a) The first pathway is condensation polymerization of the L(+) and D(−) isomers of lactic acid which produce low molecular weight PLA, (b) the second route involves ring-opening polymerization of the lactide ring (Figure 4.5). Cargill Dow LLC developed an alternative pathway of melt polycondensation and the use of a tin catalyst to obtain commodity PLA applicable for packaging

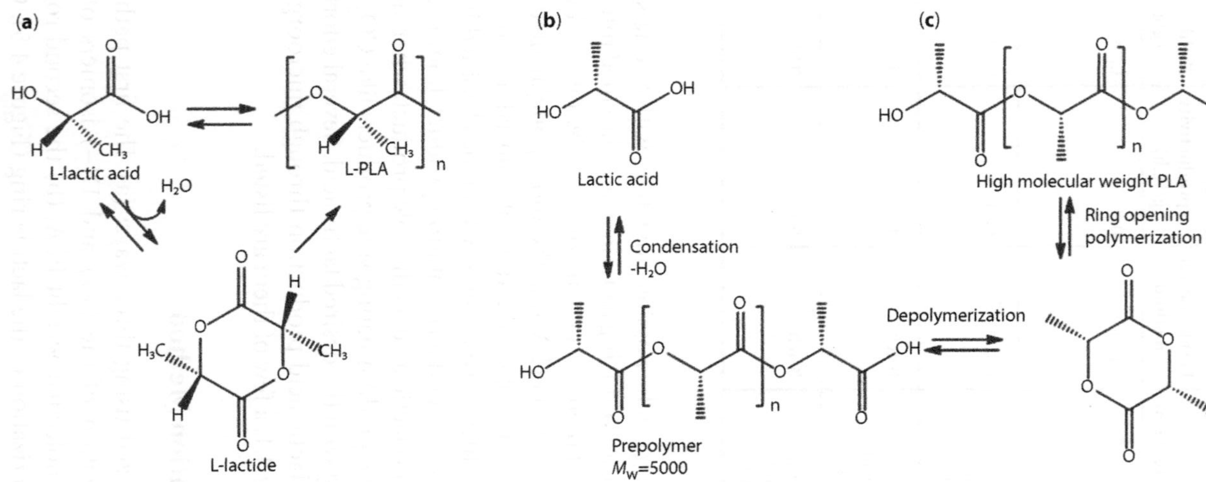

Figure 4.5 (a) Conventional polymerization of polylactic acid, (b) High M_w PLA preparation through catalytic lactide formation method (c) The third route involves azeotropic dehydrated condensation of lactic acid to obtain high M_w PLA without the chain extenders' addition.

industries. This pathway involves the pre-polymer formation of aqueous lactic acid and converting into lactide stereoisomer through intramolecular cyclization. In the next step, lactide ring-opening produces high M_w PLA [21] using stannous octoate $Sn(Oct)_2$ as a catalyst [22].

4.3.3 Biodegradability of Polylactic Acid

PLA attains researchers' keen attention among all the biodegradable plastics due to its complete application in the agriculture and packaging industry, as PLA films are mechanically stable and biodegradable. Biodegradation of PLA occurs in the following two pathways: (a) Fragmentation of the polymer leading chains occurs via hydrolysis in the presence of acidic or basic conditions. Moisture plays a vital role in lowering M_n values below 40000, (b) followed by bio-assimilation of disintegrated oligomers by environmental microorganisms to carbon dioxide, water, etc. Biodegradability depends on the chemical structure as well as the polymer sources. Proteinase K was a widely used microorganism for PLA bio-degradation [23]. However, they are able to degrade oligomer products or lactic acids but not PLA itself. Degradability can be controlled in various synthetic ways, and a few popular methods are discussed as follows.

4.3.4 Copolymerization Method

The degree of crystallization is directly connected to the rate of bio-degradability and amorphous copolymers between L-lactide and glycolic acid monomers. The prepared poly(lactide-*co*-glycolide) shows faster degradation than PLA, and increasing glycolide contribution rate further increases [24]. Grafting copolymerization between L-lactide into chitosan (high content) using a tin catalyst increases the thermal stability and degradability rate [25].

4.3.5 Blending Method

Biodegradability rate can be increased in the blending method in between PLA and lactic acid (0%–5% of lactic acid content), and the changes were observed in chemical or physical properties. Lactate is an available substrate for different bacterial species to facilitate PLA degradation by providing carbon and energy source [26].

4.3.6 Nanocomposite Formation

PLA nanocomposites with montmorillonites (nanoclay) can enhance the degradation rate because hydroxyl groups belonging to the silicate layers facilitate the hydrolysis process. Nanoclays' effect on PLA biodegradability is enhanced by their excellent dispersion over the polymer surface and depending upon their chemical structures and affinity toward bacterium [27].

4.3.7 Summary

Along with conventional uses, exciting PLA or its stereocomplex PLA applications are possible due to its favorable mechanical properties, tunable degradation rates, and high biocompatibility. These specific properties are possible with its copolymer such as Poly(lactic-*co*-glycolic acid) (PLGA) and which can be further utilized in periodontal regenerative medicine.

4.4 Polyhydroxyalkanoates

Definition

PHAs are the family of bio-polyesters and are among well-known biodegradable plastics and well recognized as entirely biosynthetic and biodegradable with almost zero toxic waste be recycled into organic waste [30–32]. PHAs act as microbial reserve compounds for energy [33] and carbon [34] and hold a great potential to replace the petroleum-based compounds in the plastic market are termed as "green plastics" [35]. They show a wide range of properties that can be accessed biosynthetically by selected prokaryotes, and this opens the potential market for substituting petroleum-based products such as elastomers, thermoplastics by PHAs.

Rising concern for greenhouse gas emissions facilitates bio-based materials and promotes the PHA market in the future. PHAs commercial use increased from an estimated value of 10,000 metric tons (MT) to 34,000 MT in 2018, with a CAGR of 27.7% [36]. PHA biopolymer is expensive as compared to PP and PE. This high price is due to the high purity of substrates such as glucose, its production in various batches, and a large amount of solvents [2]. With the increasing availability of renewable raw material and increasing demand to use biodegradable polymers for biomedical use and food, applications are beneficial to the PHA market, and its market is expected to US$93.5 million by 2021 from US$73.6 million in 2016.

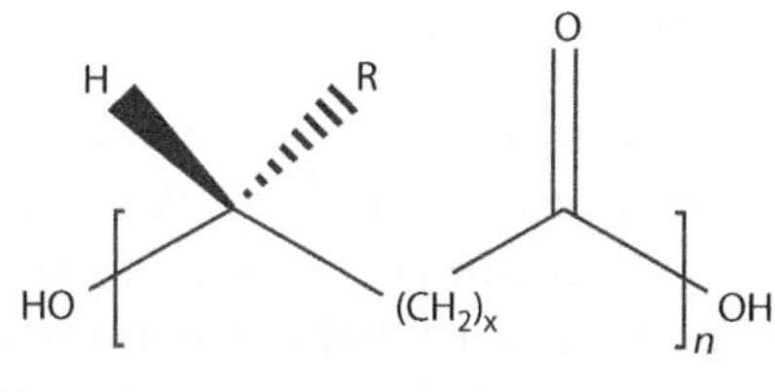

Figure 4.6 Structure of polyhydroxyalkanoates (x = number of methylene groups in the backbone; n = 1000-10000; R = alkyl groups, C_1-C_{13}).

4.4.1 Biosynthesis of Polyhydroxyalkanoates

PHA (Figure 4.6) biosynthesis begins from the feedstocks like hexoses, pentoses, lactose, maltose, lipids, alcohols, organic acids, or gases like carbon dioxide or methane under undesirable growth conditions due to imbalanced nutrient supply [2, 3]. PHAs are unique among biopolymer families whose production and degradation depend on the living cells. Hydroxyl groups of PHAs are produced in recombinant *Escherichia coli* JM109 in the presence of glycolate as the only carbon source. The propionate-CoA transferase (*pct*) gene from *Megasphaera elsdenii* and the β-ketothiolase (*bktB*) gene and *phaCAB* operon from *Ralstonia eutropha* H16 were introduced into *E. coli* JM109. Another alternative and convenient synthetic approach to synthesize PHAs is a chemical method that utilizes a ring-opening polymerization mechanism of β-lactones, including anionic, coordination-insertion, organo-catalyzed, enzymatic, and cationic processes.

In addition to PHA's biosynthesis, an alternative and convenient synthetic approach to obtain PHAs is via the ring-opening polymerization (ROP) of β-lactones, including anionic, "coordination-insertion," organo-catalyzed, enzymatic, and cationic processes.

4.4.2 Application of PHAs

PHAs can be used in various applications [37], such as

- Packaging films for foods, containers, and bags.
- Precursors for different chiral compounds.
- Act as a probe for drug delivery, herbicides, and insecticides.
- Disposable products like utensils, diapers, cups, etc.
- Medical applications such as surgical pins, staples, swabs, and wound dressings.

4.4.3 Biodegradability of PHAs

Biodegradation can be defined as the breaking down material when exposed to bacteria, fungi, or by other biological means, whether anaerobically or aerobically [38]. It can also be stated that the polymer degradation in biological space via enzymatic and non-enzymatic hydrolysis and not via thermal oxidation, radiolysis, or photolysis. The remarkable ability of PHAs to degrade biologically has made it an interesting and promising material for various applications [2, 3]. Increasing amounts of chemical waste pose a significant threat to the biosphere and damage the environment to a greater extent. So, it is a great matter of concern for the environment and materials having biospheric cycling are becoming important these days. PHAs are one of the polymeric materials synthesized by microorganisms under particular growth conditions and find a special place as biodegradable natural polyesters in the biosphere recycling [2, 3]. Biodegradation of PHAs is accelerated by microorganisms that reside in a specific natural environment such as soil. In this natural environment, PHA has the most exceptional capacity for degradation.

However, studies show that PHA degradation in soil was carried out in the laboratory, and different isolated cultures of microorganisms for PHA degradation and very few data are available on PHA biodegradation in land under field conditions. In one of the examples, PHA degradation under natural conditions showed that it too four weeks to degrade in the ground for a golf tee made of the polymer, but unfortunately, exact requirements for degradation of PHA were not mentioned yet. However, data suggest that the type of soil is an essential factor affecting PHA degradation [39–41].

4.4.4 Degradability Methods

Intracellular Degradation

Intracellular degradation takes place when carbon limitation conditions are stressed upon the bacterium. Accumulated PHA in the cells undergoes hydrolysis as carbon and energy sources. Further, it breaks down to 3-hydroxy alkanoic acid, a monomeric component by PHA depolymerase and oligomer hydrolase [42]. If the PHA is made up of one kind of monomer, such as 3-hydroxybutyrate, the resulting PHA is called poly(3-hydroxybutyrate) [P(3HB)] homopolymer, and it is the most common type of PHA which is synthesized by various bacteria naturally. P(3HB) is further degraded to 3-hydroxybutyric acid, which is oxidized by a dehydrogenase to acetoacetyl-CoA, converted into acetyl-CoA by beta ketothiolase

[43]. All these breakdown products of PHA are naturally found in animals. So, biodegradation of PHA does not lead to toxic products and can be termed biocompatible material [44, 45].

Extracellular Degradation

Extracellular depolymerase hydrolyzes partially crystallized P(3HB). These depolymerases comprises of a single peptide (22–58 amino acids) and three functional domains, catalytic domain (320–400 amino acids), linker domain (50–100 amino acids), and substrate-binding domain (40–60 amino acids) from N-terminal to C-terminal. [46, 47] The catalytic domain is further classified into two types of depolymerases, i.e., Type I and Type II, differing on the order of the sequential order of active amino acids forming a catalytic triad. Apart from these depolymerases, any lipases also possess the ability to hydrolyze poly(ω-hydroxyalkanoates) such as poly(6-hydroxyhexanoate) [P(6HHx)] and poly(4-hydroxybutyrate) [P(4HB)].

Specific enzymes, PHA depolymerases, present in the soil and aquatic microorganisms degrade the PHAs. Until this time, the identification of 600 PHA depolymerases from the wide society of microorganisms has been made. Various microorganisms in the soil, fresh waters, compost, and marine environments help in the degradation of PHAs. Bacteria present in marine environments such as *Pseudoalteromonas* sp. NRRL B-30083, *Marinobacter* sp. NK-1, *Alcaligenes faecalis* AE122, actinobacteria *Nocardiopsis aegyptia*, and *Streptomyces* sp. SNG9 are the few microorganisms that are known to be PHA degraders [46–48].

4.4.5 Summary

PHA is bio-based polyester that has a low softening temperature, and degradation occurs in the presence of the microorganism. The PHA is a promising material, and key parameters for degradation are microorganism secrets depolymerize. The industry can design suitable PHA materials for their needs because of the advantage of biocompatibility and degradability.

4.5 Conclusion and Future Development

The development of bio-derived polymer from renewable resources favors establishing a sustainable society, but non-degradability/biodegradability was a big issue for the world. In this chapter, we reviewed the recent trend for the development of bio-based materials, biodegradation, and biocompatibility, which increased fuel efficiency and reduced the threats over the plastics waste problem. The demand for bio-based plastics such as PA, PLA, and PHA has been raised in various fields and used in the automobile industry, packaging materials, and biomedical applications. The non-degradable thermoplastics especially Nylon^{TM} are used as packaging materials, fibers, film forms, and some of the PA, especially nylon 4, nylon 6, and nylon 6,6 degraded *in vivo* by using some microbes. Itaconic acid-based PAs minimize the plastics waste problem because of compostable nature inside the soil and photo-solubilization behavior in the soil. Therefore, much more work is needed to study a degradable route for predicting the development of PAs. The low softening temperature-based PLA are degraded with excellent tunable properties because of degradation with high biocompatibility. These specific properties are possible with its copolymer PLGA and can be further utilized in periodontal regenerative medicine. Restenosis is a common problem that occurs after a few months of angioplasty, and the reasons behind this might be blood vessel injury, inflammatory reaction and endothelial cell proliferation, etc., due to the use of metallic stents. Biodegradable drug-eluting stents with flexibility, high mechanical property, and specific drug-releasing features can replace the conventional ones to prevent restenosis. This book chapter has also covered the significant components of PHA degradation. All the factors affecting PHA degradation are interrelated. The key player of PHA degradation is the microorganisms that can secrete extracellular depolymerase enzymes. Besides, the degradation of PHA also plays a significant role in determining PHA application in various areas. PHA is a promising material for sustainable developments; understanding the mechanism of PHA degradation and the factors that affect its degradation will help the researcher design suitable materials according to the industrial needs. Novel synthetic pathways are still being developed to degrade the amide or ester bond of polymer, covering the polymer field's broad spectrum. Under these circumstances, much more research is needed to solve the plastic waste problem, which reduces cost and global warming.

References

1. Roopan, S.M., Surendra, T.V. *et al.*, Preparation and Properties of Biopolymers: A Critical Review, in: *Handbook of Polymers for Pharmaceutical Technologies: Structure and Chemistry*, V.K. Thakur and M.K. Thakur (Eds.), vol. 3, pp. 541–553, Wiley-Scrivener, Hoboken, NJ, USA, 2015.

2. Lendlein, A. and Sisson, A. (Eds.), *Handbook of Biodegradable Polymers: Isolation, Synthesis, Characterization and Applications*, Wiley-VCH, Weinheim, Germany, 2011.
3. Kutz, M. (Ed.), *Applied Plastics Engineering Handbook: Processing, Materials, and Applications*, 2 edition, Elsevier, William Andrew, Norwich, New York, 2016.
4. Parthiban, A., Monomers and Polymers Derived from Renewable or Partially Renewable Resources, in: *Synthesis and Applications of Copolymers*, A. Parthiban, (Ed.), pp. 101–124, Wiley-VCH: Weinheim, Germany, 2014.
5. a) Iwata, T., Biodegradable and bio-based polymers: Future prospects of eco-friendly plastics. *Angew. Chem.*, 54, 3210–3215, 2015. b) Ali, M.A., Kaneko, T., Syntheses of Aromatic/Heterocyclic Derived Bioplastics with High Thermal/Mechanical Performance. *Ind Eng Chem Res.*, 58, 15958, 2019.
6. Ali, M.A. and Kaneko, T., Microbe-Derived Itaconic Acid: Novel Route to Biopolyamides. Microbial Applications, in: *Microbial Applications*, vol. 2, V. Kalia (Ed.), pp. 279–289, Springer, Cham, 2017.
7. Yamano, N., Kawasaki, N., Ida, S., Nakayama, Y., Nakayama, A., Biodegradation of polyamide 4 *in vivo*. *Polym. Degrad. Stab.*, 137, 281, 2017.
8. van Es, D.S., van der Klis, F., Knoop, R.J.I., Molenveld, K., Sijtsma, L., van Haveren, J., Other Polyesters from Biomass Derived Monomers, in: *Bio-Based plastics: Materials and Applications*, S. Kabasci (Ed.), pp. 241–274, John Wiley & Sons Ltd., Sussex, United Kingdom, 2013.
9. Kobayashi, S. and Mullen, K., Encyclopedia of polymeric nanomaterials, in: *Polyamide Syntheses*, M.A. Ali (Ed.), pp. 1750–1762, Springer Berlin, Heidelberg, 2015.
10. Ali, M.A., Tateyama, S., Kaneko, T., Synthesis of rigid-rod but degradable biopolyamides from itaconic acid with aromatic diamines. *Polym. Degrad. Stab.*, 109, 367, 2014.
11. Ali, M.A., Tateyama, S., Oka, Y., Okajima, M., Kaneko, D., Kaneko, T., High-performance biopolyamides derived from itaconic acid and their environmental corrosion. *Macromolecules*, 46, 3719, 2013.
12. Zheng, Y., Yanful, E.K., Amarjeet, S., Bassi, A.S., A Review of Plastic Waste Biodegradation. *Crit. Rev. Biotechnol.*, 25, 243, 2005.
13. Tokiwa, Y., Calabia, B.P., Ugwu, C.U., Aiba, S., Biodegradability of Plastics. *Int. J. Mol. Sci.*, 10, 3722, 2009.
14. Dubois, P., Coulembier, O., Raquez, J.M. (Eds.), *Handbook of Ring-Opening Polymerization*, Wiley-VCH, Weinheim, Germany, 2011.
15. Misra, M. and Panday, J.K. (Eds.), *Biocomposites: Design and Mechanical Performance*, woodhead publishing, Cambridge, UK, 2015.
16. Lima, U.A., Aquarone, E., Borzani, W., Schmidell, W., *Biotechnologia Industrial Processos Fermentativos e Enzimaticos*, pp. 101–124, Editora Edgard Blücher Ltda, São Paulo, 2001.
17. Komesu, A., Oliveira, J.A.R.d., Martins, L.H.d.S., Wolf Maciel, M.R., Maciel Filho, R., Lactic acid production to purification: A review. *BioRes.*, 12, 4364, 2017.

18. Wee, Y., Kim, J., Ryu, H., Biotechnological production of lactic acid and its recent applications. *Food Technol. Biotechnol.*, 44, 163, 2006.
19. Abdel-Rahman, M.A. and Sonomoto, K., Opportunities to overcome the current limitations and challenges for efficient microbial production of optically pure lactic acid. *J. Biotechnol.*, 236, 176, 2016.
20. Wee, Y., Kim, H., Yun, J., Ryu, H., Pilot-scale lactic acid production *via* batch culturing of *Lactobacillus sp.* RKY2 using corn steep liquor as a nitrogen source. *Food Technol. Biotechnol.*, 44, 293, 2006.
21. Drumright, R.E., Gruber, P.R., Henton, D.E., Polylactic acid technology. *Adv. Mater.*, 23, 1841, 2000.
22. Puaux, J., Banu, I., Nagy, I., Bozga, G., A study of L-Lactide ring-opening polymerization kinetics. *Macromol. Symp.*, 259, 318, 2007.
23. Williams, D.F., Enzymatic hydrolysis of polylactic acid. *Eng. Med.*, 10, 5, 1981.
24. Vroman, I. and Tighzert, L., Biodegradable polymers. *Materials*, 2, 307, 2009.
25. Luckachan, G.E. and Pillai, C.K.S., Chitosan/oligo L-lactide graft copolymers: Effect of hydrophobic side chains on the physic-chemical properties and biodegradability. *Carbohyd. Polym.*, 64, 254, 2006.
26. Lee, J.C., Moon, J.H., Jeong, J., Kim, M.Y., Kim, B.M., Choi, M.C., Kim, J.R., Ha, C.S., Biodegradability of poly(lactic acid) (PLA)/Lactic acid(LA) blends using anaerobic disaster sludge. *Macromol. Res.*, 24, 741, 2016.
27. Fukushima, K., Abbate, C., Tabuani, D., Gennari, M., Camino, G., Biodegradation of poly (lactic acid) and its nanocomposites. *Polym. Degrad. Stab.*, 94, 1646, 2009.
28. Sun, X., Xu, C., Wu, G., Ye, Q., Wang, C., Poly(lactic acid-co-glycolic acid): applications and future prospects for periodontal tissue regenerations. *Polymers*, 9, 189, 2017.
29. Tamai, H., Igaki, K., Kyo, E., Kosuga, K., Kawashima, A., Matsui, S., Komori, H., Tsuji, T., Motohara, S., Uehata, H., Initial and 6-months results of biodegradable poly-l-lactic acid coronary stents in humans. *Circulation*, 102, 399, 2000.
30. Rydz, J., Sikorska, W., Kyulavska, M., Christova, D., Polyester-Based (Bio) degradable Polymers as Environmentally Friendly Materials for Sustainable Development. *Int. J. Mol. Sci.*, 16, 564, 2015.
31. Amass, W., Amass, A., Tighe, B., A review of biodegradable polymers: use, current developments in the synthesis and characterization of biodegradable polyesters, blends of biodegradable polymers and recent advances in biodegradable studies. *Polym. Int.*, 47, 89, 1998.
32. Amos, D.A. and McInerney, M.J., Poly-β-hydroxyalkanoates in synthrophomonas wolfei. *Arch. Microbiol.*, 152, 172, 1989.
33. Dunn, B., Kamath, H., Tarascon, J.M., Electrical energy storage for the grid: a battery of choices. *Science*, 334, 928, 2011.
34. Koller, M., Salerno, A., Braunegg, G., Polyhydroxyalkanoates: Basics, Production and Applications of Microbial Biopolyesters, in: *Bio-Based*

plastics: Materials and Applications, S. Kabasci (Ed.), pp. 137–170, John Wiley & Sons Ltd., Sussex, United Kingdom, 2013.
35. Anushri, S. and Archana, T., Polyhydroxyalkonates: Green Plastics of the Future. *Int. J. Biomed. Adv. Res.*, 2, 356, 2011.
36. Braunegg, G., Bona, R., Schellauf, F., Wallner, E., Polyhydroxyalkanoates (PHAs): Sustainable Biopolyester Production. *Polimery*, 47, 13, 2002.
37. Kai, D. and Loh, X.J., Polyhydroxyalkanotes: Chemical Modifications Toward Biomedical Applications. *Acs Sustain Chem Eng.*, 2, 106, 2014.
38. Ikada, Y. and Tsuji, H., Biodegradable polyesters for medical and ecological applications. *Macromol. Rapid Commun.*, 21, 117, 2000.
39. Jendrossek, D. and Handrick, R., Microbial degradation of Polyhydroxyalkanoates. *Annu. Rev. Microbiol.*, 56, 403, 2002.
40. Knoll, M., Hamm, T.M., Wagner, F., Martinez, V., Pleiss, J., The PHA depolymerase engineering database: a systematic analysis tool for the diverse family of polyhydroxyalkanoate (PHA) depolymerases. *BMC Bioinformatics*, 10, 89, 2009.
41. Mergaert, J. and Swings, J., Biodiversity of microorganisms that degrade bacterial and synthetic polyesters. *J. Ind. Microbiol.*, 17, 463, 1996.
42. Madison, L.L. and Huisman, G.W., Metabolic engineering of poly(3-hydroxyalkanoates): From DNA to plastic. *Microbiol. Mol. Biol. Rev.*, 63, 21, 1999.
43. Eggers, J. and Steinbüchel, A., Poly (3-hydroxybutyrate) degradation in Ralstonia eutropha H16 is mediated stereoselectively to (S)-3-hydroxybutyryl coenzyme A (CoA) via crotonyl-CoA. *J. Bacteriol.*, 195, 3213, 2013.
44. Handrick, R., Reinhardt, S., Focarete, M.L., Scandola, M., Adamus, G., Kowalczuk, M., Jendrossek, D., A new type of thermoalkalophilic hydrolase of *Paucimonas lemoignei* with high specificity for amorphous polyesters of short chain-length hydroxyalkanoic acids. *J. Biol. Chem.*, 276, 36215, 2001.
45. Tseng, C.L., Chen, H.J., Shaw, G.C., Identification and characterization of the Bacillus thuringiensis phaZ gene, encoding new intracellular poly-3-hydroxybutyrate depolymerase. *J. Bacteriol.*, 188, 7592, 2006.
46. Leathers, T.D., Govind, N.S., Greene, R.V., Biodegradation of poly(3-hydroxybutyrateco-3-hydroxyvalerate) by a tropical marine bacterium, Pseudoalteromonas sp. NRRL B-30083. *J. Polym. Environ.*, 8, 119, 2000.
47. Kasuya, K., Mitomo, H., Nakahara, M., Akiba, A., Kudo, T., Doi, Y., Identification of a marine benthic P(3HB)-degrading bacterium isolate and characterization of its P(3HB) depolymerase. *Biomacromolecules*, 1, 194, 2000.
48. Kita, K., Mashiba, S., Nagita, M., Ishimaru, K., Okamoto, K., Yanase, H., Cloning of poly(3-hydroxybutyrate) depolymerase from a marine bacterium, *Alcaligenes faecalis* AE122, and characterization of its gene product. *Biochim. Biophys. Acta.*, 1352, 113, 1997.

49. Patel, M., Ou, M., Ingram, L. *et al.* Fermentation of sugar cane bagasse hemicellulose hydrolysate to l(+)-lactic acid by a thermotolerant acidophilic Bacillus sp.**. *Biotechnology Letters,* 26, 865, 2004.
50. Bischoff, K.M., Liu, S., Hughes, S.R. *et al.* Fermentation of corn fiber hydrolysate to lactic acid by the moderate thermophile Bacillus coagulans. *Biotechnol Lett.,* 32, 823, 2010.
51. Ouyang, J., Cai, C., Chen, H. *et al.* Efficient Non-sterilized Fermentation of Biomass-Derived Xylose to Lactic Acid by a Thermotolerant Bacillus coagulans NL01. *Appl Biochem Biotechnol.,* 168, 2387, 2012.
52. Yun, J. -S., Wee, Y. -J., Ryu, H.-W., Production of optically pure l(+)-lactic acid from various carbohydrates by batch fermentation of Enterococcus faecalis RKY1, *Enzyme and Microbial Technology*, 33, 4, 416, 2003.

5

A Review on PHAs: The Future Biopolymer

S. Mohapatra[1,2*], K. Vishwakarma[1], N. C. Joshi[1], S. Maity[3], R. Kumar[4], M. Ramchander[5], S. Pattnaik[6] and D. P. Samantaray[6]

[1]*Amity Institute of Microbial Technology, Amity University, Amity University, Uttar Pradesh, India*
[2]*Department of Infection Biology, School of Medicine, Wonkwang University, South Korea*
[3]*Centre for Converging Technologies, University Innovation Cluster-Biotechnology (UIC-B), University of Rajasthan, Jaipur, India*
[4]*Department of Environmental Sciences, H.N.B. Garhwal University (a Central University), Uttarakhand, India*
[5]*Department of Biochemistry, Mahatma Gandhi University, Anneparthy, Nalgonda, Telangana State, India*
[6]*Department of Microbiology, Orissa University Agriculture Technology, Odisha, India*

Abstract
PHAs have shared an immense impact on the replacement of LDPE. In order to control the environmental pollution without compromising the application, the only polymer having more than 150 varieties is polyhydroxyalkanoates (PHAs); hence, one biopolymer that can easily replace the plastic is PHAs. As the need of the hour is making our environment green and healthy, so a single replacement that comes to mind is PHAs. Here, in this book chapter, we have compiled various types of biopolymer and biodegradable polymers and also discussed about the comparison. Still, a lot of research needs to be conducted to find out the potential microbe that can synthesize more than three varieties of monomeric polymer in presence of different substrate. Moreover, economic production of PHAs still is major challenges although around more than hundred companies are working with PHAs as its base material.

***Keywords*:** Polyhydroxyalkanoates, green polymer, green energy, municipality waste

**Corresponding author*: swatimohapatraiitr@gmail.com

Bibhuti Bhusan Mishra, Suraja Kumar Nayak, Swati Mohapatra, and Deviprasad Samantaray (eds.) Environmental and Agricultural Microbiology: Applications for Sustainability, (83–100) © 2021 Scrivener Publishing LLC

5.1 Introduction

Pollution is a process of contamination that affects our environment such as land, water, and air. Due to these effects, it becomes harmful for the human or not safe and suitable to use. These days, especially, solid waste is becoming surplus in the urban area, and with the due period, it is becoming out of control in India. Generally, 31% of the population of India is located in metropolitan zones, 337 million (Census of India, 2011) creates a colossal 143,449 metric lots. Due to the loss of neighborhood groups on the preparation, plan, application, and tracking of municipal solid waste control structures, it is becoming too much difficult to manage. Municipality solid waste ma nagement spends lots of money on collection to disposal and degradation of the waste. The expenditure includes environmental, transportation, and manpower expenditure costs that include investment, operation expenditure, mecilineous expenditure, many more. Due to improper rules, regulation, and understanding, it leads to poor or uncontrolled management of polluted environment.

Although our environment perpetually involves in the maintenance of a healthy ecosystem, the occurrence of any interruption in the ecosystem makes a bumpy effect during the balancing of the ecosystem. In our ecosystem, different types of pollutants contribute to different types of pollution. Among all types of pollutant, plastic pollutants are the most dangerous as these are never degrading pollutants which need proper disposal and management [1]. This tiny plastic becoming a choice for many applications especially low-density polyethylene (LDPE) having stability, durability, permeability, and apposite motorized and thermal properties. After monotonous use, dumping of the

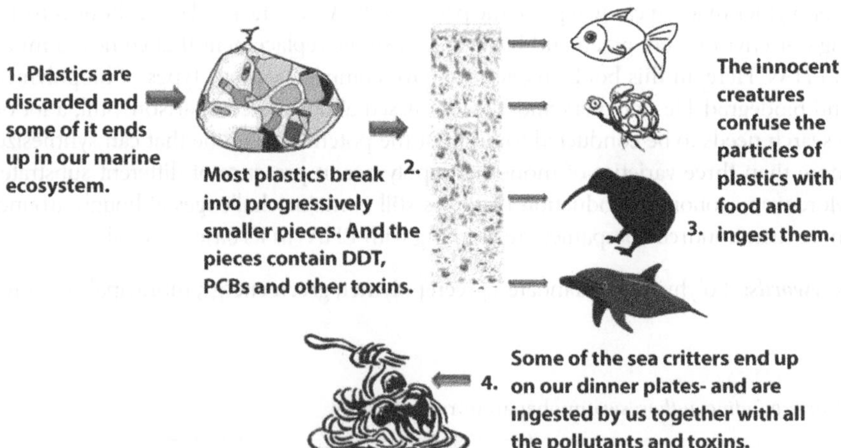

Figure 5.1 Environmental pollution: Contamination begins with us-and it ends with us.

LDPE here and there which flows to the drainage system is generally connected to water bodies that flow and meets to marine environments which subsequently threatening marine life habitats. This accumulation leach toxic chemicals in our soil, and sometimes, wildlife can also entangle the plastic for food which leads to blockage or starve to death (Figure 5.1), hence use of these plastic destroying precious bio-diversity.

So to get rid of this "poison pill" named plastic, the production of bioplastic is becoming an eco-friendly alternative and unique access to worldwide [2]. It will decrease the environmental toxic waste as well as decrease the risk factor of life; therefore, it can be called in one term that bioplastic seems to be a remedy to save this earth. It is consequently basic to believe that any step toward the widespread extinction of plastic bags would be praised and more prominently, buttressed by deed. Hence, global awareness is a solution to the problem of plastic bags [3]. Fragile life of the environment can be saved by counter-intuitive ideas to stop using plastic bags.

5.2 Green Plastic: Biodegradable Polymer Used as Plastic

Some biopolymers as mention below are the best examples of naturally synthesized polymers, such as carbohydrates and proteins that are appropriate examples of biopolymers [4, 5]. Most of the biopolymers are previously being manufactured commercially on large-scale basis, having plastic property but in small percentage; however, it would significantly decrease our dependence on manufactured and non-renewable resources (Figure 5.2). These are having different origins and different applications also having different degradation processes and time [6]. Some are described below.

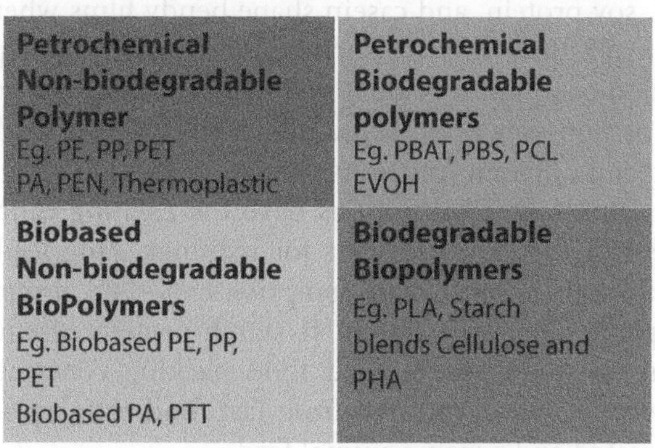

Figure 5.2 Different types of polymer available in our society.

i. Cellulose is the most plentiful carbohydrate in the world; 40% of all organic matter is made up of cellulose. Easy availability of cellulose enhances the possibility of using cellulose as an affordable biopolymer.
ii. Starch is present in potatoes, corn (maize), tapioca (cassava), wheat, and certain other plants. Annual global production of starch is properly over 70 billion pounds, with a good deal of it getting used for non-meals purposes, such as manufacturing paper, cardboard, textile sizing, and adhesives. Starch-based bioplastics are critical and not fully biodegradable biopolymer; however, it could be processed easily as synthetic polymers, like film extrusion and injection molding, dining utensils, plates, cups, and other merchandise, had been made with starch-based plastics
iii. Protein: Collagen is the maximum considerable amount protein found in mammals. Gelatin, a mutate collagen, is employed in sausage casings and drugs for tablets.

- Casein, commercially manufactured in particular from cow's skimmed milk, is utilized in adhesives, binders, defensive coatings, and other merchandise.
- Soya protein and zein (from corn) are ample plant proteins. They are used for manufacturing adhesives and coverings for paper and cardboard. Concentration in soyabeans has been revived, evoking Ford's early efforts. In research laboratories, it has been proven that soy protein, with and without cellulose extenders, can be prepared with cutting-edge extrusion and injection molding methods.
- Numerous water-soluble biopolymers along with starch, gelatin, soy protein, and casein shape bendy films when accurately plasticized. While such films have appeared especially as food coatings, it is identified that they have got potential use as non-supported stand-alone sheeting for food wrapping and different functions.
- Starch-protein compositions have the exciting function of assembly dietary necessities for livestock. Hog feed, as an instance, is suggested to comprise 13%–24% protein, and supplemented with starch. If starch-protein plastics have been commercialized, used food packing containers and service ware accumulated from fast food eating places can be treated and become animal feed.

iv. Polyesters are produced by microorganisms like bacteria and may be made commercially on huge scales via fermentation technique. They are being used in biomedical applications. Polyesters at the moment are comprised of natural sources-like starch and sugars-through massive-scale fermentation processes and used to manufacture water-proof bottles, consuming utensils, and other products. Apart from the above polymers, some different natural materials are also replaced by conventional plastic which are biodegradable.

v. Lactic acid is currently commercially produced on massive scales by the fermentation of sugar feed-stocks obtained from sugar beets or sugar cane, or from the conversion of starch from corn, potato peels, or other starch sources. It can be polymerized to produce poly-lactic acid, which has been already gained commercial applications in drug encapsulation and biodegradable medical devices. Poly-lactic acid has become a significant commercial polymer these days. The transparency property makes it reusable. Most of the place its gaining the market like storage of food, packing and serving of food, as a raping agent. The best use till date has been observed is in biomedical use example preparation of sutures surface, prosthetic materials, and tissue engineering.

vi. Triglycerides create a big a part of the lipids in animal and plant cells. Over 16 billion pounds of vegetable oils are produced in the United States each year, mainly from soybean, flax, and rapeseed. Triglycerides are another promising raw material for producing plastics. Triglycerides have recently become the basis for a new family of sturdy composites. With glass fiber reinforcement, they can be made into long-lasting durable materials with applications in the manufacture of agricultural equipment, in the automotive industry, construction, and other areas. Fibers other than glass can also be used in the process, like fibers from jute, hemp, flax, wood, and even straw or hay. If straw could replace wood in composites now used in the construction industry, it would provide a new use for an abundant, rapidly renewable agricultural commodity and at the same time conserve less rapidly renewable wood fiber.

These natural raw substances are ample, renewable, and biodegradable, making them appealing feed-stocks for bioplastics, a brand new era of environmentally pleasant plastics. The nondestructive, sizeable use of those new plastics will depend upon newly rising technologies that may be a success inside the market [7]. Mainly acceptance of these technologies and the commitment toward the society plays vital role upon resource conservation, environmental upkeep, and sustainable technologies to make this idea and technology successful. Humans certainly want to stay in extra concord with nature so bioplastics may be a suitable replacement in the cutting-edge age of plastics.

5.3 Difference Between Biopolymer and Bioplastic

Biopolymers and bioplastics are exclusive terms but frequently they are harassed with one another. However, these are one-of-a-kind materials. Biopolymers are polymers located inside the living bodies or synthesized by using living organisms. Additionally, these polymers involve in fabricating bioplastics with the aid of polymerization.

Bioplastics are the plastics that are formed by using biodegradable polymers. The splendid vehicle manufacturer Henry Ford devised a manner of producing bioplastic vehicle sections from soybeans. However, the start of the Second World War stopped the manufacturing of bioplastic motors. It is simplest and currently that bioplastic cars have made a return because of the improvement of recent production techniques through biotechnology [8].

Thus, the prominence interest in pollution free environment has shifted for the development of biodegradable plastics which is an important polymer used in our everyday life. Thereby, the use of microbes (products of microorganisms) for the production of bioplastic is becoming an alternative and a unique way worldwide [9]. It will decrease environmental pollution in addition to decrease the risk factor of life, therefore it may be said in one phrase that bioplastic is green or eco-friendly.

Although there are different types of biopolymer in form of bioplastic that are available in the market, still most of them are losing interest from the customer in some aspect such as diversified application and use of plants is a destructive method [10]. However, one polymer which has been identified in recent years that is polyhydroxyalkanoates (PHAs) which is best in all aspects and full filling all the criteria of petroleum-based plastic property which is described in detail below.

5.4 Polyhydroxyalkanoates

History of PHAs: The innovation of this newly emerging bio-based biodegradable polymer was first discovered by the eminent French scientist Lemogine in 1926 while he was working on *Bacillus megaterium*. Initially, he stated its inclusion body made up of lipids found in the cytoplasm of bacterial cells. His further finding leads to the development of the best bio-based biodegradable biopolymer to the world of material science when he found these granules mimic the properties of petroleum-derived plastic. Since the discovery of PHAs, till now various studies have been conducted and this research leads to the discovery of new polymers having diversified properties. Never the less this new monomeric composition of the polymer opening the door for new applications having tremendous social and economic desire.

PHAs have shared an immense impact on the replacement of all types of polymer. In controlling a healthy aquatic environment as well as soil, water, and air pollutants, PHAs are the hope for today's life. Future application of the PHAs polymers depends on the selection of the potential microbes by a polyphasic approach. As cytosolic deposition of the microorganisms within a stress-prone environment and types of nutrition play a key role in PHAs production [11]. Low-cost production is a major challenge, but the curiosity of researchers can economically enhance the productivity of polymers using low-cost substrate and efficient microbes.

PHAs have shared an immense impact on the replacement of LDPE. In controlling a healthy aquatic environment as well as soil, water, and air pollutants, PHAs are the hope for today's life [12, 13]. Future application of the PHAs polymers depends on the selection of the potential microbes by polyphasic approach. As cytosolic deposition of the microorganisms within a stress-prone environment and types of nutrition play key role in PHAs production. Low-cost production is a major challenge, but the curiosity of researchers can enhance the productivity of polymers in an economical fashion.

5.5 Polyhydroxyalkanoates and Its Applications

The endocellular PHAs are composed of biosynthesized hydroxy fatty acids and saved as lipid inclusions; when the carbon supply is considerable and growth conditions become limited through exhaustion of some key nutrient such as nitrogen, phosphorus, oxygen, or sulfur, many prokaryotes can synthesis intracellular storage compounds [14, 15]. These granules act as carbon and energy reserves which can be utilized when the growth of microbes are resumed. PHAs normally can be identified through the use of iodine and lipophilic dyes, respectively.

PHAs accumulation is one of the responses toward pressure practiced by means of microbes dwelling at unique ecological niches inclusive of estuarine sediments, marine habitat, rhizosphere, groundwater sediments,

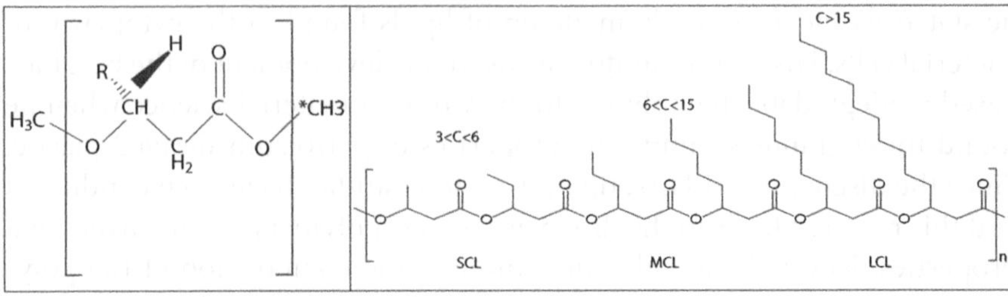

Figure 5.3 Structure and types of PHAs monomer.

waste, and sludge. When present environments are frequently rich in organic content and less rich in nitrogen content that enhance PHAs accumulation by microbes to meet metabolic energy source necessities through carbon famishment condition [16]. PHAs can be divided into three broad classes such as short (up to C5 carbon atom), medium (C6 to C14 carbon atom), and long-chain length PHAs (more than C14), based on the number of carbon atoms in the polymer chain (Figure 5.3). More than 150 different monomer of PHAs has been reported [17] and molecular weight of these polymers range between 200 and 300 kDa depending on the microbes and the fermentation conditions [18].

5.6 Microorganisms Producing PHAs

The PHAs extracted from various bacteria are widely used for the preparation of plastics materials, medical implants, drug delivery carriers, printing and photographic materials, nutritional supplements, drugs, and fine chemicals [19]. However, the broadly spread substitution of conservative plastics has been restricted due to their high manufacturing cost, which holds back PHAs successful commercialization [20]. Thus, more efforts are needed for making this method economically achievable by analyzing the inherent mechanism of the PHAs accumulation method and civilizing its manufacture.

Utilization of various inexpensive carbon sources such as palm oil mill effluent, pulp industry waste, cane molasses, whey lactic acid, olive-mill wastewater, waste lipids like cooking oil, non-edible acidic oil cake, agro waste like lignocellulosic hydrolysates [21], soya waste, malt waste, date syrup [22], animal-derived waste, sugar cane molasses, beet molasses, soya molasses [23], and industrial by-products like olive-mill wastewater, wastes from pulp industry [24], vinasse from the ethanol industry, whey from dairy and cheese industry [16], and waste glycerol from biodiesel industry [22, 25] has proved to have cost affordable for PHAs production. However, very few reports are available regarding PHAs production by inexpensive carbon sources [26]. Process economics of industrial PHAs production study reveals that use of cheap renewable carbon source like agricultural, municipal, and industrial wastes as well as their by-products as a feedstock can reduce the production cost as much as 40%–50% which can serve as a potential alternative carbon source to synthetic carbon sources such as glucose [24]. In light of the above, here, we have enlightened on microbes involved in production of this novel eco-friendly polymer PHAs, its raw material, productivity, advantages, disadvantages, and its possible applications (Tables 5.1 to 5.3).

A. Examples of potential Gram-negative marine bacteria

- A halophilic Gram-negative culture of *Halomonas boliviensis* LC1 (DSM 15516) could form P(3HB) (scl-PHA) from starch hydrolysate under reasonably salty environments.
- Under constrained nitrogen circumstances using acetate, glucose, methanol, pentothal, propionic acid, or valeric acid as substrates, strains of *Methylobacterium* and *Paracoccus* are correct for producing copolymers, together with poly (R-3-hydroxybutyrate-co-hydroxyvalerate) (PHBV) [27, 28].
- A marine isolate SM-P-3M was originated hopeful for PHA production. PHA accumulation for SM-P-3M was 75% PHA/cell dry weight (CDW) and recognized as *Halomonas hydrothermalis* by MTCC [29].
- PHAs were started to be accumulated by *Vibrio* spp. strain M11, M14, M20, and M31, marine isolates. Strain M11 amassed PHB in concentrations as excessive as 41% of cellular dry weight while grown in medium containing 4% of sodium chloride [30].

B. Examples of potential Gram-positive marine bacteria
Gram-positive bacteria play a beneficial role over Gram-negative bacteria owing to their deficiency in lipopolysaccharide (LPS). Till the studies, lack of LPS has made PHAs production in Gram-positive bacteria a progressed basis of raw material for biomedical applications.

- PHA production in Gram-positive bacteria has been mentioned in genera *Bacillus, Caryophanon, Clostridium, Corynebacterium, Micrococcus, Microlunatus, Microcystis, Nocardia, Rhodococcus, Staphylococcus,* and *Streptomyces.*

Table 5.1 Bacteria are used for the production of PHA from wastes.

Strains	Substrates (waste material)	Types of PHA	Productivity	References
Pseudomonas guezennei	biovar. tikehau, Coprah oil	mcl PHA	63%	[32]
Cupriavidus necator H16	Crude palm kernel oil, olive oil, sunflower oil, palm kernel oil, cooking oil, palm oli, crude palm oil, coconut oil + sodium propionate	P(3HB-co3HV)	65%–90%	[17]
B. thuringiensis EGU45	1%–10% CG [v/v] and nutrient broth	P[3HB-co-3HV]	1.54 g/L to 1.83 g/L	[33]
Cupriavidus necator	Bagasse hydrolysates	P(3HB)	54%	[25]
B. thuringiensis IAM12077	-	PHB	4 g/L	
IFO3924	Palm oil	mcl PHA	39%	[25]
B. subtilis OK2.	agro-industrial waste, orange peel	PHB	1.24 g/L	[25]
Bacillus cereus	-	PHB	0.436 g/L	[34]
Recombinant *Escherichia coli*	Soybean oil	P(3HB-co3HHx-co-3HO)	6	[35]
Comamonas testosterone	Castor oil, coconut oil, mustard oil, cottonseed oil, groundnut oil, olive oil, sesame oil	MCL-PHA	79%–88%	[36]

(Continued)

Table 5.1 Bacteria are used for the production of PHA from wastes. (*Continued*)

Strains	Substrates (waste material)	Types of PHA	Productivity	References
Recombinant Cupriavidus necator	Palm kernel oil, palm olein, crude palm oil, palm acid oil	P(3HB-co3HHx)	40%–90%	[17]
Bacillus megaterium	Beet molasses, date syrup	P(3HB)	~50%	[37]
Burkholderia sp. USM (JCM 15050)	Palm oil derivatives, fatty acids, glycerol	P(3HB)	22%–70%	[33]
Cupriavidus necator DSM 545	Waste glycerol	P(3HB)	50%	[33]
Thermus thermophilus HB8	Whey	P(3HV-co3HHp-co-3HNco-3HU)	36%	[38]
Alcaligenes latus DSM 1124	Soya waste, malt waste	P(3HB)	33.71%	[39]
Pseudomonas aeruginosa NCIB 40045	Waste frying oil	mcl PHA	29%	[40]
Bacillus cereus PHA 008	Palm oil mill effluent [POME]	PHA	64.09% of DCW	[17]

- Compared to Gram-negative bacteria, Gram-positive bacteria have been mostly produce scl-PHA. High scl-PHA content of 82% cell dry mass (CDM) has been formerly produced from *Streptomyces* sp. (ATCC 1238) and the growth was on glucose.
- New *Bacillus* spp. ND153, ND97, and QN194 synthesized poly (3-hydroxybutyrate) from glucose, isolated from the Vietnamese mangrove [24, 31].
- Report utters, halophilic archaeal species also produce PHAs, and the genera include *Haloferax*, *Haloarcula*, *Natrialba*, *Haloterrigena*, *Halococcus*, *Haloquadratum*, *Halorubrum*, *Natronobacterium*, *Natronococcus*, and *Halobacterium* [24].

Table 5.2 Bacteria used for production of PHA.

Strains	Substrates	Productivity	Types of PHA	References
Bacillus sp. ND153	Glucose and propionate	3.6 g/L	PHBV	[41, 42]
Yangia pacifica QN271	Glucose and propionate	5.1 g/L	PHB	[41]
Bacillus megaterium A9	Activated sludge	74.00	PHB n12888	[30]
B. thuringiensis	Glucose	11.30	PHB	[43]
Bacillus sp. AS 3-2	Yeast extract	59.90	2-methyl-3-HB	[44]
Bacillus cereus	Glucose	13.77	PHB-3HHX	[34]
Bacillus mycoides DFC1	Glucose	76.32	PHB	[34]
Bacillus sp. SW1-2	Glucose	36.00	PHB	[45]
Bacillus megaterium uyuni S29	Glucose	70.00	PHB	[45]
Bacillus thuringiensis IAM12077	Glucose	64.16	PHB	[46]
Paenibacillus durus BV-1	Fructose	0.93g/L	PHB	[46]
Bacillus sp. Ti3	Starch	58.73	PHB	[46]
Bacillus aryabhattai	Sucrose & trace glucose and fructose	57.62	PHA	[32]
Bacillus sp.	Sucrose	51.49	PHA	[46]
Lysinibacillus sp. CH-N5	Glucose	80.94	PHB	[15]
Bacillus licheniformis	Glucose	53.01	PHB	[14]
Bacillus subtilis	Cashew fruits drink	4.40	PHB	[30]
Bacillus sp.	Date syrup	70.50	PHA 5433	[30]
B. cereus PHA 008	Palm oil mill effluent	64.09	P(3HB) 2345	[30]

Derivation: MCL-PHA, medium-chain-length polyhydroxyalkanoate; P(3HB), poly(3-hydroxybutyrate); P(3HB-co-3HV), poly(3-hydroxybutyrate-co-3-hydroxyvalerate); P(3HB-co-3HHx), poly(3-hydroxybutyrate-co-3-hydroxyhexanoate); P(3HV-co-3HHp-co3HN-co-3HU), poly(3-hydroxyvalerate-co-3-hydroxyheptanoate-co-3-hydroxynonanoate-co-3-hydroxyundecanoate).

Table 5.3 Advantages, application, and disadvantages of PHAs.

Advantages	Application	Disadvantages
Biodegradable	Domestic plastic, fibers, ruber products	Short polymer chains cause to let through oxygen
Non-toxic	Biomedical devices	Costly affair: small production high cost
Non-immunogenic	Pharmaceuticals	Instability of PHAs granules
Non-carcinogenic	Adhesives	Secondary metabolites production during fermentation
Non-thrombogrnic	Cosmetics	Low physical, chemical, and mechanical resistance
Sustainable materials	Oil industry	Expensive than conventional plastic including the green plastic: poly-lactic acid (PLA)
Produced from renewable resources	Textiles and clothing	
Control water and soil pollution	Automotive	
Enhances agricultural productivity		
Condense global warming		
Create a healthy aquatic ecosystem		

5.7 Advantages

Besides being available on a sustainable basis, biopolymers have several financial and environmental benefits.

- Biopolymers can also show an asset to waste processing. For instance, changing the polyethylene used in covered papers with the aid of a biopolymer definitely assist get rid of plastic scraps occurring in compost.
- Consumers have an active interest in biopolymers too, as conventional plastics are frequently seen as environmentally unfavorable.
- Sustainable plastics could therefore offer lots off advantages.
- The foremost benefit of the biodegradable container is that it could be composted.
- The biodegradability of raw materials now does not essentially suggest that the product or bundle crafted from them (Lined paper) is itself compostable.
- Biopolymers also can have blessings for waste processing.

5.8 Conclusion and Future Prospective

PHAs is the best green polymer having biological origin, is biodegradable, and can replace most of the conventional plastic. It has been observed that PHAs served lots of properties to replace LDPE. In order to control the environmental pollution without compromising the application, the only polymer having more than 150 varieties is PHAs; hence, one biopolymer that can easily replace the plastic is PHAs. As the need of the hour is make our environment green and healthy so a single replacement is PHAs. Moreover, economic production of PHAs still is major challenges although around more than hundred companies are working with PHAs as its base material. Till date, in all the laboratory condition, PHAs have shown its most flexible and adaptable polymer and easy to use.

References

1. Chanprateep, S. Buasri, K. Muangwong, A. Utiswannakul, P., Biosynthesis and biocompatibility of biodegradable poly(3-hydroxybutyrate-co-4-hydroxybutyrate). *Polym. Degrad. Stab.*, 95, 2003, 2010.

2. Mohapatra, S., Pattnaik, S., Maity, S., Sharma, S., Akhtar, J., Pati, S., Samantaray, D.P., Varma, A., Comparative analysis of PHAs production by Bacillus megaterium OUAT 016 under submerged and solid-state fermentation. *Saudi J. Biol. Sci.*, 27, 1242, 2020.
3. Mohapatra, S., Mohanta, P.R., Sarkar, B., Daware, A., Kumar, C., Samantaray, D.P., Production of Polyhydroxyalkanoates (PHAs) by Bacillus Strain Isolated from Waste Water and Its Biochemical Characterization. *Proc. Natl. Acad. Sci. India Sect. B Biol. Sci.*, 87, 459, 2017.
4. Maity, S., Das, S., Mohapatra, S., Tripathi, A.D., Akthar, J., Pati, S., Pattnaik, S., Samantaray, D.P., Growth associated polyhydroxybutyrate production by the novel *Zobellellae tiwanensis* strain DD5 from banana peels under submerged fermentation. *Int. J. Biol. Macromol.*, 153, 461, 2020.
5. Pati, S., Maity, S., Dash, A. Jema, S., Mohapatra, S., Das, S., Samantaray, D.P., Biocompatible PHB Production from Bacillus Species Under Submerged and Solid-State Fermentation and Extraction Through Different Downstream Processing. *Curr. Microbiol.*, 77, 1203, 2020.
6. Mohapatra, S., Maity, S., Dash, H.R., Das, S., Rath, C.C., Samantaray, D.P., Bacillus and biopolymer: Prospects and challenges. *Biochem. Biophys. Reports*, 12, 206, 2017.
7. Mohapatra, S., Sarkar, B., Samantaray, D.P., Daware, A., Maity, S., Pattnaik, S., Bhattacharjee S., Bioconversion of fish solid waste into PHB using *Bacillus subtilis* based submerged fermentation process. *Environ. Technol.*, 38, 3201, 2017.
8. Mohapatra, S., Samantaray, D.P., Samantaray, S.M., Study on Polyhydroxyalkanoates Production using Rhizospheric Soil Bacterial Isolates of Sweet Potato. *Indian J. Sci. Technol.*, 8, 57, 2015.
9. Mohapatra, S., Samantaray, D.P., Samantaray, S.M., Mishra, B.B., Das, S., Majumdar, S., Pradhan, S.K., Rath, S.N., Rath, C.C., Akthar, J., Achary, K.G., Structural and thermal characterization of PHAs produced by *Lysinibacillus* sp. through submerged fermentation process. *Int. J. Biol. Macromol.*, 93, 1161, 2016.
10. Nayak, S.K., Mishra, B.B., (Ed.), *Frontiers in Soil and Environmental Microbiology*, CRC Press, Boca Raton, US, 2020.
11. Mohapatra, S., Samantaray, D.P., Samantaray, S.M., Impact of process recovery on pha production by *Bacillus thuringiensis RKD-12. Pollut. Res.*, 34, 395, 2015.
12. Nayak S.K., Dash B., Baliyarsingh B., Microbial Remediation of Persistent Agro-chemicals by Soil Bacteria: An Overview, in: Microbial Biotechnology, J.K. Patra, G. Das, H.S., Shin, (Eds.) Vol-II, pp. 275–301, Springer, Singapore, 2018.
13. Baliyarsingh, B., Nayak, S.K., Mishra, B.B., Soil Microbial Diversity: An Ecophysiological Study and Role in Plant Productivity, in: *Advances in Soil Microbiology: Recent Trends and Future Prospects, Microorganisms for Sustainability, vol 4*, T.K. Adhya, B.B. Mishra, K. Annapurna, D.K. Verma, U. Kumar (Eds.), vol 4, pp. 1–17, Springer Nature, Singapore, 2017.

14. Dash, S., Mohapatra, S., Samantaray, D.P., Sethi, A.K., Production of Polyhydroxyalkanoates by Sugar Cane Rhizospheric Soil Bacterial Isolates. *J. Pure Appl. Microbiol.*, 8, 4889, 2014.
15. Samantaray, D.P., and Samantaray, S.M., Mohapatra, S., Phylogenetic heterogeneity of the rhizospheric soil bacterial isolates producing PHAs revealed by comparative analysis of 16s-rRNA. *Int. J. Curr. Microbiol. Appl. Sci.*, 3, 680, 2014.
16. Koller, M., Gasser, I., Schmid, F., Berg, G., Linking ecology with economy: Insights into polyhydroxyalkanoate-producing microorganisms. *Eng. Life Sci.*, 11, 222, 2011.
17. Lee, W.-H., Loo, C.-Y., Nomura, C.T., Sudesh, K., Biosynthesis of polyhydroxyalkanoate copolymers from mixtures of plant oils and 3-hydroxyvalerate precursors," *Bioresour. Technol.*, 99, 6844, 2008.
18. Keshavarz, T., Roy, I., Polyhydroxyalkanoates: bioplastics with a green agenda, *Curr. Opin. Microbiol.*, 13, 321, 2010.
19. Chen, G.G.-Q. (Ed.), *Plastics from Bacteria: Natural Functions and Applications*, vol. 14, Springer, Berlin, Heidelberg, 2010.
20. Pati, S., Maity, S., Samantaray, D.P., Mohapatra, S., Bacillus and Biopolymer with Special Reference to Downstream Processing. *Int. J. Curr. Microbiol. Appl. Sci.*, 6, 1504, 2017.
21. Cesário, M.T., Raposo, R.S., de Almeida, M.C.M.D., van Keulen, F., Ferreira, B.S., da Fonseca, M.M.R., Enhanced bioproduction of poly-3-hydroxybutyrate from wheat straw lignocellulosic hydrolysates, *N. Biotechnol.*, 31, 104, 2014.
22. Chee, J.-Y., Tan, Y., Samian, M.-R., Sudesh, K., Isolation and Characterization of a *Burkholderia* sp. USM (JCM15050) Capable of Producing Polyhydroxyalkanoate (PHA) from Triglycerides, Fatty Acids and Glycerols. *J. Polym. Environ.*, 18, 584, 2010.
23. Du, C., Sabirova, J., Soetaert, W., Ki Carol Lin, S., Polyhydroxyalkanoates production from low-cost sustainable raw materials, *Curr. Chem. Biol.*, 6, 14, 2012.
24. Queirós, D., Rossetti, S., Serafim, L.S., PHA production by mixed cultures : a way to valorize industrial waste. *Bioresour. Technol.*, 157, 197, 2014.
25. Chee, J.Y., Yoga, S.S., Lau, N.S., Ling, S.C., Abed, R.M., Sudesh, K., Bacterially Produced Polyhydroxyalkanoate (PHA): Converting Renewable Resources into Bioplastics, in: *Current Research, Technology and Education Topics in Applied Microbiology and Microbial Biotechnology*, A. Méndez-Vilas (Ed.), pp. 1395–1404, Formatex Research Center, Spain, 2010.
26. Liau, C.P., Bin Ahmad, M., Shameli, K., Yunus, W.M.Z.W., Ibrahim, N.A., Zainuddin, N., Then, Y.Y., Preparation and Characterization of Polyhydroxybutyrate/Polycaprolactone Nanocomposites, 2014, 572726, 2014.

27. Mohanty, S., Samantaray, D.P., Mohapatra, S., Rath, S.N., Mohanty, D.P., Rath, C.C., Studies on Biodegradation of Medium Chain Length PHAs by Soil Bacterial Isolates, *J. Adv. Microbiol.*, 1, 350, 2014.
28. Poli, A., Di Donato, P., Abbamondi, G.R., Nicolaus, B., Synthesis, Production, and Biotechnological Applications of Exopolysaccharides and Polyhydroxyalkanoates by Archaea. *Archaea*, 2011, 693253, 2011.
29. Shrivastav, A., Mishra, S.K., Shethia, B., Pancha, I., Jain, D., Mishra, S., Isolation of promising bacterial strains from soil and marine environment for polyhydroxyalkanoates (PHAs) production utilizing Jatropha biodiesel byproduct. *Int. J. Biol. Macromol.*, 47, 283, 2010.
30. Chookietwattana, K., Khonsarn, N., Biotechnological conversion of wastewater to polyhydroxyalkanoate by Bacillus in a sequencing batch reactor. *World Appl. Sci. J.* 15, 1425, 2011.
31. Mohapatra, S., Rath, S.N., Pradhan, S.K., Samantaray, D.P., Rath, C.C., Secondary Structural Models (16S rRNA) of Polyhydroxyalkanoates Producing Bacillus Species Isolated from Different Rhizospheric Soil : Phylogenetics and Chemical Analysis Phylogenetic analysis. *Int. J. Bio-Automation*, 20, 329, 2016.
32. Simon-Colin, C., Raguénès, G., Crassous, P., Moppert, X., Guezennec, J., A novel mcl-PHA produced on coprah oil by *Pseudomonas guezennei* biovar. *tikehau*, isolated from a 'kopara' mat of French Polynesia. *Int. J. Biol. Macromol.*, 43, 176, 2008.
33. Chee, J.-Y., Yoga, S.-S., Lau, N.-S., Ling, S.-C., Abed, R.M.M., Sudesh, K., Bacterially produced polyhydroxyalkanoate (PHA): converting renewable resources into bioplastics, *Curr. Res. Technol. Educ. Top. Appl. Microbiol. Microb. Biotechnol.*, 2, 1395, 2010.
34. Vijaya Abinaya, R., Balasubramanian, V., Ramesh, N., Natrajan, K., Rajeshkannan, V., Exploration of polyhydroxyalkanoates production from rhizosphere soil bacteria, *ENVIS Newsl.*, 10, 1, 2012.
35. Fonseca, G.G., Antonio, R.V., Polyhydroxyalkanoates Production by Recombinant *Escherichia coli* Harboring the Structural Genes of the Polyhydroxyalkanoate Synthases of *Ralstonia eutropha* and *Pseudomonas aeruginosa* Using Low Cost Substrate. *J. Appl. Sci.*, 6, 1745, 2006.
36. Thakor, N., Trivedi, U., Patel, K.C., Biosynthesis of medium chain length poly(3-hydroxyalkanoates) (mcl-PHAs) by *Comamonas testosteroni* during cultivation on vegetable oils. *Bioresour. Technol.*, 96, 1843, 2005.
37. Omar, S., Rayes, A., Eqaab, A., Voß, I., Steinbüchel, A., Optimization of cell growth and poly(3-hydroxybutyrate) accumulation on date syrup by a *Bacillus megaterium* strain, *Biotechnol. Lett.*, 23, 1119, 2001.
38. Pantazaki, A.A., Papaneophytou, C.P., Pritsa, A.G., Liakopoulou-Kyriakides, M., Kyriakidis, D.A., Production of polyhydroxyalkanoates from whey by *Thermus thermophilus* HB8, *Process Biochem.*, 44, 847, 2009.

39. Yu, P.H., Chua, H., Huang, A.L., Ho, K.P., Conversion of industrial food wastes by *Alcaligenes latus* into polyhydroxyalkanoates, *Appl. Biochem. Biotechnol.*, 78, 445, 1999.
40. Fernández D., Rodríguez, E., Bassas, M., Viñas, M., Solanas, A.M., Llorens, J., Marqués, A.M., Manresa, A., Agro-industrial oily wastes as substrates for PHA production by the new strain *Pseudomonas aeruginosa* NCIB 40045: Effect of culture conditions, *Biochem. Eng. J.*, 26, 159, 2005.
41. Van-Thuoc, D., Huu-Phong, T., Thi-Binh, N., Thi-Tho, N., Minh-Lam, D., Quillaguamán, J., Polyester production by halophilic and halotolerant bacterial strains obtained from mangrove soil samples located in Northern Vietnam, *Microbiologyopen.*, 1, 395, 2012.
42. Mohanty, S., Samantaray, D.P., Mohapatra, S., Rath, S.N., Mohanty, D.P., Rath, C.C., Studies on Biodegradation of Medium Chain Length PHAs by Soil Bacterial Isolates, *J. Adv. Microbiol.*, 1, 350, 2015.
43. Patel, S.K.S., Singh, M., Kalia, V.C., Hydrogen and Polyhydroxybutyrate Producing Abilities of *Bacillus* spp. From Glucose in Two Stage System, *Indian J. Microbiol.*, 51, 418, 2011.
44. Shah, K.R., FTIR analysis of polyhydroxyalkanoates by a locally isolated novel Bacillus sp. AS 3-2 from soil of Kadi region, North Gujarat, India. *J. Biochem. Tech.*, 3, 380, 2012.
45. Narayanan, A., Ramana, K.V., Polyhydroxybutyrate production in *Bacillus mycoides* DFC1 using response surface optimization for physico-chemical process parameters. *3 Biotech.*, 2, 287, 2012.
46. Hungund, B., Shyama, V.S., Patwardhan, P., Saleh, A.M., Production of polyhydroxyalkanoate from *Paenibacillus durus* BV-1 isolated from oil mill soil. *J. Microb. Biochem. Technol.*, 5, 13, 2013.

6

Polyhydroxybutyrate as an Eco-Friendly Alternative of Synthetic Plastics

Shikha Sharma, Priyanka Sharma, Vishal Sharma and Bijender Kumar Bajaj*

School of Biotechnology, University of Jammu, Jammu, India

Abstract

Synthetic plastics though have vast application spectrum but their recalcitrance to biodegradation is the major cause of environmental pollution. Polyhydroxybutyrate (PHB), an intracellular lipid reserve material produced by several bacteria, has recently emerged as an alternative for synthetic plastics as its structural properties are similar to polypropylene, and yet, it is fully degradable. PHB due to its biocompatibility may have potential for application in tissue engineering/transplantology and for food, pharmaceutical, agriculture, and several other industries. Thus, PHB is a green biomaterial as it can be produced from renewable resources unlike the conventional plastics which are fossil-fuel based products. Several bacterial species are known to accumulate PHB. The major barrier in the production and commercialization of PHB is its high production cost. High PHB-producing bacterial strains could be obtained from microbial diversity or by genetic engineering. Furthermore, the production cost could be lowered by exploiting agro-industrial residues as medium components for PHB production. Process optimization by statistical design of experiments may be used for enhancing product yield. Biochemical characterization of PHB is crucial not only for the structural elucidation but also from application view point like characterizing the structure of mixed polymers developed from blends of two or more monomers. The current chapter presents the recent developments on PHB production from bacteria, its characterization, and application potential in different industries.

***Keywords*:** Polyhydroxybutyrate, PHB production, optimization, characterization, application

**Corresponding author*: bajajbijenderk@gmail.com; bijenderbajaj@yahoo.com

Bibhuti Bhusan Mishra, Suraja Kumar Nayak, Swati Mohapatra, and Deviprasad Samantaray (eds.) Environmental and Agricultural Microbiology: Applications for Sustainability, (101–150) © 2021 Scrivener Publishing LLC

6.1 Introduction

Synthetic plastics due to their vast application spectrum ranging from small domestic setups to huge industries have become one of the most integral parts of human lives. However, plastics due to their recalcitrance to biodegradation have become one of the major causes of environmental pollution and it sensitized the researchers, policy-makers, and civil society equally around the globe [1]. Nowadays, there is an overpowering demand for plastic products with the rising affluence and public embracement of western consumerism. Plastics have made a doorway in modern life and became indispensable part of humankind [2]. Synthetic plastic (polymer) has become essential because of its magnificent properties like mechanical and thermal stability, durability, and resistance to degradation. It is an imperative gift of modern technology to society [3]. At this stage of industrialization and urbanization, present era cannot be imagined without plastic due to its diverse applications in all the sectors including medicine, health, agriculture, food, household, automobiles, and other industries. Plastic is replacing glass, wood and other constructional materials [4]. Synthetic polymer's structures can be chemically manipulated and easily moulded into variety of shapes and range of strengths [5]. They are resistant to chemical and biological degradation hence used for manufacture of durable goods, packaging material and disposal goods.

Conventional plastics are presenting a big threat to our atmosphere; firstly, because of its non-biodegradable nature, and secondly, the raw material used for synthetic plastic fabrication is petroleum, the resources of which are rapidly diminishing. Therefore, there is an intense focus on its research globally to find suitable alternatives. Considering dearth of biodegradable plastics, and unfortunately, most of the substitutes compete with food/feed resources, production of bioplastic from microbial sources especially using low-cost agro-wastes can be very important and significant. Large molecular size (50,000 to 1,000,000 Da) seems to be mainly responsible for the resistance to biodegradation and their persistence in soil for a long time [6]. The acquisition of non-degradable plastic in the environment leads to detrimental effects on the flora and fauna, water bodies, quickly fills up natural domains, and greatly affects exquisite quality of the region [7].

Nowadays, plastic bags are one of the major causes of the pollution because of their long life, light weight, and persistence [8]. The incineration of synthetic plastic waste generates potential hazards and the economy of disposal process makes waste management a problem [9]. Worldwide increasing concern of non-renewable resources not only influences the energy industry but also changes the chemical industry. At present, majority of the production of plastic is from petroleum or from petroleum-derived products. The depletion of petroleum and hydrocarbon resources leads to the increases in production cost of synthetic/conventional plastics. In view of non-biodegradable nature of synthetic plastic and shortage of non-renewable energy resources, there is a dire need to develop eco-friendly and cost-effective technologies for production of suitable substitutes of the conventional plastics.

Bacterial polyhydroxyalkanoates (PHAs) are being considered as most appropriate substitutes of petrochemical-based plastics due to close resemblance of their material properties with various thermoplastics [10]. PHAs have advantageous features including apt molecular weight, biodegradability, and biocompatibility. PHAs represent aliphatic bacterial polyesters that are accumulated by many bacteria intracellularly as carbon and energy sources under unfavorable conditions [11]. Poly-β-hydroxybutyrate (PHB) is one of the most prevalent PHA and represents a suitable candidate for bioplastic production. PHB is a non-toxic, biocompatible, biodegradable, and recyclable thermoplastic. Furthermore, distinctive properties like insolubility in water, resistance to hydrolytic degradation, impermeability to oxygen, resistance to UV, and others make it most appropriate candidate for wide range of applications in industrial, agricultural, and biomedicine sectors [12]. Figure 6.1 depicts the various properties of PHB. The current chapter presents the recent developments on PHB production from bacteria, its characterization and application potential in different industries.

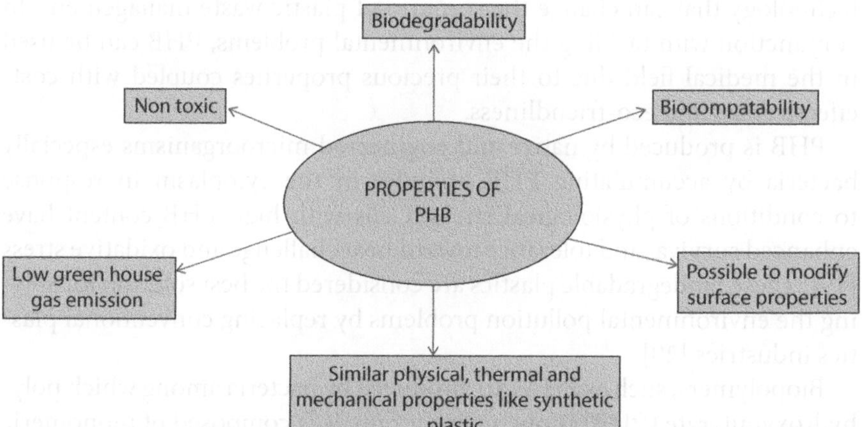

Figure 6.1 Various properties of PHB.

6.2 Bioplastics

Growth and application of bioplastics is still in its infancy stage but holds great assurance for sustainable and eco-friendly future. Considering scarcity of biodegradable plastics and, unfortunately, most of the options compete with food/feed assets, so production of bioplastic from microbial sources particularly using low-cost agro-wastes would be very useful and significant [13].

Bio-based polymers are the best replacements for synthetic polymers. The term biopolymer is used for polymers that are bio based, i.e., either produced by living organisms or derived from biomass. Bioplastics are partly or completely biodegradable. This can be done with the help of and/or in presence of microorganisms in the environment that convert bio-based materials (bioplastics) into natural substances such as water, carbon dioxide, and compost. This can be achieved in both natural aerobic and anaerobic environments [14]. Biopolymer produced using starch, sugars, or cellulose, vegetable oil, or microbiota is bio-degradable and derived from sustainable biomaterials making it an environmentally benevolent process [15]. Some microorganisms have the capability to accumulate polymeric material as carbon and energy storage materials in the form of mobile, amorphous, and lipid granules under stressful condition [16].

The presence of biopolymers in the bacterial cells has been known since 1920s. Biopolymers were first commercially introduced in 1980s. Initially, traditional polymers like polyolefins blend with starch or some organic substance. In 1926, Lemoigne first reported the formation of poly(3-hydroxybutyrate) (PHB) inside bacteria [17, 18]. PHB production is a promising technology that can change the scenario of plastic waste management. In conjunction with tackling the environmental problems, PHB can be used in the medical field due to their precious properties coupled with cost-effectiveness and eco-friendliness.

PHB is produced by native and engineered microorganisms especially bacteria by accumulating PHB granules in the cytoplasm in response to conditions of physiological stress. Cells with high PHB content have enhanced survival and tolerance toward heat challenge and oxidative stress [19]. These biodegradable plastics are considered the best solution for solving the environmental pollution problems by replacing conventional plastics industries [20].

Biopolymers, such as PHAs are produced by bacteria among which polyhydroxybutyrate (PHB) is one major group. It is composed of monomeric units of β-hydroxybutyrate and is polyester of the PHAs family. It possesses similar thermo-mechanical properties like that of synthetic plastics [8]. Therefore, it can be used as a substitute to the present day traditional plastic and may lead to sustainability due to its complete biodegradability. In response to imbalanced nutrient conditions, PHB is accumulated as granules in the cytoplasm of bacterial cells in excess of carbon and in limitation of oxygen or nitrogen [21].

The concept of biodegradable plastics appeared as a solution for this problem when a wide variety of poly-β-hydroxyalkanoates (PHAs) natural biopolymers were found as an intracellular storage compound in diverse taxonomic group of prokaryotes. The homopolymer PHB is the best known example of the PHA family and prokaryotic organisms are known to accumulate PHB amounting to as much as 80% of their cellular weight [22]. This polymer is completely biocompatible and biodegradable in nature and possess thermoplastic like properties and has high tensile strength, inertness, and high melting point [23]. These properties are of great interest to several industries such as packaging material, long-term dosage of drugs, medicines, insecticides, herbicides, fertilizers cosmetic world, and disposable items such as razors, utensils, diapers, feminine hygiene products, cosmetics containers, shampoo bottles, and cups [24, 25]. Furthermore, bioplastics finds its application in the field of pharmaceutical and biomedical areas like surgical implants, scaffolds for tissue engineering or as wound dressing and blood vessel replacements [26].

PHB is produced by more than 300 bacterial species and are present in both aquatic and terrestrial environments. It includes *Ralstonia eutropha*, *Alcaligenes latus*, *Aeromonas hydrophila*, *Pseudomonas putida*, *Burkholderia sacchari*, *Azotobacter vinelandii*, *B. megaterium*, and recombinant *Escherichia coli* [15, 26]. The major hurdle in large-scale production of PHB is its high cost. In order to make the economically feasible process, many issues have to be addressed simultaneously including use of recombinant microbial strains that are able to achieve both a high substrate conversion rate and production of biopolymer with good strength. Another way is to reduce the substrate cost used for production. The use of renewable carbon substrates and lignocellulosic biomass in biopolymers production can reduce the production cost by 40%–50% [24]. Hence, the use of waste residues like agricultural waste, whey, molasses, soymeal, and dairy waste can appreciably minimize the substrate cost which in turn cut down the production costs [15]. Other factors which also affect the total production costs are bacterial strains, fermentation strategies, and recovery processes [7].

6.3 Bioplastics vs. Petroleum-Based Plastics

Bioplastics confer several advantages in order to be used as commodity products over petroleum-derived plastics. Table 6.1 represents comparison of bioplastics with petroleum-derived plastics as related to sustainability.

Biodegradable plastics act as an essential alternative to petroleum-derived plastics. These biopolymers are obtained from renewable resources and can be produced by different life forms, e.g., plants and microbes. Microbes synthesize the biopolymers are mostly lipid in nature and gathered in the form of granules in the cell. These mobile granules help microbes to survive under stress conditions [27]. Presently, scientific research is not only concentrated on finding alternative to petroleum-based plastic, but the attention is toward the consequences of biodegradability of the plastics. Numerous scientific research groups tried to exploit the various options of making bio-based plastics photodegradable. In the past few years, society's point of view had now changed due to plastics never ending effects on the whole world. Since, plastics are derived from non-renewable resources like petroleum and are not even friendly to nature because of their non-degradable properties, so to overwhelm this situation, the production and use of eco-friendly materials such as bioplastics is obligatory.

Considering the consequences associated with synthetic plastic, the present chapter deals with the production of biodegradable plastic using agro-industrial waste residues, resulting in valorization of biomass to value

Table 6.1 Properties of bioplastics and petroleum-derived plastics.

Properties	Bioplastics	Petroleum-based plastics
Production	From renewable sources	From non-renewable sources
Biodegradability	Yes	No or very low
Biocompatibility	Yes	No
Sustainable	Yes	No
Toxicity	No	Yes
Range of polymers	Biopolymers (starch, lipids, proteins), Bacterial polymers	Extensive
Green house gases emission	Low	High

added products. This chapter further deals with optimized production of PHB, production by genetic manipulated microbes, its application potential and biodegradability. Hence, bioplastic production would replace the deteriorating effects of synthetic plastic on environment and dependence on fossil reserves.

6.4 Classification of Biodegradable Polymers

Biodegradable polymers are derived from biomass. These are macromolecules of biological origin, susceptible to modifications on exposure to environment, and converted to molecules of lower molar mass. They can be classified on the basis of chemical composition of biopolymer, origin and method of synthesis, processing method, economic importance, application, etc. Mainly, biodegradable polymers are classified on the basis of origin into two groups as natural and synthetic polymers as shown in Figure 6.2.

The biodegradable polymers can be produced by various ways. Polymers produced in nature are modified and then processed before use as bioplastic. It is produced by recasting of plant and animal polymers like polysaccharides, proteins, and lipids. The various biomass derived biodegradable polymers are starch, rice, wheat, zein, and casein. In response to nutrient starving condition, microorganisms produced biopolymers (PHAs) using complex metabolic processes within the cells and are completely degraded by living organisms.

Fermentation of bio-based substrates causes chemical polymerization of monomers and synthesized biodegradable polymers (e.g., polylactic acid derived from starch). Polymerization of synthetic monomers derived from oil products produce biodegradable polymers like polycaprolactones and polyesteramides.

Bioplastics are also classified on the basis of degradability. The various classes include photodegradable, semi-biodegradable, and completely biodegradable. Photodegradable plastics get easily degraded in presence of light due to the presence of light sensitive groups present in the backbone of the polymer as additives. Extensive ultraviolet radiation for long period of time can disrupt the polymeric structure and make them available for bacterial degradation [28]. Despite that, deep areas or landfills lack sunlight where it remains non-degradable. Semi-biodegradable plastics are partially degraded form of plastic. This form of bioplastic incorporates starch into the short fragments of polyethylene. On discarding, starch will be attacked by soil bacteria leaving behind the small fragments of

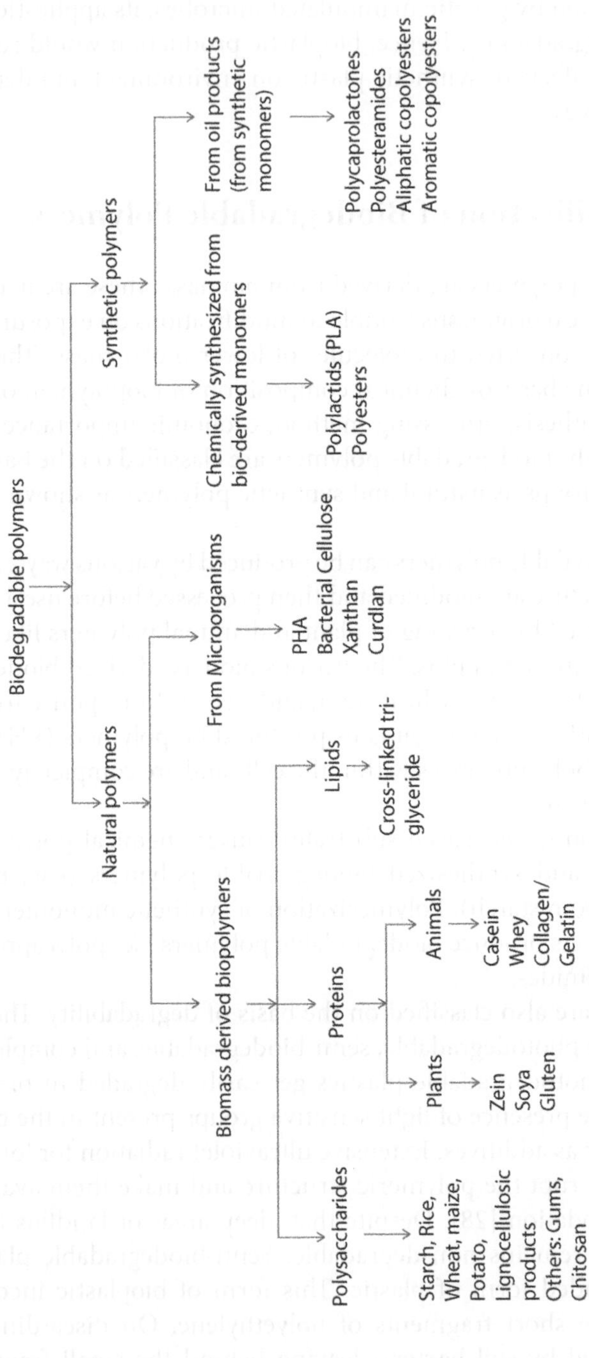

Figure 6.2 Different categories in which biodegradables polymers are classified.

polyethylene [29]. The third type of bioplastic is completely degradable form of plastic. This form is relatively new and encouraging because of its full utilization by bacteria. It includes PHAs, polylactides (PLA), aliphatic polyesters, polysaccharides, copolymers, and/or mixture of the above [6].

On the basis of application biodegradable polymers are classified and include biomedical polymers and ecological polymers. Those polymers that have biomedical relevance and contribute to the medical care of patients by bone replacements, drug delivery, and for the preparation of materials supporting surgical operation (e.g., suture, wound dressings, and sealant) are biomedical polymers. The various polymers used are poly(acid anhydride), poly(acyanoacrylate), etc. On the other hand, ecological polymers are those polymers that maintain and sustain the earth environments clean such as poly(butylene succinate) and poly(3-hydroxybutyrate). These biomedical polymers and ecological polymers are generally used for either of the purpose, but few of them are available in both the categories, for example, chitin and starch.

6.5 PHB-Producing Bacteria

Currently, all the necessities of an individual to the small and large-scale industries are fulfilled by petrochemical based plastics, covering all the spheres of life. By permeating the daily needs, plastic became the most popular material of this world. But this requisite material possesses several limitations like production from non-renewable resources and highly resistant to biological and chemical degradation. Owing to the fact, the research focus shifts toward production of eco-friendly bio-based plastic. Lately, PHA and various members of their family procured attention because of its highly similar properties to that of synthetic plastic, production from renewable resources and complete degradation in nature by microbes to CO_2 and H_2O [30]. The best studied form of PHA is PHB, naturally accumulated in the bacterial cell cytoplasm. During imbalanced growth conditions such as excess of carbon and limiting concentrations of nitrogen, phosphorus, sulfate, and oxygen, PHB accumulation by microorganisms can be stimulated. It is produced in the form of distinct lipid granules as an intracellular storage material and utilized these granules as energy reserve during unfavorable conditions [30].

The natural diversity is very extensive and it is claimed that few living organisms are able to synthesize PHAs. For the biological production of PHAs, plants and prokaryotic microorganisms like bacteria are eligible. Prokaryotic microorganisms can generate wide range of extracellular and intracellular polymers having multiple functions [31]. Plant cells yield <10% (w/w) of dry weight of PHA production and higher yield of 10%–40% (w/w) have a negative effect on development of plant. In contrast, bacteria yield PHAs as high as 90% (w/w) of the dry cell mass [18]. Microbes were reported as the potential producers of PHB because of their high versatile nature and adaptation to the various extreme environmental conditions. These polyesters are synthesized by many gram-positive and gram-negative bacteria from at least 75 different genera [6].

Many studies have been attempted on the isolation and characterization of PHB producers from various natural sources [32]. Bacterial species producing PHB have been isolated from varied environments like activated sludge [33], hypersaline lake [34], soil from contaminated urban and hilly areas [35], oil contaminated soils [36], rhizosphere [37], and from different soil types [32]. Panigrahi and Badveli [38] collected soil samples from vegetable soil, paddy field soil, sunflower field soil, red soil, and Hussein Sagar lake (India) soil for screening of potential PHB-producing bacteria. Soil samples from contaminated sites like sewage water were found to be more valuable for the isolation of PHB producers than non-contaminated sites [39]. High yield of PHB depends upon the physiology of microbial communities. The microbes present in contaminated areas regularly experience occurrence of unbalanced growth conditions and it leads to higher PHB production. Thus, PHB-producing bacteria should be isolated from diverse ecological niches.

Numerous bacterial species were known to produce PHBs but only a few bacterial species have been used for the production of PHBs like *Ralstonia eutropha*, *Alcaligenes latus*, *Azotobacter beijerinckii*, *Pseudomonas oleovorans*, *Bacillus megaterium*, *Micrococcus luteus*, *Halomonas campisalis*, *Haloferax mediterranei*, *Halomonas halophila*, *Halomonas hydrothermalis* MTCC 5445, and a halophilic bacterium [40–45].

For the intended application of PHB in medicine and living tissues, PHB from Gram-positive bacteria may be preferred. The production from Gram-negative bacteria produces endotoxins along with PHB, which may necessitate extra purification steps for removing potential contaminants. Hence, no such purification is required for PHB produced by Gram-positive bacteria [46]. Various Gram-positive bacterial sp. reported for PHB production are *Clostridium*, *Corynebacterium*, *Nocardia*, *Bacillus*, *Rhodococcus*, *Streptomyces*, and *Staphylococcus* [47]. Among all, *Bacillus* species provide numerous advantages over other and have been studied for variety of industrial products including PHB [48, 49]. They show comparatively faster growth and have potential to use range of agro-industrial wastes as substrates [50]. Furthermore, it acts as a model system for the heterologous expression of foreign genes associated with PHA production and several other chemicals [51, 52].

Among all the microbes, *Bacillus* sp., *Pseudomonas* sp., and *Vibrio* sp. are more effective for PHB production due to their more stable and reproducible nature under extreme environmental conditions. *Bacillus* sp., *Pseudomonas* sp., *Cupriavidus* sp., and *Aeromonas* sp. were studied for their higher capacities to produce PHA on large scale. Few bacterial species like *Ralstonia eutropha* and *Bacillus megaterium* have attained more attention from the researchers. Production potential of *Bacillus megaterium* was reported to be about 84%. Various other species of bacteria like *Actinobacillus, Azotobacter, Agrobacterium, Rhodobacter*, and *Sphaerotilius* also found to possess ability of converting organic waste to biopolymers [8]. Different group of microorganisms producing PHB are shown in Figure 6.3. *Bacillus* species found in soil sample of Al-Kharj produced highest concentration of PHB by using date palm syrup as a carbon source that can considerably reduce substrate and production costs of PHB [20].

Halomonas halophila produces PHAs using inexpensive substrates and have ability to accumulate high intracellular fractions of poly(3-hydroxybutyrate) up to 82% of cell dry mass [44]. Hsiao *et al.* [53] studied *Caldimonas manganoxidans*, a thermophilic bacterium producing high concentration of PHB using glycerol as raw material and fermentation conditions of unbuffered initial pH of 7 and 50°C. It enhanced the yield of PHB concentration by 8.4 ± 1.5g/L and PHB content of 71 ± 7 wt%. In another study, *Cupriavidus necator* PTCC 1615 synthesized PHB using

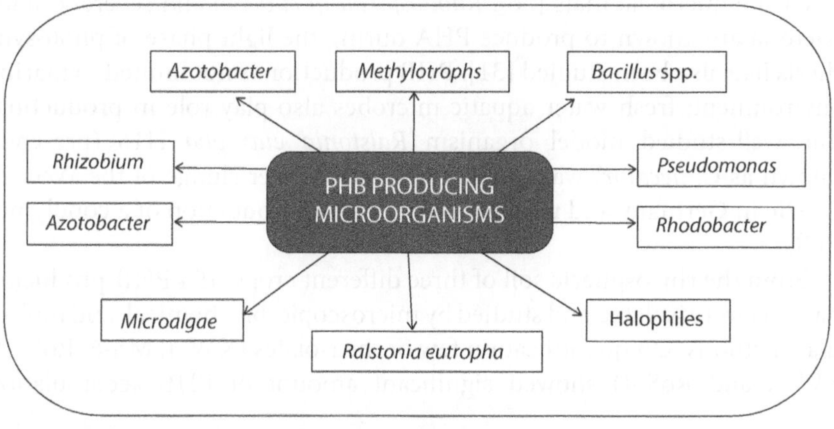

Figure 6.3 Various microorganisms producing PHB.

brown seaweed *Sargassum* sp. as biomass [23]. *Nocardia* species capable of generating PHB was identified during screening for PHB-producing bacteria. The use of alternative substrates and novel extraction methods makes PHB more commercially attractive. The aim of the research was to determine PHB production by actinobacteria and to test further PHB production in different agro-waste as carbon and nitrogen sources [16].

Short chain length PHAs (scl-PHAs) composed of 3–5 carbon atoms are produced by wide range of bacteria like *Cupriavidus necator*; on the other hand, medium chain length PHAs (mcl-PHAs) comprise of 6–14 carbon atoms and produced mainly by *Pseudomonas* species. However, on the basis of mixed substrates used by bacteria as raw material produced copolymers like poly (3-hydroxybutyrate-co-3-hydroxyvalerate). The composition of product formed depends upon the type of carbon supplied and the strain of bacteria used for production [54]. The side of biopolymers can be saturated, branched, halogenated, or aromatic. For example, bromide and aromatic group was extracted from polymer produced by *Pseudomonas putida* [55]. By using fed-batch production strategies, *Azotobacter vinelandii* OPNA produces large quantities of PHB, having ultra-high molecular weight [26]. *Methylobacterium extorquens* AM1 and *Cupriavidus necator* H16 produced PHB in an integrated one-pot electromicrobial setup and studied the influence of different stress conditions, like high salinity, nutrient limitation, coexisting electrolysis, and starvation, on the production of PHB [19].

Nutrient limitation is one of the factors required for biopolymer production from bacteria. Numerous studies have been done on production, the amount and the composition of PHA produced by different species in marine environment. *Halomonas* and *Labrenzia* have been isolated from estuarine microbial mats [56]. *Rhodospirillum rubrum* and several cyanobacteria are known to produce PHA during the light phase of photosynthesis had also been studied [31]. PHB production is not limited to marine environment; fresh water aquatic microbes also play role in production. The well-studied model organism *Ralstonia eutropha* H16 (presently known as *C. necator*) was isolated from freshwater sludge of the Weende Quelle in Germany and used for the industrial production of a copolymer [57].

From the rhizospheric soil of three different crops, 194 PHB-producing bacteria were isolated and studied by microscopic, biochemical, and molecular methods. On quantification few of the isolates (KW-4, MS-6, RoW-1, AW-1, and RoS-4) showed significant amount of PHB accumulation

(120–132 mg/ml). PHB granules detection by transmission electron microscopy (TEM) confirmed that isolates are PHB producers [30].

6.6 Methods for Detecting PHB Granules

PHBs are produced in many bacterial cells as inclusion bodies in their cytoplasm. The potential of the microorganisms to produce bioplastic can be detected by various screening methods and quantified by different methods [58]. PHB in the form of granules in cell cytoplasm can be visualized with a phase contrast light microscope (1,000×) due to their high refractivity. With the help of lipophilic dyes, intracellular polymers are usually visible. Various phenotypic detection methods can be used for screening PHB producers are Nile blue A staining [58, 59]. Nile red [60] and other method is by using sudan black dye for direct staining of bacterial colonies [49]. Another way is growing bacteria on plates containing Nile blue A or Nile red and then observed fluorescence in plates under UV illumination in case of PHB producers [61]. The presence of this polymer can be seen as black-blue granule in a clear or light pink background. Different bacterial isolates were screened for accumulation of PHB using Nile blue A and Sudan black B dyes [22].

Nile Red is the oxidized form of Nile Blue A, and on staining PHB granules showed bright fluorescence. It is soluble in neutral lipids and therefore absorbed by PHB granules. Various other cellular organelles and cell membranes containing lipid do not absorb ample dye to give a detectable fluorescence. At excitation wavelength of 460 and 546 nm, stained PHB granules showed fluorescence. With the increase in PHB concentration, the fluorescent response also increases and this helps in quantification of PHB. Balaji *et al.* [62] analyzed the PHB accumulation in strains of *Anabaena* sp., *Synechocystis* sp. and *Spirulina* sp. by cell staining with Nile red and observed under fluorescent microscope as red intracellular PHB in cell cytoplasm. After Nile red staining of living cells of *R. eutropha*, quantification of PHB content was done by flow cytometry and spectrofluorometry. Cells show fluorescence maximum between 590 and 630 nm, when excited between 520 and 550 nm [63]. Nile Blue A staining excited between 540 and 560 nm and show a clear fluorescence maximum between 570 and 605 nm. This shows a direct correlation between intensity used for fluorescence and PHB concentration. Gabr [20] used Nile Red staining approach for screening of PHB production by the isolated strains. Schlegal *et al.* [64]

developed alternative staining method using sudan black staining and results in dark blue granules.

Fourier transform infrared (FTIR) spectroscopy was used to carry out routine and low-cost rapid qualitative analysis of PHB content in cells [65–67]. In spite of many methods used for screening of PHB, all are insufficient to distinguish between different monomers, and therefore, they cannot be used to ascertain the composition of PHB copolymers. Shamala *et al.* [68] used PCR technique for the identification of PHA-producing *Bacillus* sp. Hsiao *et al.* [53] quantified the PHB content using gas chromatography (GC). To measure PHA in intact cells, flow cytometry and spectrofluorometry was used by Degelau *et al.* [69] and Vidal-Mas *et al.* [70]. Another methods used were two-dimensional fluorescence spectroscopy, flow cytometry, FTIR spectroscopy, and Raman spectroscopy [63, 67, 71]. HPLC can also be used as one of the analytical methods for determination of PHAs [26, 72].

6.7 Biochemical Pathway for Synthesis of PHB

Bioplastics are produced by microbes and plants either naturally or by genetic manipulations. The microbial polyesters are produced by microorganisms using the biosynthetic pathway and easily degraded by microorganisms. It is also degraded in the body of higher animals, including humans.

The metabolic pathway observed in organisms like *Azotobacter beijerinckii* and *Zoogloea ramigera* includes the enzyme ketothiolase, which catalyses the conversion of acetyl coenzyme A (acetyl CoA) to acetoacetyl CoA. The intermediate thus formed is reduced to D-(-)-P3-hydroxybutyryl-CoA by an NADPH-dependent acetoacetyl CoA reductase [73, 74]. Then, the last step involves the head to tail polymerization of the monomer to PHB and is catalyzed by the enzyme PHB synthase [8].

In some organisms like *Rhodospirillum rubrum*, synthetic pathway of PHB synthesis is carried out in five steps. L-(+)-3-hydroxybutyryl CoA is formed from acetoacetyl-CoA with the help of an NADH-dependent acetoacetyl CoA reductase enzyme. The product formed is further converted to D-P-hydroxybutyryl-CoA, by using two stereospecific enoyl-CoA hydratases. Then, it undergoes the process of polymerization to produce PHB [8].

The chemolithoautrophic bacterium *Alcaligenes eutrophus* H16 possesses the property of producing PHB granules in the metabolic pathway

[74]. This organism has a magnificent property of producing 70%–80% dry cell weight of PHB, under limiting conditions for nitrogen or phosphate.

PHAs are produced by *A. eutrophus* and many other species of bacteria in the excess of carbon and nitrogen or phosphorus limitation. PHB is produced in the form of granules of size 1–2 μm and accumulates as nearly 80% in the form dry weight. By the successive action of three enzymes 3-ketothiolase, acetoacetyl-CoA reductase, and PHB synthase on acetyl-CoA, PHB is synthesized (Figure 6.4). These enzymes are encoded by the *phbA*, *phbB*, and *phbC* genes, respectively [57]. This pathway was also observed in *Cupriavidus nectar*, *Aeromonas hydrophila*, and *Pseudomonas stutzeri* [54]. For the synthesis of mcl-PHAs, three types of pathways are associated. In the first pathway, 3-hydroxyacyl-CoA is produced by the *β*-oxidation of carbon sources. R-specific enoyl-CoA hydratase PhaJ is an enzyme that generates 3-hydroxyacyl-CoA precursors for the production of PHA when provided with fatty acids as a substrate [75]. In the second pathway, 3-hydroxyacyl-CoA is generated using non-related substrates. In the last pathway chain, elongation reaction takes place where 3-hydroxyacyl-CoA is produced by elongation of acetyl-CoA moieties. PhaG reported as 3-hydroxyacyl-acyl-carrier protein (ACP) CoA transferase acts as crucial link for the production of PHA from unrelated substrates [76].

The key enzymes used for PHA biosynthesis are PHA synthases (PhaC), which polymerizes the substrate 3-hydroxyacyl CoA (3HACoA) into PHA polymer. Based on the primary structures and specificities toward substrate, PHA synthases have been cloned and divided into four classes [77].

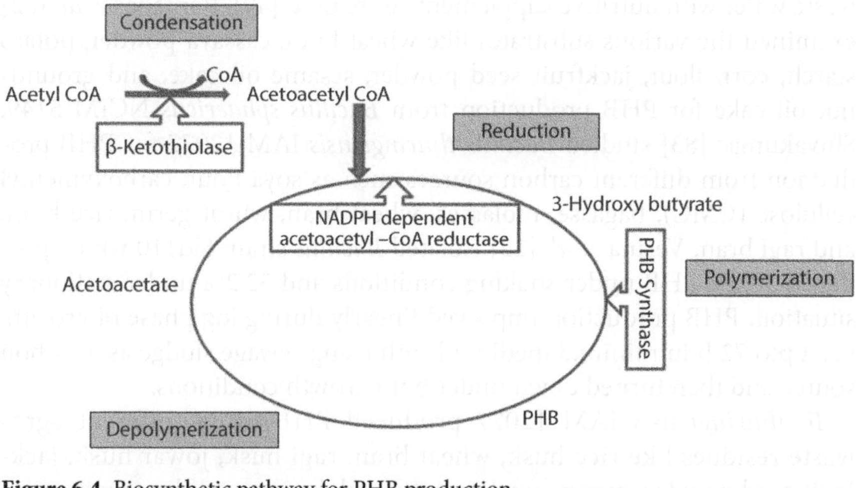

Figure 6.4 Biosynthetic pathway for PHB production.

In *Ralstonia eutropha*, Class I of PHA synthases is composed of only one subunit and represented by PhaCRe. The substrates for this class of enzyme are (R)-3-hydroxybutyryl-CoA and C3–C5 carbon chain length of hydroxyacyl CoA thioesters. Class II of PHA synthases was cloned from *Pseudomonas aeruginosa* and represented by PhaCPa, and use 6-14 carbon atoms hydroxyalkanoic acids. Class I and class II of PHA synthases possess molecular mass between 61 and 73KDa [78]. PHA synthases for both class III and class IV show nearly similar substrate specificity and contain two non-identical subunits PhaE and PhaR. From *Allochromatium vinosum*, PhaCAv is PhaE subunit and from *Bacillus megaterium* PhaCBm is PhaR subunit are typical class III and class IV PHA synthase, respectively [54, 79].

6.8 Production of PHB

Numerous findings are available on production of plastic using non-renewable resources [15]. But the most recommendable and worthwhile attribute is to produce bioplastic by bacteria using cost-effective renewable resources. The cost of any industrial process depends mainly on the substrate used. Bulk production of PHB using agro-industrial waste as carbon source reduce the production cost by 50%. Several bacterial species were reported for PHB production using inexpensive carbon sources [80, 81]. Moreover, easily available low-cost carbon sources possess appreciable content of nutrients like amino acid and peptides, which further improves the growth rate and helps in cost-effective production of PHB.

For the production of PHB from *Bacillus subtilis* NG220, sugar industry waste water with nutritive supplements were used [49]. Ramadas *et al.* [82] examined the various substrates like wheat bran, cassava powder, potato starch, corn flour, jackfruit seed powder, sesame oil cake, and groundnut oil cake for PHB production from *Bacillus sphaericus* NCIM 5149. Shivakumar [83] studied *Bacillus thuringiensis* IAM 12077 for PHB production from different carbon sources such as soya flour, carboxymethyl cellulose (CMC), bagasse, molasses, wheat bran, wheat germ, rice bran, and ragi bran. Verma *et al.* [22] isolated *Bacillus* strain Sld110 which produced 38.6% PHB under shaking conditions and 52.2% under stationary situation. PHB production improved linearly during log phase of growth, i.e., upto 72 h in minimal medium broth using sewage sludge as a carbon source and then turned down under both growth conditions.

B. thuringiensis IAM 12077 produced PHB using different agro-waste residues like rice husk, wheat bran, ragi husk, jowar husk, jackfruit seed powder, mango peel, potato peel, bagasse, and straw. Among all carbon sources used, mango peel yielded the highest production [24]. *Cupriavidus necator* used oil extracted from spent coffee grounds for the production of PHAs [84]. Bhattacharyya *et al.* [43] studied the production of PHAs from polluting waste of ethanol industry—Vinasse using halophilic archaeon and *Haloferax mediterranei.* In another study, methanol and saponified palm kernel oil was used for enhanced production of PHAs [85].

Azizi *et al.* [23] studied *Cupriavidus necator* PTCC 1615 for PHB production using brown seaweed *Sargassum* sp. and produced PHB concentration of 3.93 ± 0.24g/L. Besides, *Cupriavidus taiwanensis* 187 cause 100% phenol degradation and generate appreciable quantities of PHB [86]. In another study, mixed microbial culture was used by Colombo *et al.* [87] to produce PHAs from two fermented cheese whey FCW1 and FCW2 comprised of lactic, acetic, butyric acids and acetic, propionic, butyric, lactic, and valeric acid, respectively.

The model strain *Paraburkholderia sacchari* IPT 101 utilized softwood hydrolysate as a substrate for the production of PHB [88]. Naranjo *et al.* [89] produced PHB using raw glycerol and cause 60% aggregation of PHB at laboratory scale. Additionally, the two model strains *M. extorquens* AM1 and *C. necator* H16 were investigated for production of PHB using an integrated electromicrobial system where electrochemically carbon dioxide is reduced to formate and its successive conversion into PHB [19].

Diverse kinds of substrates were used for production of PHB. Variety of renewable resources such as agricultural wastes and industrial wastes utilized either directly or on hydrolysis produced sugars and fatty acids which are further used as carbon and energy reserves for PHB production.

6.8.1 Process Optimization for PHB Production

Higher cost of production of bioplastic is the major factor that limits its use. Improvement in strategy of production can reduce the cost and imply its usage in day-to-day life [90]. This generates the interest of researchers in cost-effective production of PHB. Bioprocess optimization is one such method of that may substantially reduce the cost of production and further improves the product yield [91]. Exploring low-cost agro-waste residues along with bioprocess optimization significantly reduce the production cost of PHB. It is of utmost importance for commercial ventures.

Conventional optimization of process variables by one variable at a time (OVAT) approach suffers with severe shortcomings like time consuming and laborious, does not study the effect of various variables interaction [92]. However, this can be prevented by using statistical design of experiment (DoE) approach which provides numerous advantages including enhanced process economy. Plackett Burman (PB) design and response surface methodology (RSM) are two powerful tools under DoE approach. PB design helps in screening of the various process variables that have outstanding impact on outcome of the process. RSM is very efficient tool for investigation of earmarked variables by PB and optimization of multiple variable processes [93]. There are numerous reports available on the optimization process for PHB to make it a cost-effective process. Various organisms studied for biopolymers production using cost-effective substrates are shown in Table 6.2.

6.8.2 Optimization of PHB Production by One Variable at a Time Approach

For the enhanced production, the various physicochemical parameters of medium should be optimized. The various parameters includes composition of the medium, the carbon and nitrogen sources, pH, minerals, temperature, agitation, aeration, and inoculum size and age. The process of optimization can increase the yield and further reduce the production cost of any process [115]. In one factor at a time approach, one factor is altered and all other factors are kept constant. Traditionally, this approach was used, but nowadays, various other statistical approaches are used for optimization as this method is laborious and time consuming.

Bacillus megaterium R11 was studied for PHB production using glucose and xylose as carbon source and supplemented with different nitrogen sources such as peptone, yeast extract, tryptone, $(NH_4)SO_4$, and NH_4Cl. Then, the selected nitrogen source tryptone was further investigated in five different concentrations keeping the carbon concentration constant. The selected medium composition was supplemented with oil palm empty fruit bunch leading to the appreciable production of PHB [102]. Furthermore, Sharma and Bajaj [104] studied *Bacillus cereus* PS10 for PHB production using low-cost agro-based residues, *viz.*, maize bran, rice husk, wood waste, molasses, whey, walnut shell powder, almond shell powder, corn steep liquor, soy bean bran, and mustard cake. Molasses supported maximum PHB production of 9.5 g L^{-1} after 48 h of fermentation at pH 7. Few cost-effective substrates used for PHB production are shown in Figure 6.5. OVAT helps in selection of the parameters but do not study the interaction among different parameters. Hence, process should be optimized statistically to enhance the rate of production.

Table 6.2 Various organisms producing PHB and other biopolymers using cost-effective substrates.

Organism	Biomass used	Product	Yield	References
Bacillus megaterium B2	Raw glycerol obtained from Colombian biodiesel plant	PHB	1.20 g/L	[94]
Cupriavidus necator	CO_2	PHB	0.26 g/g cell/H	[95]
Bacillus megaterium	Glycerol	PHB	4.8 g/L	[89]
Crop switchgrass (*Panicum virgatum* L.)	–	PHB	3.72% leaf tissues and 1.23% in whole tillers	[96]
Rhodococcus equi	Crude palm kernel oil	PHB	38%	[97]
Bacillus megaterium ATCC 6748	Corn steep liquor and sugarcane molasses	PHB	43%	[98]
Bacillus species	soy molasses oligosaccharides	PHAs	90% of CDW	[99]
Bacillus megaterium strain	sugarcane molasses and corn steep liquor	PHB	46.2%/mg CDW	[100]
Bacillus thuringiensis IAM 12077	rice husk, wheat bran, ragi husk, jowar husk, jackfruit seed powder, mango peel, potato peel, bagasse and straw	PHA	0.96–8.03 g/L	[24]
Bacillus megaterium	sugarcane molasses, urea and trace elements	PHB	1.27/g/L/h	[101]
Bacillus sp.Strain COL1/A6	Hydrolysed wafer residue	PHAs	62.41 ± 1.04%	[7]
Bacillus megaterium R11	Oil palm empty fruit bunch	PHB	9.32 g/L	[102]
Cupriavidus necator PTCC 1615	Brown sea weed *Sargassum sp.*	PHB	3.93 ± 0.24g/L	[23]

(*Continued*)

Table 6.2 Various organisms producing PHB and other biopolymers using cost-effective substrates. (*Continued*)

Organism	Biomass used	Product	Yield	References
Ralstonia eutropha MTCC 8320 sp.	P. hysterophorus and E. crassipes	PHB	8.1–21.6% CDW	[103]
Bacillus cereus PS 10	Rice straw hydrolysate	PHB	10.61 g/L	[1]
Bacillus cereus PS 10	Molasses	PHB	57.5%	[104]
Methylosinustrichosporium OB3b	Methane, methanol and Nitrate	PHB	52.5 ± 6.3% CDW	[105]
Alcaligenes sp.	Cane molasses and urea	PHB	8.8±0.4 g/L	[4]
Bacterial consortium	Molasses	PHAs	0.37–0.5 Cmol/ Cmol VFA	[106]
Pseudomonas corrugate	Soy molasses	PHAs	5%–17%	[107]
Methylobacterium sp. ZP24	Whey	PHA	2.6–5.9 g/L	[90]
Burkholderia cepacia ATCC 17759	Glycerol	PHB	23.6 g/L	[108]
Burkholderia cepacia ATCC 17759	Hemicellulosichy-drolysates	PHB	2.0 g/L	[109]
Pseudomonas sp. strain DR2	Waste vegetable oil	PHA	23.5% CDW	[110]
Halomonas hydrothermalis	*Jatropha* biodiesel byproduct	PHA	75%/CDW	[111]
Cupriavidus necator	*Glycerol and Rapeseed*	PHB	0.21 g L^{-1} h^{-1}	[112]
Bacillus subtilis	*Rice Bran*	PHB	30.4% CDW	[113]
Cupriavidus necator strain A-04	*Hydrolyzed pineapple*	PHB	35.6±0.1% (w/w)	[114]
Caldimonas manganoxidans	*Glycerol*	PHB	8.4±1.5 g/L	[53]

6.8.3 Statistical Approaches for PHB Optimization

Statistical optimization helps to study all the process parameters simultaneously. For the optimization of significant factors and to maximize the response, RSM plays a significant role [4]. The various statistical designs have been used for PHB production. *Bacillus* sp. possesses an inherent

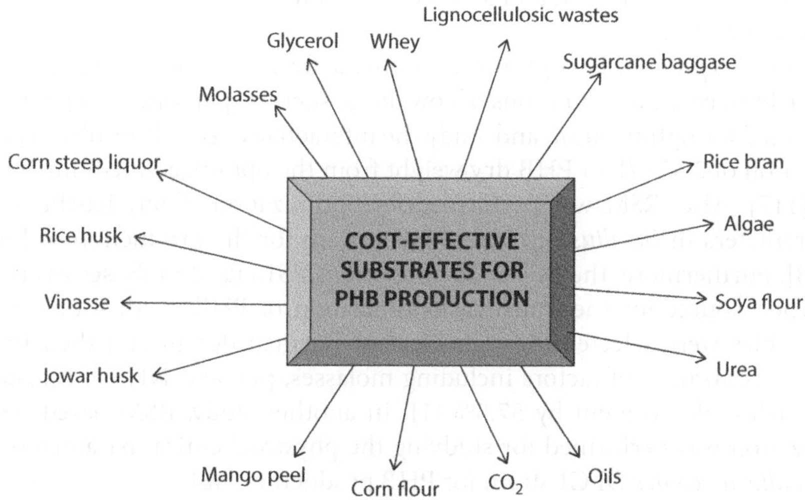

Figure 6.5 Various cost effective substrates used for PHB production.

capacity to produce wide variety of industrial products including PHB. It has tendency to explore numerous agro-industrial wastes as substrates [50]. It has potential to grow faster and act as expression system for foreign genes such as genes for PHA production and various other chemicals [51, 52]. Plenty of reports are available from *Bacillus* sp. for the production of PHB. Moreno *et al.* [94] studied *B. megaterium* B2 for the PHB production using raw glycerol from biodiesel production as the carbon source. It has the ability to accumulate PHB using statistical designs like PB and central composite designs (CCDs). Various variables effecting PHB production are temperature, glycerol concentration, and Na_2HPO_4 in shake flask. After 14 h of fermentation, 0.43 g L^{-1} of PHB was produced with 34% accumulation in the cells. The maximum PHB concentration of 1.20 g L^{-1} was reached at 11 h using the same conditions in the bioreactor.

In another study, *Bacillus megaterium* used cheapest substrate like cane molasses and corn steep liquor for biopolymer production [100]. Highest PHB production was obtained with cane molasses and glucose as carbon substrate (40.8 and 39.9 per mg of cell dry matter, respectively). The maximum yield of PHB was obtained with 2% molasses, i.e., 46.2% per mg cell dry matter. For the synthesis of PHB, i.e., 32.7 mg per cell dry matter corn steep liquor was considered as best nitrogen source [100]. Pandian *et al.* [116] also studied *B. megaterium* for optimized production of PHB using RSM design for four variables including concentration of dairy waste, rice bran, sea water and pH, and showed maximum PHB production of 11.32 g/L at 36th hour.

Bacillus pumilus H9 optimized the media for PHB production. The four variables, *viz*., concentrations of cow dung, sucrose, peptone, and pH were selected for optimization and study the interactive effects. It results in production of 2.47 g/L of PHB dry weight from the optimized medium at pH 7 [117]. Also, RSM was performed for optimization of physicochemical parameters in *Bacillus mycoides* DFC1 strain for the production of PHB [48]. Furthermore, the isolate *Bacillus cereus* PS10 used molasses as crude carbon source for the statistical optimization of PHB production. First, variables were selected through Plackett-Burman design and then RSM for optimization of factors including molasses, pH, and NH_4Cl. It results in yield enhancement by 57.5% [1]. In another study, RSM-based optimization was performed for studying the physicochemical parameters in *Bacillus mycoides* DFC1 strain for PHB production [48].

Johar *et al.* [118] studied *Comamonas* sp. EB171 for PHA production. The media was optimized using statistical design from mixed organic acids under anaerobically treated POME. Moreover, Berwig *et al.* [119] used whey after protein precipitation for production of PHB using *Alcaligenes latus* in a 4-L bioreactor having temperature of 35°C, 750 rpm, 7 L/min air flow and pH of 6.5 having polymer yield of 1.08g/g. The three physical factors, *viz*., pH, temperature, and agitation speed were optimized by central composite rotatable design for increasing the PHB production by *Alcaligenes* sp. using cane molasses and urea as carbon and nitrogen source. It results in PHB mass fraction yield of 76.80% on dry molasses. Same media on scale up produces maximum yield and productivity of 0.78 and 0.19 g L^{-1} h^{-1}, which was higher than previous reports [4]. *Paraburkholderia sacchari* IPT 101 reached maximum concentration of PHB to 5.72 g/L. The strain converted all sugars, sugar mixtures of glucose, mannose, galactose, xylose, and arabinose simultaneously to reach a PHB concentration of 80.5% PHB after 51 h [88].

Methylobaccterium organophilum produced PHB under potassium limited condition and methanol concentration in the range of 2–3 g/L so that it did not show any inhibitory effect on growth. PHB contents were produced in the range of 52% to 56% of dry cell weight having a yield factor of 0.19 g-PHB/g-methanol [120]. *Methylobacterium* sp. ZP24 utilized processed cheese supplemented with whey and ammonium sulfate under limiting dissolved oxygen condition leads to 4.58 fold increase in PHB production [90]. The two invasive weeds *Parthenium hysterophorus* and *Eichhornia crassipes* were used as biomass for PHB production using *Ralstonia eutropha* MTCC 8320 sp. the content of PHB obtained from dry cell mass was 8.1%–21.6% w/w and yield was 6.85×10^{-3} to 36.41×10^{-3} w/w raw biomass [103].

6.9 Production of PHB Using Genetically Modified Organisms

Genetic engineering is the way of causing genetic modifications in organisms' genome using various technologies. It has opened a new path for maximizing the production of various industrially important products such as bioplastics (PHB). The production of PHA can be enhanced by widening the use of economically feasible substrates range, and it also leads to production of novel PHAs [121]. Construction of recombinant strains for increase in PHAs yield involves the cloning of the PHA synthase genes from diverse bacterial species.

Escherichia coli, a model organism in the field of research, do not produce biopolymers normally, but genetic engineering made it feasible by transferring the complete operon for PHAs synthase [54]. Wang *et al.* [122] reported that on genetic modification *E. coli* produced upto 90% of P(3HB). The main reason for using recombinant *E. coli* as PHB producers is that recombinant strains produce appreciable quantity of PHB and eases its extraction. As *E. coli* is not a natural producer of PHA, so it was supposed to be best possible host for production due to the lacking of intracellular depolymerization system [123]. The first metabolic pathway in which whole PHB gene was cloned in *E. coli* for the synthesis of PHAs was described by Schubert *et al.* [74]. Various organisms producing PHB naturally transferred the genes responsible for their productions into *E. coli* are *Cupriavidus necator* [124], *Pseudomonas aeruginosa* [125], *Alcaligenes latus* [126, 127], *Streptomyces aureofaciens* [128], and *Thiocapsa pfennigii* [129].

Advantages of using recombinant organisms over natural producers are that it produces various types of copolymers and terpolymers. Furthermore, recombinant strains did not need specific conditions like nutrient limitations for production. Biosynthetic genes from *Aeromonas* sp. were used to construct recombinant *E. coli* for the production of terpolymers using dodecanoic acid and odd number carbon fatty acids as carbon sources [130]. Genetically modified *E. coli* (K24K) strain bears biosynthetic gene from *Azotobacter* sp. strain FA8 utilized wide variety of waste substrates like whey and corn steep liquor and reached the P(3HB) productivity to 2.13 g/l/h [131]. Recombinant *E. coli* possessing phaC1 gene of *Pseudomonas* sp. LDC-5 utilized molasses and produced PHAs to the level of 3.06 g/l [132].

Many reports were available on using *C. necator* as ideal host for the biosynthesis of PHA. *C. necator* Re2160/Pcb113 possessing PHA synthase gene from *Ralstonia aetherivorans* produce copolymers using crude palm kernel oil as substrate [133]. *Cupriavidus necator* H16 receiving PHA synthase gene from *Aeromonas caviae* accumulates 78% of the copolyester using crude palm kernel oil and supplemented with butyrate as cosubstrate [136]. The wild gene of PHA synthase from *Pseudomonas* sp. 61-3 was transferred to *C. necator* for production of terpolymers using soyabean oil as substrate [135]. Chung *et al.* [136] demonstrated the production of pure extracellular 3HA polymers using the fadBA knockout mutant of *P. putida* KT2442 possessing tesB gene and medium was supplemented with dodecanoic acid. By engineering the β oxidation cycle of reversed fatty acids in *E. coli*, 6.62% CDW of mcl-PHAs heteropolymers were produced [137].

Jin and Nikolau [138] studied the bioengineered *Rhodospirillum rubrum* for the effect of overexpression of PHAs gene and found that *phaC1* and *phaC2* significantly contributes in the production of PHA; on the other hand, *phaC3* showed very little impact on production. The mutant strain produced 30% of PHA which is nearly 2.5% higher than the production from wild type.

Snell and Peoples [139] studied the production of PHA from plant-based systems over the last few years. All the three enzymes can express (β-ketothiolase, acetoacetyl-CoA reductase and PHB synthase) in the plastid of plant (transgenic *Arabidopsis*) for production of PHB [142]. For the large-scale production in plants, there is no need of external organic carbon source which reduces the production cost maximally [141, 142]. In spite of that, plant cells produce lower yields [<10% (w/w)] of the polymer, and if higher yields are obtained, it inhibits plant growth and development. Contrarily, bacterial cells possess the enormous potential of aggregating PHAs, i.e., 90% (w/w) of their dry cell mass [18]. Another limitation of producing plant expression systems is prolonging growth rates resulting in reduced profits. Even though these expression systems can be developed for improvement in yield, the spread of transgenic plants is hard to control and causes an ethical issue which further leads to strict regulatory controls of transgenic plants in many countries. This makes them unable to compete with presently available bacterial systems. There is still much research to be done for the establishment of commercial plant-based systems. Few examples available are the developed Metabolix technologies in transgenic plants like tobacco (*Nicotiana tabacum* L.) [143] and switchgrass [96] for the production of PHB.

6.10 Characterization of PHB

Due to the diversity in structure of PHAs and their close resemblance with plastic, they have gained major attention as bioplastics. Microorganisms produce biopolymers generally in the stress environment. Nearly, 150 different kinds of PHAs have already been reported [144]. The structure of PHA is shown in Figure 6.6a in which R could be a hydrogen or hydrocarbon chain of length up to 13 carbon. The x can be in range of 1 to 3 or more. If R is CH3, the polymer formed is called polyhydroxybutyrate or polyhydroxybutyric acid (PHB), and if R is C3H7, the formed polymer is called polyhydroxyoctanoate (PHO) and so on are further shown in Table 6.3. Variation in structure due to the changes in number of x and R provides a wide range of physical and mechanical properties. For example, glass transition temperature (*Tg*), hydrophobicity, melting point (T_m), and level of crystallinity which can range from very low to around 70%, giving both stiffness and elasticity, accordingly [145]. It is formed by polycondensation of carboxylic acids with hydroxyl alcohol and is the only waterproof plastic which is fully biodegradable in both aerobic and anaerobic environments.

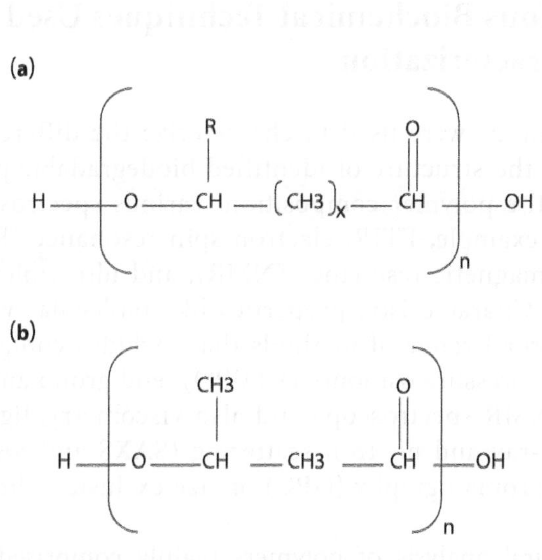

Figure 6.6 (a) Structure of polyhydroxyalkanoates (PHAs). (b) Structure of poly-3-hydroxybutyrate (PHB).

Table 6.3 Structure of PHAs and various representative members of its family.

X	R	Polymers
0	Methyl	Poly(lactic acid)
	Hydrogen	Poly(3-hydroxypropionate)
	Methyl	Poly(3-Hydroxybutyrate)
1	Ethyl	Poly(3-Hydroxyvalerate)
	Propyl	Poly(3-Hydroxyhexanoate)
	Pentyl	Poly(3-Hydroxyoctanoate)
	Nonyl	Poly(3-Hydroxydodecanoate)
2	Hydrogen	Poly(4-Hydroxybutyrate)
	Methyl	Poly(4-Hydroxyvalerate)

Most studied member of PHA family is poly-3-hydroxybutyrate (PHB) and is usually used for bioplastic production. It is composed of 3-hydroxybutyrate (3HB) repeating units and has the linear polyester structure with the general formula as shown in Figure 6.6b.

6.11 Various Biochemical Techniques Used for PHB Characterization

Different methods were used to characterize the different properties and elucidate the structure of identified biodegradable polymers. For determining the polymer composition, various spectroscopic studies are used, for example, FTIR, electron spin resonance (ESR), 1H and 13C nuclear magnetic resonance (NMR), and ultraviolet/visible light spectroscopy. Characteristic properties like molecular weight is analyzed using broad range of methods that includes colligative properties like vapor pressure osmometry (VPO), end group analysis such as quantitative NMR spectroscopy and also viscometry, light scattering, small-angle x-ray and neutron scattering (SAXS and SANS), and gel permeation chromatography (GPC) or size exclusion chromatography (SEC).

The structural analysis of polymers mainly comprised of measurement of surface properties in the solid phase. Although the crystallization studies are from the melted or from solution, and the degradation analysis of polymer is from the liquid or from gas phase [146].

6.11.1 Fourier Transform Infrared Spectroscopy

FTIR spectroscopy is analytical technique used for identification of unknown materials by creating an infrared absorption spectrum which helps in the identification of chemical bond existing in the molecule. It also helps in determination of various kind of functional groups and types of bending and stretching. Liau *et al.* [147] confirmed the presence of functional groups in the polymer finding IR spectra in the range 4,000–400 cm^{-1} at a resolution of 4 cm^{-1}. The FTIR studies for *Bacillus subtilis* NG220 found that the region of 1,675–1,735 cm^{-1} was in association with the C-O stretching of the ester carbonyl bond [49]. The two strong absorption peaks of PHB was obtained at 1,724.2 cm^{-1} and 1,280.3 cm^{-1}, showing resemblance with -C=O and -C-O stretching groups [148]. The absorption bands emerged in the spectrum associated with the side chains from the ester C=O stretching vibration at 1,727 cm^{-1}, the CH^{3-} deformation peak at 1,286 cm^{-1} and the ester C-O-C at 1,072 cm^{-1} which evidently put the extracted polymer sample obtained from isolate G-4 in the class of PHB [20].

Bacillus megaterium uyuni S29 produced PHA was characterized using IR transmission spectrum and showed bands at 1,726, 2,960–2,850, 1,390–1,370, and 1,230–1,050 cm^{-1} which corresponds to different groups like carbonyl, methyl and methylene, methyl, and the ester group, respectively [34]. Because of the stretching of the C–O bond of the ester group most of the bands are located at 1,000–1,200 cm^{-1}. Balaji *et al.* [62] studied the *Spirulina* strain and found absorption spectra at 3450 cm^{-1} showing correspondence to the terminal OH group of PHB.

6.11.2 Differential Scanning Calorimetry

DSC is a standard thermo-analytical technique for determining the thermal nature of semi-crystalline samples of polymers and studies the enthalpy changes during melting. It is the method used to establish polymer miscibility and helps in finding the glass transition temperature (Tg) or the depression of the melting temperature [149]. The degree of crystallinity (Xc) calculated from melting enthalpy helps to maintain the mechanical properties of the substance. The thermal properties of polymer like melting temperature (T_m) are critical for processing of polymer [49]. ΔH obtained by DSC helps in estimation of crystallinity of PHB [150]. DSC technique was used for PHB characterization including its nanocomposites [151, 152].

6.11.3 Thermogravimetric Analysis

Thermogravimetric analysis (TGA) is an analytical technique used for studying the decomposition pattern and thermal stability of the polymers. It helps in detecting various changes in physical and chemical properties of polymers with the increase in temperature (at constant heating rate) or time (with constant temperature and/or constant mass loss) [150]. It also provides information about chemical events like chemisorptions, desolvation, decomposition, and solid gas reactions. This is an extremely useful method for the study of materials like thermoplastics, thermosets, elastomers, composites, plastic films, fibers, coatings, and paints [153]. Sindhu *et al.* [150] prepared PHB biofilm and its blends and then TGA thermograms of both biofilm and its blends were recorded.

6.11.4 X-Ray Powder Diffraction (XRD)

This is the analytical technique used for examining the phase identification of a crystalline material and atomic spacing. It is based on the constructive interference of monochromatic X-rays and a crystalline sample. Single crystal diffraction studies find its application in semi-crystalline polymers, but not able to study stretched fibers and films. Small-angle X-ray scattering (SAXS) and wide-angle X-ray scattering (WAXS) or diffraction (WAXD) are the techniques used to study the crystalline structure of solids [146]. Kiran *et al.* [154] demonstrated the crystalline structure of PHB produced by *Brevibacterium casei* MSI04. The peaks (2θ) obtained were at 17.1°, 27.6°, 32.7°, 46.5°, and 58.2° which shows resemblance with the reflections of the orthorhombic crystalline lattice. The XRD study verifies that, with the increase in the hydroxyhexanoate content, the crystallinity of PHB decreases. It also showed that sample of PHB was nearly same to homopolymer of PHB. Sato *et al.* [155] studied PHB using WAXD patterns which are dependent on temperature.

6.11.5 Nuclear Magnetic Resonance Spectroscopy

The spectroscopic technique used for the identification of the structure of unknown compounds is NMR. It consists of two types of proton (1H) and carbon (13C) spectra. Bhattacharyya *et al.* [43] studied *Haloferax mediterranei* for poly-3-(hydroxybutyrate-*co*-hydroxyvalerate) (PHBV) production and then analyzed by 1H NMR and found multiplet (m) proton signal at 0.86–0.95 region and doublet (d) proton sign at 1.26–1.28 areas which shows presence of peaks of methyl (CH_3) from the HV unit and HB unit, respectively. The γ-CH_2 proton of HV unit of the polymer is limited to the multiplet proton signal at 1.586 ppm. Deepa and Vidhya [16] examined the structure of polymer produced by *Nocardia* sp. RD13 by using 1H NMR spectroscopy. The spectra of the PHB extracted from strain RD13 illustrated three groups of signals characteristic of PHB. A doublet at 1.53 ppm signifies the methyl (CH3) group coupled to one proton. The second signal was doublet of quadruplet at 2.5 ppm credited to methylene (CH2) group adjacent to asymmetric carbon bearing single proton and multiplet at 5.2 ppm represents the methyne group.

Chaijamrus and Udpuay [98] analyzed the PHB spectra produced by *Bacillus megaterium* ATCC 6748 and got peaks that corresponds to the various types of carbon atoms present in the PHB structure [-O-CH-(CH_3) -CH_2 -(C=O)-]$_n$. Characterization of PHB produced by *Brevibacterium casei* MSI04 by 1H NMR showed major peaks at 1.23, 2.5, and 5.2 ppm, mainly due to resonance assimilation of methyl (CH_3), methylene (CH_2), and methane (CH) groups, respectively, in 3-hydroxybutyrate (3-HB) [156]. Pan *et al.* [156] studied *Burkholderia cepacia* for PHA and analyzed it using 1H NMR which demonstrates the characteristic chemical shifts of hydrogens of methyl, methylene, and methine groups of PHB at 1.25, 2.55, and 5.25 ppm in a ratio of 3:2:1, respectively. Carbon (13C) spectrum is again of two types: solid state and high resolution NMR spectroscopy. Solid state spectra allowed examination of specific region with in solid structure. On the other hand, high resolution determines the tacticity of various polymers [146].

6.11.6 Microscopic Techniques

Various microscopy-based techniques are used to study the structure of polymer and erosion pattern during degradation. The different microscopic methods used are optical microscopy, transmission and scanning electron microscopy, atomic force microscopy, scanning probe microscopy, and confocal laser scanning reflection microscopy. Blends of different polymers can be studied through optical microscopy. In aqueous solution, the erosion of biodegradable polymer is easily studied by atomic force microscopy with good resolution. Scanning probe microscopy is used for investigating the drug delivery system. The technique used to visualize and quantify the surface properties of copolymers after degradation like the porous matrix of PLA-co-GA.

6.11.7 Elemental Analysis

Elemental analysis (EA) is a process where a sample is thoroughly studied for its elemental and isotopic composition. It refers to the analysis of mass fraction of carbon, hydrogen, nitrogen, and heteroatoms (X) (halogens, sulfur) in the sample. Results of EA help in structural elucidation as well as determination of the purity of the compound. Kulkarni *et al.* [40] studied the composition of PHA extracted from *Halomonas campisalis* MCM B-1027 analyzed and observed the absence of nitrogen in all the samples. It was also found that carbon and hydrogen content was 50.48% and 6.10%, respectively, for standard PHB, and carbon and hydrogen in PHB-*co*-PHV is 58.22% and 7.67%, and in extracted PHA the amount is 54.47% of carbon and 6.91% of hydrogen. The elemental studies of PHA accumulated in bacterium *R. eutropha* was done by Lutke-Eversloh *et al.* [157].

6.11.8 Polarimetry

Polarimetry is a technique used for measuring the optical activity of compounds including polymers. It is the quantification and exposition of the polarization of transverse waves. Major feature used in optical applications of organic polymers is their specific rotation. The optical properties of polymers depend on the dimensions, chain structure of macromolecule, or its conformational state [158].

6.11.9 Molecular Size Analysis

GPC is one of the well studied and widely used techniques for molecular size analysis. It is a type of chromatographic methods which separates particles on the basis of their molecular size in solution. It involves the use of organic solvent as a mobile phase to distinguish the *Mw* distribution of organic-soluble polymers. Kshirsagar *et al.* [159] performed the analysis of standard PHB and PHB-*co*-PHV (9:1) for molecular weight determination. PHB produced by *Burkholderia cepacia* ATCC 17759 is determined by GPC method [108].

The use of various analytical methods has been illustrated in the literature. These include the use of DSC, dynamic mechanical testing and analysis, and high performance liquid chromatography (HPLC)/GC. All these techniques help in the structural elucidation and analysis of the polymer.

6.12 Biodegradation of PHB

One of the main properties that distinguish bioplastic from petroleum-based plastics is its biodegradable nature. The property of biodegradation without any toxic effects makes PHB an appealing candidate for use in both conventional medical devices and tissue engineering [160]. As it is eco-friendly in nature, it easily degrades upon exposure to environments like soil, compost, landfill, or aquatic systems. The process of degradation is based on various environmental factors, *viz.*, moisture, pH, temperature, exposed surface area, polymer composition, and microbes present in that environment [161]. Various bacteria, fungi, and algae possess PHB degrading enzyme in the environment. Degradation of PHB by microbes eventually results into water, carbon dioxide, and methane biomass. This valuable property of biodegradation was studied in laboratory conditions as well as in simulated environment [45, 162].

The degradation of polymer in environment is a very complex process and can occur by following ways like hydrolysis, oxidation, mechanical, photochemical destruction, and biodegradation. Plastic can also be categorized as photo-degradable, oxidatively degradable, hydrolytically degradable, or those that may be composted. In view of the American Society for Testing of Materials (ASTM) and the International Standards Organization (ISO), plastic to be degraded underwent noteworthy changes in its chemical structure under specific environmental conditions [8].

PHA hydrolases and PHA depolymerases are enzymes secreted by microbes and capable to degrade PHAs [163, 164]. Rate of degradation of polymer may vary and depends upon the activities of these enzymes, framework of the polymer, and environmental conditions. It also depends upon the bulkiness of side chain of polymer, which can cause steric hindrance to the enzymatic attack on the ester bond. The enzymes, lipases, and PHA depolymerase are of the endo-type that attacks the main chain of the polymer randomly [165].

The industrial applications of PHB have been hindered because of its low thermal stability and immoderate brittleness [166]. These properties of PHB can be improved by incorporation of a second monomer unit, i.e., addition of 3-hydroxyvalerate (3HV) into structure produces poly-(3-hydroxybutyrate-*co*-3-hydroxyvalerate) [P(3HB-*co*-3HV)]. The product formed is tougher and flexible in nature, and gets easily degraded in the environment. It shows outstanding properties of biocompatibility, biodegradability, and thermoplasticity. Hence, PHB can be considered as subject to chemical and physical modifications [167].

The frequency of PHB degradation follows the sequential degradation pattern from few months (in anaerobic sewage) to years (in seawater) [5]. Shangguan *et al.* [168] described that UV light speed up the rate of degradation. The radiations of UV trigger such bond in polymeric structure to form free radicals, which then reacts with oxygen present in the environment to produce carbonyl group in the structure, even though these fragments will not degrade anymore. The degradation of biodegradable plastic has no toxicological effect on the environment [169]. In another study, it is proved that PHAs as biocompatible having no harmful effects on the living beings. In *in vivo* studies, it was found that polymer mass loss was even less than 1.6% (w/w) after a period of six month implantation. This showed the slow pace of degradation [171].

6.13 Application Spectrum of PHB

Biopolymers produced by microbes possess a wide domain of applications at commercial level in industrial, agricultural, and medical sector. The wide appliance of PHB is mainly due to two primary properties, i.e., biodegradability and biocompatibility. It is able to replace the existing material or compliment the previous one.

These properties provide an extensive range of potential functions. Previously, the uses of PHAs are limited to packaging industry mainly in bags, containers, and paper coatings and also in production of disposable items like razors, utensils, diapers, feminine hygiene products, cosmetic items, containers, and cups. But nowadays, it finds its application in various sectors like medicine, automotive, and agricultural industries. Few basic properties must be possessed by biodegradable plastics and these include being biosafe, i.e., non-toxic, durable in nature, sterilizable, economic viability, processability, biodegradation, and biocompatible. These properties broaden its application spectrum in all the fields especially medical and pharmacology fields as shown in Figure 6.7.

The putative applications of PHB in diverse fields are as follows:

- Biomedical uses: Its use in medical field is mainly due to its biodegradable nature because of this property it can be placed into human body and does not necessarily require to be removed from the body. As the product of degradation 3-hydroxy butyric acid is normally present in blood in the concentration of 0.3–1.3 mmol/l [18] making it biocompatible. PHAs in pure form or in blends produced implanted

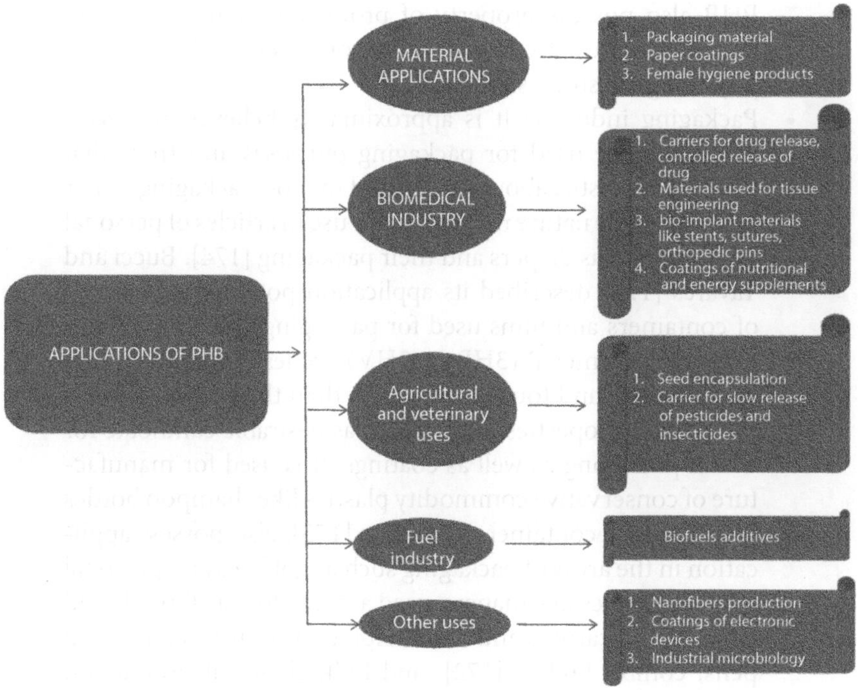

Figure 6.7 Application potential of PHB.

medical devices for dental, orthopedic, hernioplastic, and skin surgery. The number of promising medical devices were prepared using PHAs are sutures, stents, adhesion barriers, repair patches, nerve guides, orthopedic pins, biodegradable screws, and bone marrow scaffolds [83]. The property of thermo-processability along with non-toxicity makes PHB an appealing candidate for medical devices and tissue engineering [160]. The research proved that materials like PHA can be useful in bone healing processes. PHA along with hydroxyapatite (HA) acts as a bioactive and biodegradable composite and has application in tissue replacement and regeneration [172].

- The stereoregular behavior of PHA makes it a chiral precursor used for the synthesis of optically active compounds, specifically for synthesis of certain drugs or insect pheromones [173]. The very important and versatile property of these polymers is to find application in controlled drug delivery. Active ingredient in the drug released slowly by the ultimate degradation of carrier material [146]. Additionally,

PHB also possess property of producing systems for sustained enzyme activators or inhibitors liberated for advance physiological studies.

- Packaging industry: It is approximately believed that 41% of plastics are used for packaging purposes and from that amount of plastic; about half is used in food packaging. Their biodegradable nature makes it fit for use in articles of personal hygiene such as diapers and their packaging [174]. Bucci and Tavares [175] described its application potential in making of containers and films used for packaging. The short chain length copolymer P (3HB-*co*-3HV) are less crystalline, easier to mould, and tougher in nature than the homopolymers [54]. These properties make PHAs as desirable candidate for use in packaging as well as coatings. It is used for manufacture of conservative commodity plastics like shampoo bottles and cosmetic containers and films [175], also possess application in the areas of packaging such as golf tees and personal hygiene articles like diapers, used as wrap for cardboards and papers, milk cartons and films, nappies moisture barriers and pens, combs, bullets [172], and bulk chemical production using depolymerized PHA [176].
- Agricultural and veterinary uses: PHB also finds its application in agriculture by encapsulation of seeds to provide protection from harsh conditions, encapsulating fertilizers for slow release, use of biodegradable plastic films for safeguarding crop and for preparation of biodegradable containers for hothouse facilities. It acts as a biodegradable carrier for long term and slow release dosage of insecticides and herbicides [177].
- Detergent applications: The polymeric carboxylic acids (poly(acrylic acid) and its copolymers) were used in detergent formulations was reviewed by Paik *et al.* [178]. It was at first introduced in the 1980s in mixture with zeolites as partial replacements for polyphosphates. As wastewater treatment plants are not able to eliminate phosphates, it leads to eutrophication in water bodies.
- Other uses: PHAs are processed to produce into fibers and used to produce materials like nonwoven fabrics. Bioplastics are used in blend in electronic devices like mobile phones (NEC Corporation and UNITIKA Ltd. 2006). Mainly produced from from P (3HB) and P (3HB-3HV) polymers.

Various molecules are hydrolyzed chemically and produced to commercially attractive molecules such as β-hydroxyacids, 2-alkenoic acids, β-hydroxyalkanols, β-acyllactones, β-amino acids, and β-hydroxyacid esters [179]. It can also be used in dairy cream substitutes or flavor delivery agents in foods and used in biofuel industry as additives.

6.14 Conclusion

Owing to a growing environmental alertness and the inadequacy of fossil resources, it is anticipated that renewable biopolymers will substitute a considerable market fraction for synthetic polymers. Certainly, demand for bacterial polymers with material properties that are purposely tailored for applications in various fields of daily life will be growing. The recent thrust area to overcome the drawbacks of conventional plastics has renewed interest in the expansion of PHB. Bacteria remain perfect production organisms for custom-made polymers owing to the availability of genetic systems and methods for engineering metabolic pathways. These extensively important biological by-products demand a dynamic establishment of industrial process contributing as key element toward high-cost production. The advent research on simultaneous production of polymeric substances (intracellularly plus extracellularly) opens the gateway to understand new aspects of the metabolic links and ecological prospects (i.e., defining role, diversity, and evolution). For concurrent production of this high-valued endopolymers and exopolymers with the same organisms under optimized conditions using domestic, industrial, agricultural, or industrial effluents waste may support us to combat the issues linked to environmental pollution, cost production, and its marketing. Thus, the real question in our minds should be: if waste is really "waste" in relation to manufacture of high value products from its dwelling microbes. By scrutinizing the above cited review, it can be concluded that bioplastics are the most promising substitute for conventional plastics. The investigational reports on production of such eco-friendly plastics established that industrial scale mass production of bioplastics could be skilful in near future, which eventually could swap the conventional plastics.

6.15 Future Perspectives

- Despite extensive research on numerous features of PHB, there is dearth of knowledge about large-scale production of

PHB and standard to be used for its higher thermostability. Decoding these points would enable us to fully utilize bacterial PHBs for their potential applications in biotechnology.

- From application viewpoint, PHB copolymers are more useful than homopolymers. So, more polymer blends with PHB could be prepared and characterized.
- Metabolic engineering principles can be used to totally understand the biosynthetic pathways for PHB production, to augment the production of PHB from bacteria, to widen various substrates utilizing ability and to produce fresh PHBs with impending better characteristics.

Acknowledgements

Dr. Bijender Kumar (Bajaj) gratefully acknowledges various organizations for providing fellowships/grants for overseas research stays, i.e., Institute of Advanced Study, Durham University, Durham, UK, for providing COFUND-International Senior Research Fellowship; Commonwealth Scholarship Commission (Aberystwyth University UK), for commonwealth fellowship; VLIR-UOS, Belgium for Short Research Stay Scholarship at University of Antwerp, Antwerpen, Belgium; ERUSMUS-MUNDUS (EU) for invited Professor Scholarship at University of Naples, Italy; INDO-US (IUSSTF)-2011 for Visiting Professorship at Ohio State University, USA. All the funding agencies, i.e., University Grants Commission (UGC), Council of Scientific and Industrial Research (CSIR), Department of Science and Technology (DST) and Department of Biotechnology (DBT), Indian Council of Medical Research (ICMR), of Government of India, and BBSRC (UK) are gratefully acknowledged for granting financial support in the form of Research Projects to Dr. Bijender Kumar Bajaj. Authors thank the Director, School of Biotechnology, University of Jammu, Jammu, for laboratory facilities.

References

1. Sharma, P. and Bajaj, B.K., Cost-effective substrates for production of poly-β-hydroxybutyrate by a newly isolated *Bacillus cereus* PS10. *J. Environ. Biol.*, 36, 1297, 2015.
2. Brydson, J.A., *Plastic materials*, 7th ed, p. 920, Butterworth-Heinemann, Oxford, United Kingdom, 2012.

3. Luckachan, G.E. and Pillai, C.K.S., Biodegradable polymers- A review on recent trends and emerging perspectives. *J. Polym. Environ.*, 19, 637, 2011.
4. Tripathi, A.D., Srivastava, S.K., Singh, R.P., Statistical optimization of physical process variables for bio-plastic (PHB) production by *Alcaligenes* sp. *Biomass Bioenerg.*, 55, 243, 2013.
5. Madison, L.L. and Huisman, G.W., Metabolic engineering of poly (3hydroxy-alkanoates): from DNA to plastic. *Microbiol. Mol. Biol. Rev.*, 63, 2153, 1999.
6. Reddy, C., Ghai, R., Rashmi, K.V., Polyhydroxyalkanoates: an overview. *Bioresour. Technol.*, 87, 137–146, 2003.
7. Santimano, M.C., Prabhu, N.N., Garg, S., PHA Production Using Low-Cost Agro-Industrial Wastes by *Bacillus* sp. Strain COL1/A6. *Res. J. Microbiol.*, 4, 89, 2009.
8. Shivam, P., Recent developments on biodegradable polymers and their future trends. *Int. Res. J. Sci. Eng.*, 4, 17, 2016.
9. Webb, H., Arnott, J., Crawford, R., Ivanova, E., Plastic Degradation and Its Environmental Implications with Special Reference to Poly(ethylene terephthalate). *Polymers*, 5, 1, 2012.
10. Ong, S.Y., Zainab-L, I., Pyary, S., Sudesh, K., A novel biological recovery approach for PHA employing selective digestion of bacterial biomass in animals. *Appl. Microbial Biotechnol.*, 102, 2117, 2018.
11. Zhu, J., Tian, Q., Zhu, Y., Yang, J., Wang, M., Factors for Promoting Polyhydroxyalkanoate (PHA) Synthesis in Bio-Nutrient-Removal and Recovery System. *IOP Conference Series: Earth Environ. Sci.*, vol. 178, p. 012021, 2018.
12. Ahmed, S., Kanchi, S., Kumar, G. (Eds.), *Handbook of Biopolymers: Advances and Multifaceted Applications*, Jenny Stanford Publishing, Singapore, 2018.
13. Sharma, V., Bhat, B., Gupta, M., Vaid, S., Sharma, S., Nargotra, P., Singh, S., Bajaj, B.K., Role of Systematic Biology in Biorefining of Lignocellulosic Residues for Biofuels and Chemicals Production, in: *Sustainable Biotechnology-Enzymatic Resources of Renewable Energy*, pp. 5–55, Springer Cham, Switzerland, 2018.
14. Emadian, S.M., Onay, T.T., Demirel, B., Biodegradation of bioplastics in natural environments. *Waste Manag.*, 59, 526, 2017.
15. Du, C., Sabirova, J., Soetaert, W., Ki Carol Lin, S., Polyhydroxyalkanoates Production From Low-cost Sustainable Raw Materials. *Curr. Chem. Biol.*, 6, 14, 2012.
16. Deepa, R. and Vidhya, A., Production and characterization of Polyhydroxybutyrate by *Nocardia* sp. RD13 isolated from agriculture rhizosphere soil. *Int. J. Res. BioSci.*, 7, 45, 2018.
17. Lemoigne, M., Produits de dehydration et de polymerisation de l'acide ßoxobutyrique. *Bull. Soc. Chim. Biol.*, 8, 770, 1926.
18. Verlinden, R.A.J., Hill, D.J., Kenward, M.A., Williams, C.D., Radecka, I., Bacterial synthesis of biodegradable polyhydroxyalkanoates; a review article. *J. Appl. Microbiol.*, 102, 1437, 2007.

19. Al Rowaihi, I.S., Paillier, A., Rasul, S., Karan, R., Grötzinger, S.W., Takanabe, K., Eppinger, J., Poly(3-hydroxybutyrate) production in an integrated electromicrobial setup: Investigation under stress-inducing conditions. *PLOS ONE*, 13, e0196079, 2018.
20. Gabr, G.A., Isolation and Identification of Bacterial Strains Able to Biopolymer Polyhydroxybutyrate (PHB) Production from Soil of Al-Kharj Probes, Saudi Arabia. *J. Pharm. Res. Int.*, 21, 1, 2018.
21. Batista, M.B., Teixeira, C.S., Sfeir, M.Z., Alves, L.P., Valdameri, G., Pedrosa, F.D.O., Sassaki, G.L., Steffens, M.B., de Souza, E.M., Dixon, R., Müller-Santos, M., PHB Biosynthesis Counteracts Redox Stress in *Herbaspirillum seropedicae*. *Front. Microbiol.*, 9, 472, 2018.
22. Verma, N., Sindhu, S.S., Goyal, S., Efficient poly-β-hydroxybutyrate production from *Bacillus* sp. *Pharma Inv. J.*, 7, 153, 2018.
23. Azizi, N., Najafpour, G., Younesi, H., Acid pretreatment and enzymatic saccharification of brown seaweed for polyhydroxybutyrate (PHB) production using *Cupriavidus necator*. *Int. J. Biol. Macromol.*, 101, 1029, 2017.
24. Gowda, V. and Shivakumar, S., Agrowaste-based Polyhydroxyalkanoate (PHA) production using hydrolytic potential of *Bacillus thuringiensis* IAM 12077. *Braz. Arch. Biol. Technol.*, 57, 55, 2014.
25. Rehm, B.H.A., Genetics and Biochemistry of Polyhydroxyalkanoate Granule Self-assembly: The Key Role of Polyester Synthases. *Biotechnol. Lett.*, 28, 207, 2006.
26. Castillo, T., Flores, C., Segura, D., Espín, G., Sanguino, J., Cabrera, E., Barreto, J., Diaz-Barrera, A., Pena, C., Production of polyhydroxybutyrate (PHB) of high and ultra-high molecular weight by *Azotobacter vinelandii* in batch and fed-batch cultures. *J. Chem. Technol. Biotechnol.*, 92, 1809, 2017.
27. Rasheed, F., Production of sustainable bioplastic materials from wheat gluten proteins, http://pub.epsilon.slu.se/8393/2/Rasheed_f_111020.pdf., 2011.
28. Kalia, V.C., Raizada, N., Sonakya, V., Bioplastics. *J. Sci. Ind. Res.*, 59, 433, 2000.
29. Johnstone, B., A throw away answer. *Far Eastern Econ. Rev.*, 147, 62, 1990.
30. Lathwal, P., Nehra, K., Singh, M., Rana, J.S., Characterization of Novel and Efficient Poly-3-hydroxybutyrate (PHB) Producing Bacteria Isolated from Rhizospheric Soils. *J. Polym. Environ.*, 26, 3437, 2018.
31. Koller, M., Gasser, I., Schmid, F., Berg, G., Linking ecology with economy: Insights into polyhydroxyalkanoate-producing microorganisms. *Eng. Life. Sci.*, 11, 222, 2011.
32. Shaaban, M.T., Attia, M., Turky, A.S., Mowafy, E.I., Production of some biopolymers by some selective Egyptian soil bacterial isolates. *J. Appl. Sci. Res.*, 8, 94, 2012.
33. Khardenavis, A.A., Kumar, M.S., Mudliar, S.N., Chakrabarti, T., Biotechnological conversion of agro-industrial waste waters into biodegradable plastic, PHB. *Bioresour. Technol.*, 98, 3579, 2007.

34. Rodriguez-Contreras, A., Koller, M., Dias, M.M., de, S., Calafell, M., Braunegg, G., Marqués-Calvo, M.S., Novel poly[(R)-3-hydroxybutyrate]-producing bacterium isolated from a bolivian hypersaline lake. *Food Technol. Biotechnol.*, 51, 123, 2013.
35. Razzaq, A., Jamil, N., Naheed, N., Hasnain, S., Bacteria from contaminated urban and hilly areas as a source of polyhydroxyalkanoates production. *Afr. J. Biotechnol.*, 9, 1919, 2010.
36. Haba, E., Vidal-Mas, J., Bassas, M., Espuny, M.J., Llorens, J., Manresa, A., Poly 3(hydroxyalkanoates) produced from oily substrates by *Pseudomonas aeruginosa* 47T2 (NCBIM 40044): Effect of nutrients and incubation temperature on polymer composition. *Biochem. Eng. J.*, 35, 99, 2007.
37. Bonatto, D., Matias, F., Lisboa, M.P., Bogdonia, H.M., Hanriques, J.A.P., Production of ssc-PHA by a newly isolated *Ralstonia pickettii* strain. *World J. Microbiol. Biotechnol.*, 20, 395, 2004.
38. Panigrahi, S. and Badveli, U., Screening, isolation and quantification of PHB producing soil bacteria. *Int. J. Eng. Sci. Inv.*, 2, 01, 2013.
39. Arshad, M.U., Jamil, N., Naheed, N., Hasnain, S., Analysis of bacterial strains from contaminated and non-contaminated sites for the production of biopolymers. *Afr. J. Biotechnol.*, 6, 1115, 2007.
40. Kulkarni, S.O., Kanekar, P.P., Jog, J.P., Patil, P.A., Nilegaonkar, S.S., Sarnaik, S.S., Kshirsagar, P.R., Characterisation of copolymer, poly (hydroxybutyrate-co-hydroxyvalerate) (PHB-coPHV) produced by *Halomonas campisalis* (MCM B-1027), its biodegradability and potential application. *Bioresour. Technol.*, 102, 6625, 2011.
41. Mohanrasu, K., Premnath, N., Prakash, G.S., Sudhakar, M., Boobalan, T., Arun, A., Exploring multi potential uses of marine bacteria; an integrated approach for PHB production, PAHs and polyethylene biodegradation. *J. Photochem. Photobiol. B.*, 185, 55, 2018.
42. Bhuwal, A.K., Singh, G., Aggarwal, N.K., Goyal, V., Yadav., A., isolation and screening of polyhydroxyalkanoates producing bacteria from pulp, paper, and cardboard industry wastes. *Int. J. Biomater.*, 2013, 752821, 2013.
43. Bhattacharyya, A., Pramanik, A., Maji, S.K., Haldar, S., Mukhopadhyay., U.K., Mukherjee, J., Utilization of vinasse for production of poly-3-(hydroxybutyrateco-hydroxyvalerate) by *Haloferax mediterranei*. *AMB Express*, 2, 1, 2012.
44. Kucera, D., Pernicová, I., Kovalcik, A., Koller, M., Mullerova, L., Sedlacek, P., Mravec, F., Nebesarova, J., Kalina, M., Marova, I., Krzyzanek, V., Characterization of the promising poly(3-hydroxybutyrate) producing halophilic bacterium *Halomonas halophila*. *Bioresour. Technol.*, 256, 552, 2018.
45. Bharadwaj, S.V., Shrivastav, A., Dubey, S., Ghosh, T., Paliwal, C., Maurya, R., Mishra, S., Draft genome sequence of *Halomonas hydrothermalis* MTCC 5445, isolated from the west coast of India. *Genome Announc.*, 3, e0141914, 2015.

46. Lopez, J.A., Naranjo, J.M., Higuita, J.C., Cubitto, M.A., Cardona, C.A., Villar, M.A., Biosynthesis of PHB from a new isolated *Bacillus megaterium* strain: outlook on future developments with endospore forming bacteria. *Biotechnol. Bioprocess. Eng.*, 17, 250, 2012.
47. Singh, M., Patel, S.K.S., Kalia, V.C., *Bacillus subtilis* as potential producer for polyhydroxyalkanoates. *Microb. Cell Fact.*, 8, 38, 2009.
48. Narayanan, A. and Ramana, K.V., Polyhydroxybutyrate production in *Bacillus mycoides* DFC1 using response surface optimization for physico-chemical process parameters. *3Biotech.*, 2, 287, 2012.
49. Singh, G., Kumari, A., Mittal, A., Yadav, A., Aggarwal, N.K., Poly--hydroxybutyrate production by *Bacillus subtilis* NG220 using sugar industry waste water. *BioMed. Res. Int.*, 2013, 952641, 2013.
50. Masood, F., Hasan, F., Ahmed, S., Hameed, A., Biosynthesis and characterization of poly (3-hydroxybutyrate-co-3-hydroxyvalerate) from *Bacillus cereus* FA11 isolated from TNT contaminated soil. *Ann. Microbiol.*, 62, 1377, 2012.
51. Law, K.H., Cheng, Y.C., Leung, Y.C., Lo, W.H., Chu, H., Yu, H.F., Construction of recombinant *Bacillus subtilis* strains for PHA synthesis. *Biochem. Eng. J.*, 16, 203, 2003.
52. Schallmey, M., Singh, A., Ward, O.P., Developments in the use of *Bacillus* sp. from industrial production. *Can. J. Microbiol.*, 50, 1, 2004.
53. Hsiao, L.J., Lee, M.C., Chuang, P.J., Kuo, Y.Y., Lin, J.H., Wu, T.M., Li, S.Y., The production of poly (3-hydroxybutyrate) by thermophilic *Caldimonas manganoxidans* from glycerol. *J. Polym. Res.*, 25, 85, 2018.
54. Możejko-Ciesielska, J. and Kiewisz, R., Bacterial polyhydroxyalkanoates: Still fabulous? *Microbiol. Res.*, 192, 271, 2016.
55. Fritzsche, K., Lenz, R.W., Fuller, R.C., Bacterial polyesters containing branched poly(β-hydroxyalkanoate) units. *Int. J. Biol. Macromol.*, 12, 92, 1990.
56. Villanueva, L., Del Campo, J., Guerrero, R., Diversity and physiology of polyhydroxyalkanoate-producing and-degrading strains in microbial mats. *FEMS microbial. Ecol.*, 74, 42, 2010.
57. Steinbüchel, A. and Schlegel, H.G., Physiology and molecular genetics of poly (β-hydroxyalkanoic acid) synthesis in *Alcaligenes eutrophus*. *Mol. Microbial.*, 5, 535, 1991.
58. Mesquita, D.P., Leal, C., Cunha, J.R., Oehmen, A., Amaral, A.L., Reis, M.A., Ferreira, E.C., Prediction of intracellular storage polymers using quantitative image analysis in enhanced biological phosphorus removal systems. *Anal. Chim. Acta.*, 770, 36, 2013.
59. Van Doan, T. and Nguyen, B.T., Polyhydroxyalkanoates production by a bacterium isolated from mangrove soil samples collected from Quang Ninh province. *J. Viet Env.*, 3, 76, 2012.
60. Sitepu, I.R., Ignatia, L., Franz, A.K., Wong, D.M., Faulina, S.A., Tsui, M., Kanti, A., Boundy Mills, K., An improved high-throughput Nile red

fluorescence assay for estimating intracellular lipids in a variety of yeast species. *J. Microbiol. Meth.*, 91, 321, 2012.
61. Spiekermann, P., Rehm, B.H., Kalscheuer, R., Baumeister, D., Steinbüchel, A., A sensitive, viable-colony staining method using Nile red for direct screening of bacteria that accumulate polyhydroxyalkanoic acids and other lipid storage compounds. *Arch. Microbiol.*, 171, 73, 1999.
62. Balaji, S., Gopi, K., Lavanya, B., Muthuvelan, B., Isolation and optimization of Poly-β- hydroxybutyrate producing cyanobacterial strains. *Int. J. Appl. Biol. Pharm. Technol.*, 3, 137, 2012.
63. Gorenflo, V., Steinbuchel, A., Marose, S., Rieseberg, M., Scheper, T., Quantification of bacterial polyhydroxyalkanoic acids by Nile red staining. *Appl. Microbiol. Biotechnol.*, 51, 765, 1999.
64. Schlegel, H.G., Lafferty, R., Krauss, I., The isolation of mutants not accumulating PHB. *Arch. Microbiol.*, 71, 283, 1970.
65. Randriamahefa, S., Renard, E., Guérin, P., Langlois, V., Fourier transform infrared spectroscopy for screening and quantifying production of PHAs by *Pseudomonas* grown on sodium octanoate. *Biomacromolecules*, 4, 1092, 2003.
66. Jarute, G., Kainz, A., Schroll, G., Baena, J., Lendl, B., On-line determination of the intracellular poly(beta-hydroxybutyric acid) content in transformed *Escherichia coli* and glucose during PHB production using stopped-flow attenuated total reflection FT-IR spectrometry. *Anal. Chem.*, 76, 6353, 2004.
67. Hong, K., Sun, S., Tian, W., Chen, G.Q., Huang, W., A rapid method for detecting bacterial polyhydroxyalkanoates in intact cells by Fourier transform infrared spectroscopy. *Appl. Microbiol. Biotechnol.*, 51, 369, 1999.
68. Shamala, T.R., Chandrashekar, A., Vijayendra, S.V.N., Kshama, L., Identification of PHA producing *Bacillus* sp. using the polymerase chain reaction (PCR). *J. Appl. Microbiol.*, 94, 369, 2003.
69. Degelau, A., Scheper, T., Bailey, J.E., Guske, C., Fluorometric measurement of poly-b-hydroxybutyrate in *Alcaligenes eutrophus* by flow cytometry and spectrofluorometry. *Appl. Microbiol. Biotechnol.*, 42, 653, 1995.
70. Vidal-Mas, J., Resina-Pelfort, O., Haba, E., Comas, J., Manresa, A., Vives-Rego, J., Rapid flow cytometry-Nile red assessment of PHA cellular content and heterogeneity in cultures of *Pseudomonas aeruginosa* 47T2 (NCIB 40044) grown in waste frying oil. *Antonie Van Leeuwenhoek*, 80, 57, 2001.
71. De Gelder, J., Willemse-Erix, D., Scholtes, M.J., Sanchez, J.I., Maquelin, K., Vandenabeele, P., De Boever, P., Puppels, G.J., Moens, L., De Vos, P., Monitoring poly(3-hydroxybutyrate) production in *Cupriavidus necator* DSM 428 (H16) with Raman spectroscopy. *Anal. Chem.*, 80, 2155, 2008.
72. Serafim, L.S., Lemos, P.C., Oliveira, R.F., Reis, M.A.M., Optimization of polyhydroxybutyrate production by mixed cultures submitted to aerobic dynamic feeding conditions. *Biotechnol. Bioeng.*, 87, 145, 2004.
73. Nishimura, T., Saito, T., Tomita, K., Purification and properties of β-ketothiolase from *Zoogloea ramigera. Arch. Microbial.*, 116, 21, 1978.

74. Schubert, P., Steinbüchel, A., Schlegel, H.G., Cloning of the *Alcaligenes eutrophus* genes for synthesis of poly-beta-hydroxybutyric acid (PHB) and synthesis of PHB in *Escherichia coli. J. Bacterial.*, 170, 5837, 1988.
75. Fiedler, S., Steinbüchel, A., Rehm, B.H., The role of the fatty acid β-oxidation multienzyme complex from Pseudomonas oleovorans in polyhydroxyalkanoate biosynthesis: molecular characterization of the fadBA operon from *P. oleovorans* and of the enoyl-CoA hydratase genes phaJ from *P. oleovorans* and *Pseudomonas putida. Arch. Microbial.*, 178, 149, 2002.
76. Rehm, B.H. and Steinbüchel, A., Biochemical and genetic analysis of PHA synthases and other proteins required for PHA synthesis. *Int. J. biol. Macromol.*, 25, 3, 1999.
77. Amara, A.A. and Moawad, H., PHAC synthases and PHA depolymerases: The enzymes that produce and degrade plastic. *IIUM Engineering J.*, 12, 21, 2011.
78. Qi, Q. and Rehm, B.H., Polyhydroxybutyrate biosynthesis in *Caulobacter crescentus*: molecular characterization of the polyhydroxybutyrate synthase. *Microbiol.*, 147, 3353, 2001.
79. Ye, Z.Q., Song, G., Chen, G.Q., Chen, J.Y., Location of functional region at N terminus of polyhydroxyalkanoate (PHA) synthase by N-terminal mutation and its effects on PHA synthesis. *Biochem. Eng. J.*, 41, 67, 2008.
80. Van-Thuoc., D., Quillaguaman, J., Mamo, G., Mattiasson, B., Utilization of agricultural residues for poly(3-hydroxybutyrate) production by *Halomonas boliviensis* LC1. *J. Appl. Microbiol.*, 104, 420, 2008.
81. Yezza, A., Fournier, D., Halasz, A., Hawari, J., Production of PHA from methanol by a new methylotropic bacterium *Methlobacterium* sp.GW2. *Appl. Microbiol. Biotechnol.*, 73, 211, 2006.
82. Ramadas, N.V., Singh, S.K., Soccol, C.R., Pandey, A., Polyhydroxybutyrate production using agro-industrial residue as substrate by *Bacillus sphaericus* NCIM 5149. *Braz. Arch. Biol. Technol.*, 52, 17, 2009.
83. Shivakumar, S., Polyhydroxybutyrate (PHB) production using agro-industrial residue as substrate by *Bacillus thuringiensis* IAM 12077. *Int. J. Chem. Tech. Res.*, 4, 1158, 2012.
84. Cruza, M.V., Paivaa, A., Lisboaa, P., Freitasa, F., Alvesb, V.D., Simõesa, P., Barreirosa, S., Reisa, M.A.M., Production of polyhydroxyalkanoates from spent coffee grounds oil obtained by supercritical fluid extraction technology. *Bioresour Technol.*, 157, 360, 2014.
85. Mokhtari-Hosseini, Z.B., Vashegahani-Farahani, E., Heidazardeh-Vazifekhoran, A., Shojaosadati, S.A., Karimzadeh, R., Darani, K.K., Statistical media optimization for growth and PHB production from methanol by a methylotrophic bacterium. *Bioresour. Technol.*, 100, 2436, 2009.
86. Chen, W.C., Chang, S.M., Chang, J.S., Chen, W.M., Chu, I.M., Tsai, S.L., Wang Li., F., Chang, Yu., K., Wei, Y.H., A process for simultaneously achieving phenol biodegradation and polyhydroxybutyrate accumulation using *Cupriavidus taiwanesis* 187. *J. Polym. Res.*, 25, 137, 2018.

87. Colombo, B., Sciarria, T.P., Reis, M., Scaglia, B., Adani, F., Polyhydroxyalkanoates (PHAs) production from fermented cheese whey by using a mixed microbial culture. *Bioresour. Technol.*, 218, 692, 2016.
88. Dietrich, K., Dumont, M.J., Schwinghamer, T., Orsat, V., Del Rio, L.F., Model Study To Assess Softwood Hemicellulose Hydrolysates as the Carbon Source for PHB Production in *Paraburkholderia sacchari* IPT 101. *Biomacromolecules*, 19, 188, 2017.
89. Naranjo, J.M., Posada, J.A., Higuita, J.C., Cardona, C.A., Valorization of glycerol through the production of biopolymers: The PHB case using *Bacillus megaterium*. *Bioresour. Technol.*, 133, 38, 2013.
90. Nath, A., Dixit, M., Bandiya, A., Chavda, S., Desai, A.J., Enhanced PHB production and scale up studies using cheese whey in fed batch culture of *Methylobacterium* sp. ZP24. *Bioresour. Technol.*, 99, 5749, 2008.
91. Vaid, S., Nargotra, P., Bajaj, B.K., Consolidated Bioprocessing for Biofuel-Ethanol Production from Pine Needle Biomass. *Environ. Prog. Sustain. Energy*, 37, 546, 2018.
92. Singh, S., Sambhyal, M., Vaid, S., Singh, P., Bajaj, B.K., Two-step sequential optimization for production of ionic liquid stable cellulase from *Bacillus subtilis* I-2. *Biocatal Biotransform.*, 33, 224, 2015.
93. Bajaj, B.K. and Wani, M.A., Purification and characterization of a novel phytase from *Nocardia* sp. MB 36. *Biocatal. Biotransfor.*, 33, 141, 2015.
94. Moreno, P., Yanez, C., Cardozo, N.S.M., Escalante, H., Combariza, M.Y., Guzman, C., Influence of nutritional and physicochemical variables on PHB production from raw glycerol obtained from a Colombian biodiesel plant by a wild-type *Bacillus megaterium* strain. *New Biotechnol.*, 32, 682, 2015.
95. Mozumder, M.S., Garcia-Gonzalez, L., Wever, H.D., Volcke, E.I.P., Poly(3-hydroxybutyrate) (PHB) production from CO_2: Model development and process optimization. *Biochem. Eng. J.*, 98, 107, 2015.
96. Somleva, M.N., Snell, K.D., Beaulieu, J.J., Peoples, O.P., Garrison, B.R., Patterson, N.A., Production of polyhydroxybutyrate in switchgrass, a value-added co-product in an important lignocellulosic biomass crop. *Plant Biotechnol. J.*, 6, 663, 2008.
97. Altaee, N., Fahdil, A., Yousif, E., Sudesh, K., Recovery and subsequent characterization of polyhydroxybutyrate from *Rhodococcus equi* cells grown on crude palm kernel oil. *J. Taibah Univ. Sci.*, 10, 543, 2016.
98. Chaijamrus, S. and Udpuay, N., Production and Characterization of Polyhydroxybutyrate from Molasses and Corn Steep Liquor produced by *Bacillus megaterium* ATCC 6748. *Agri. Eng. Int., the CIGR E J, X*,1–12, 2008.
99. Full, T.D., Jung, D.O., Madigan, M.T., Production of poly-β-hydroxyalkanoates from soy molasses oligosaccharides by new, rapidly growing *Bacillus* species. *Lett. Appl. Microbiol.*, 43, 377, 2006.
100. Gouda, M.K., Swellam, A.E., Omar, S.H., Production of PHB by a *Bacillus megaterium* strain using sugarcane molasses and corn steep liquor as sole carbon and nitrogen sources. *Microbiol. Res.*, 156, 201, 2001.

101. Kulpreecha, S., Boonruangthavorn, A., Meksiriporn, B., Thongchul, N., Inexpensive fed-batch cultivation for high poly(3-hydroxybutyrate) production by a new isolate of *Bacillus megaterium*. *J. Biosci. Bioeng.*, 107, 240, 2009.
102. Zhang, Y., Sun, W., Wang, H., Geng, A., Polyhydroxybutyrate production from oil palm empty fruit bunch using *Bacillus megaterium* R11. *Bioresour. Technol.*, 147, 307, 2013.
103. Pradhan, S., Borah, A.J., Poddar, M.K., Dikshit, P.K., Rohidas, L., Moholkar, V.S., Microbial production, ultrasound-assisted extraction and characterization of biopolymer polyhydroxybutyrate (PHB) from terrestrial (*P. hysterophorus*) and aquatic (*E. crassipes*) invasive weeds. *Bioresour. Technol.*, 242, 304, 2017.
104. Sharma, P. and Bajaj, B.K., Production and characterization of poly-3-hydroxybutyrate from *Bacillus cereus* PS 10. *Int. J. Biol. Macromol.*, 81, 241, 2015.
105. Carrillo, J.A.Z., Stein, L.Y., Sauvageau, D., Defining nutrient combinations for optimal growth and polyhydroxybutyrate production by *Methylosinus trichosporium* OB3b using Response Surface Methodology. *Front Microbial.*, 9, 1513, 2018.
106. Albuquerque, M.G.E., Eiroa, M., Torres, C., Nunes, B.R., Reis, M.A.M., Strategies for the development of a side stream process for polyhydroxyalkanoate (PHA) production from sugar cane molasses. *J. Biotechnol.*, 130, 411, 2007.
107. Solaiman, D.K.Y., Ashby, R.D., Hotchkiss, A.T., Foglia, T.A., Biosynthesis of medium-chain-length Poly(hydroxyalkanoates) from soy molasses. *Biotechnol Lett.*, 28, 157, 2006.
108. Zhu, C., Nomura, C.T., Perrotta, J.A., Stipanovic, A.J., Nakas, J.P., Production and characterization of poly-3-hydroxybutyrate from biodiesel-glycerol by *Burkholderia cepacia* ATCC 17759. *Biotechnol prog.*, 26, 424, 2010.
109. Keenan, T.M., Nakas, J.P., Tanenbaum, S.W., Polyhydroxyalkanoate copolymers from forest biomass. *J. Ind. Microbiol.*, 33, 616, 2006.
110. Song, J.H., Jeon, C.O., Choi, M.H., Yoon, S.C., Park, W., Polyhydroxyalkanoate (PHA) production using waste vegetable oil by *Pseudomonas* sp. strain DR2. *J. Microbiol. Biotechnol.*, 18, 1408, 2008.
111. Shrivastav, A., Mishra, S.K., Shethia, B., Pancha, I., Jain, D., Mishra, S., Isolation of promising bacterial strains from soil and marine environment for polyhydroxyalkanoates (PHAs) production utilizing Jatropha biodiesel byproduct. *Int. J. Biol. Macromol.*, 47, 283, 2010.
112. Salakkam, A. and Webb, C., Production of poly (3-hydroxybutyrate) from a complete feedstock derived from biodiesel by-products (crude glycerol and rapeseed meal). *Biochem. Eng. J.*, 137, 358, 2018.
113. Hassan, M.A., Bakhiet, E.K., Hussein, H.R., Ali, S.G., Statistical optimization studies for polyhydroxybutyrate (PHB) production by novel *Bacillus subtilis* using agricultural and industrial wastes. *Int. J. Environ. Sci. Technol.*, 16, 3497, 2019.

114. Sukruansuwan, V. and Napathorn, S.C., Use of agro-industrial residue from the canned pineapple industry for polyhydroxybutyrate production by *Cupriavidus necator* strain A-04. *Biotechnol Biofuels*, 11, 202, 2018.
115. Verlinden, R.A.J., Hill, D.J., Kenward, M.A., Williams, C.D., Piotrowska-Seget, Z., Radecka, I.K., Production of polyhydroxyalkanoates from waste frying oil by *Cupriavidus necator*. *AMB Express*, 1, 1, 2011.
116. Pandian, R.S., Deepak, V., Kalishwaralal, K., Rameshkumar, N., Jeyaraj, M., Gurunathan, S., Optimization and fed-batch production of PHB utilizing dairy waste and sea water as nutrient sources by *Bacillus megaterium* SRKP-3. *Bioresour. Technol.*, 101, 705, 2010.
117. Devi, N.C., Mazumder, P.B., Bhattacharjee, A., Statistical Optimization of Polyhydroxybutyrate Production by *Bacillus Pumilus* H9 Using Cow Dung as a Cheap Carbon Source by Response Surface Methodology. *J. Polym. Environ.*, 26, 3159, 2018.
118. Johar, N.A.M., Hassan, M.A., Zakaria, M.R., Yee, P.L., Shirai, Y., Ariffin, H., Evaluation of factors affecting polyhydroxyalkanoates production by *Comamonas* sp. EB172 using central composite design. *Malays. J. Microbiol.*, 8, 184, 2012.
119. Berwig, K.H., Baldasso, C., Dettmer, A., Production and characterization of poly (3-hydroxybutyrate) generated by *Alcaligenes latus* using lactose and whey after acid protein precipitation process. *Bioresour. Technol.*, 218, 31, 2016.
120. Kim, S.W., Kim, P., Lee, H.S., Kim, J.H., High production of poly-β-hydroxybutyrate (PHB) from *Methylobacterium organophilum* under potassium limitation. *Biotechnol. Lett.*, 18, 25, 1996.
121. Lee, S.Y., Plastic bacteria? Progress and prospects for polyhydroxyalkanoate production in bacteria. *Trends. Biotechnol.*, 14, 431, 1996.
122. Wang, Y., Ruan, L., Chua, H., Yu, P.H.F., Cloning and expression of the PHA synthase genes phaC1 and phaC1AB into *Bacillus subtilis*. *World J. Microbiol. Biotechnol.*, 22, 559, 2006.
123. Li, R., Zhang, H., Qi, Q., The production of polyhydroxyalkanoates in recombinant *Escherichia coli*. *Bioresour. Technol.*, 98, 2313, 2007.
124. Vandamme, P. and Coenye, T., Taxonomy of the genus Cupriavidus: a tale of lost and found. *Int. J. Syst. Evol. Microbial.*, 54, 2285, 2004.
125. Langenbach, S., Rehm, B.H., Steinbüchel, A., Functional expression of the PHA synthase gene pha C1 from *Pseudomonas aeruginosa* in *Escherichia coli* results in poly (3-hydroxyalkanoate) synthesis. *FEMS Microbial. Lett.*, 150, 303, 1997.
126. Choi, J.I., Lee, S.Y., Han, K., Cloning of the *Alcaligenes latus* polyhydroxyalkanoate biosynthesis genes and use of these genes for enhanced production of poly (3-hydroxybutyrate) in *Escherichia coli*. *Appl. Enviro. Microbial.*, 64, 4897–4903, 1998.
127. Seo, M.C., Shin, H.D., Lee, Y.H., Transcription level of granule-associated phaP and phaR genes and granular morphogenesis of poly-β-hydroxyalkanoate granules in *Ralstonia eutropha*. *Biotechnol. Lett.*, 26, 617, 2004.

128. Mahishi, L.H., Tripathi, G., Rawal, S.K., Poly (3-hydroxybutyrate)(PHB) synthesis by recombinant *Escherichia coli* harbouring *Streptomyces aureofaciens* PHB biosynthesis genes: effect of various carbon and nitrogen sources. *Microbiol. Res.*, 158, 19, 2003.
129. Liu, S.J. and Steinbüchel, A., A novel genetically engineered pathway for synthesis of poly (hydroxyalkanoic acids) in *Escherichia coli. Appl. Environ. Microbiol.*, 66, 739, 2000.
130. DeMarco, S., Advances in polyhydroxyalkanoate production in bacteria for biodegradable plastics. *Basic Biotechnol.*, 1, 1, 2005.
131. Nikel, P.I., Almeida, A.D., Melillo, E.C., Galvagno, M.A., Pettinari, M.J., New recombinant *Escherichia coli* strain tailored for the production of Poly (3Hydroxybutyrate) from agroindustrial by-products. *Appl. Environ. Microbiol.*, 72, 3949, 2006.
132. Saranya, V. and Shenbagarathai, R., Production and characterization of PHA from recombinant E. coli harbouring phaC1 gene of indigenous *Pseudomonas* sp. LDC-5 using molasses. *Braz. J. Microbiol.*, 42, 1109, 2011.
133. Wong, Y.M., Brigham, C.J., Rha, C., Sinskey, A.J., Sudesh, K., Biosynthesis and characterization of polyhydroxyalkanoate containing high 3-hydroxyhexanoate monomer fraction from crude palm kernel oil by recombinant *Cupriavidus necator. Bioresour. Technol.*, 121, 320, 2012.
134. Sato, S., Maruyama, H., Fujiki, T., Matsumoto, K., Regulation of 3-hydroxyhexanoate composition in PHBH synthesized by recombinant *Cupriavidus necator* H16 from plant oil by using butyrate as a co-substrate. *J. Biosci. Bioeng.*, 120, 246, 2015.
135. Tsuge, T., Yamamoto, T., Yano, K., Abe, H., Doi, Y., Taguchi, S., Evaluating the Ability of Polyhydroxyalkanoate Synthase Mutants to Produce P (3HB-co-3HA) from Soybean Oil. *Macromolecul. Biosci.*, 9, 71, 2009.
136. Chung, A.L., Jin, H.L., Huang, L.J., Ye, H.M., Chen, J.C., Wu, Q., Chen, G.Q., Biosynthesis and characterization of poly (3-hydroxydodecanoate) by β-oxidation inhibited mutant of *Pseudomonas entomophila* L48. *Biomacromolecules*, 12, 3559, 2011.
137. Zhuang, Q., Wang, Q., Liang, Q., Qi, Q., Synthesis of polyhydroxyalkanoates from glucose that contain medium-chain-length monomers via the reversed fatty acid β-oxidation cycle in *Escherichia coli. Metab. Eng.*, 24, 78, 2014.
138. Jin, H. and Nikolau, B.J., Evaluating PHA productivity of bioengineered *Rhodosprillum rubrum. PloS one*, 9, e96621, 2014.
139. Snell, K.D. and Peoples, O.P., PHA bioplastic: a value-added coproduct for biomass biorefineries. *Biofuels Bioprod. Biorefin.*, 3, 456, 2009.
140. Nawrath, C., Poirier, Y., Somerville, C., Targeting of the polyhydroxybutyrate biosynthetic pathway to the plastids of *Arabidopsis thaliana* results in high levels of polymer accumulation. *Proc. Natl. Acad. Sci.*, 91, 12760, 1994.
141. Lee, W.H., Loo, C.Y., Nomura, C.T., Sudesh, K., Biosynthesis of polyhydroxyalkanoate copolymers from mixtures of plant oils and 3hydroxyvalerate precursors. *Bioresour. Technol.*, 99, 6844, 2008.

142. Yunus, A.M.M., Ghulam-Kadir, A.P., Chai-Ling, H., Transgenic Plants Producing Polyhydroxyalkanoates. *Asia Pac. J. Mol. Biol. Biotechnol.*, 16, 1, 2008.
143. Bohmert-Tatarev, K., McAvoy, S., Daughtry, S., Peoples, O.P., Snell, K.D., High levels of bioplastic are produced in fertile transplastomic tobacco plants engineered with a synthetic operon for the production of polyhydroxybutyrate. *Plant Physiol.*, 155, 1690, 2011.
144. Zhang, B., Carlson, R., Srienc, F., Engineering the monomer composition of polyhydroxyalkanoates synthesized in *Saccharomyces cerevisiae. Appl. Environ. Microbial.*, 72, 536, 2006.
145. Zhang, M., *Development of polyhydroxybutyrate based blends for compostable packaging*, vol. 190, Ph.D. Thesis. Department of Materials, University of Loughborough., Loughborough, 2010.
146. Amass, W., Amass, A., Tighe, B., A review of biodegradable polymers: uses, current developments in the synthesis and characterization of biodegradable polyesters, blends of biodegradable polymers and recent advances in biodegradation studies. *Polym. Intl.*, 47, 89, 1998.
147. Liau, C.P., Ahmad, M.B., Shameli, K., Yunus, W.M.Z.W., Ibrahim, N.A., Zainuddin, N., Then, Y.Y., Preparation and characterization of polyhydroxybutyrate/polycaprolactone nanocomposites. *Scientific. W. J.*, 2014, 572726, 2014.
148. Muralidharan, R. and Radha, K.V., Growth and biological production of polyhydroxy butyrate (PHB) by *Bacillus megaterium* MTCC 8075. *J. Chem. Biol. Phys. Sci.*, 4, 1271, 2014.
149. Al-Rawajfeha, A.E., Al-Salahb, H.A., Al-Rhaelc, I., Miscibility, crystallinity and morphology of polymer blends of polyamide-6/poly (β-hydroxybutyrate). *Jordan J. Chem.*, 1, 155, 2006.
150. Sindhu, R., Ammu, B., Binod, P., Deepthi, S.K., Ramachandran, K.B., Soccol, C.R., Pandey, A., Production and characterization of poly-3-hydroxybutyrate from crude glycerol by *Bacillus sphaericus* NII 0838 and improving its thermal properties by blending with other polymers. *Braz. Arch. Biol. Technol.*, 54, 783, 2011.
151. Chang, L. and Woo, E.M., Effects of molten poly (3-hydroxybutyrate) on crystalline morphology in stereocomplex of poly (L-lactic acid) with poly (D-lactic acid). *Polymer*, 52, 68, 2011.
152. Prakalathan, K., Mohanty, S., Nayak, K.S., Reinforcing effect and isothermal crystallization kinetics of poly (3-hydroxybutyrate) nanocomposites blended with organically modified montmorillonite. *Polym. Composit.*, 35, 999, 2014.
153. Tikhonov, N.A., Arkhangelsky, I.V., Belyaev, S.S., Matveev, A.T., Carbonization of polymeric nonwoven materials. *Thermochim. Acta.*, 486, 66, 2009.
154. Kiran, G.S., Lipton, A.N., Priyadharshini, S., Anitha, K., Suárez, L.E.C., Arasu, M.V., Choi, K.C., Selvin, J., Al-Dhabi, N.A., Antiadhesive activity

of poly-hydroxy butyrate biopolymer from a marine *Brevibacterium casei* MSI04 against shrimp pathogenic vibrios. *Microb. Cell. Fact.*, 13, 114, 2014.
155. Sato, H., Murakami, R., Zhang, J., Ozaki, Y., X-Ray diffraction and infrared spectroscopy studies on crystal and lamellar structure and CHO hydrogen bonding of biodegradable poly (hydroxyalkanoate). *Macromol. Res.*, 14, 408, 2006.
156. Pan, W., Perrotta, J.A., Stipanovic, A.J., Nomura, C.T., Nakas, J.P., Production of polyhydroxyalkanoates by *Burkholderia cepacia* ATCC 17759 using a detoxified sugar maple hemicellulosic hydrolysate. *J. Ind. Microbiol. Biotechnol.*, 39, 459, 2012.
157. Lütke-Eversloh, T., Bergander, K., Luftmann, H., Steinbüchel, A., Identification of a new class of biopolymer: bacterial synthesis of a sulfur-containing polymer with thioester linkages. *Microbiol.*, 147, 11, 2001.
158. Mallakpour, S. and Zamanlou, M.R., Synthesis of new active poly (amide-imide)s containing epiclon and L-phenylalanine in the main chain by microwave irradiation and classical heating. *J. Appl. Polym. Sci.*, 91, 3281, 2004.
159. Kshirsagar, P.R., Kulkarni, S.O., Nilegaonkar, S.S., Niveditha, M., Kanekar, P.P., Kinetics and model building for recovery of polyhydroxyalkanoate (PHA) from *Halomonas campisalis*. *Sep. Purif. Technol.*, 103, 151, 2013.
160. Jain, R., Kosta, S., Tiwari, A., Polyhydroxyalkanoates: A way to sustainable development of bioplastics. *Chron. Young Sci.*, 1, 10, 2010.
161. Boopathy, R., Factor limiting bioremediation technologies. *Bioresour. Technol.*, 74, 63, 2000.
162. Saharan, B.S., Grewal, A., Kumar, P., Biotechnological production of polyhydroxyalkanoates: A review on trends and latest developments. *Chin. J. Biol.*, 2014, 802984, 2014.
163. Jendrossek, D. and Handrick, R., Microbial degradation of polyhydroxyalkanoates. *Annu. Rev. Microbiol.*, 56, 403, 2002.
164. Choi, G.G., Kim, H.W., Rhee, Y.H., Enzymatic and non-enzymatic degradation of poly (3-hydroxybutyrate-co-3-hydroxyvalerate) copolyesters produced by *Alcaligenes* sp MT-16. *J. Microbiol.*, 42, 346, 2004.
165. Ikada, Y. and Tsuji, H., Biodegradable polyesters for medical and ecological applications. *Macromol. Rapid. Commun.*, 21, 117, 2000.
166. Huey, C., Polyhydroxybutyrate (PHB) *production from cafeteria wastes under anoxic and aerobic conditions in sequencing batch reactor*, B.Sc. Thesis. Civil Eng Env Faculty Civil Eng, Universiti Teknologi, Malaysia, 2006.
167. Weng, Y.X., Wang, X.L., Wang, Y.Z., Biodegradation behaviour of PHAs with different chemical structures under controlled composting conditions. *Polym. Test.*, 30, 372, 2011.
168. Shangguan, Y.Y., Wang, Y.W., Wu, Q., Chen, G.Q., The mechanical properties and *in vitro* biodegradation and biocompatibility of UV-treated poly(3-hydroxybutyrate-co3-hydroxyhexanoate). *Biomaterialsv*, 27, 2349, 2006.

169. Bagheriasl, S., *Development and characterisation of polyhydroxybutyrate from selected bacterial species*, vol. 209, Ph.D. Thesis. School of Metallurgy and Materials, University of Birmingham, Birmingham, 2012.
170. Volova, T.G., Shishatskaya, E.I., Sevastianov, V.I., Efremov, S., Mogilnaya, O., Results of biomedical investigations of PHB and PHB/PHV fibers. *Biochem. Eng. J. Biopolym.*, 16, 125, 2003.
171. Pouton, C.W. and Akhtar, S., Biosynthetic polyhydroxyalkanoates and their potential in drug delivery. *Adv. Drug Deliv. Rev.*, 18, 133, 1996.
172. Chen, G.Q. and Wu, Q., The application of polyhydroxyalkanoates as tissue engineering materials. *Biomaterials*, 26, 6565, 2005.
173. Zinn, M., Witholt, B., Egli, T., Occurrence, synthesis and medical application of bacterial polyhydroxyalkanoate. *Adv. Drug Deliver. Rev.*, 53, 5, 2001.
174. Noda, I., *Films comprising biodegradable PHA copolymers*, vol. 6174990, United States Patent, Bainbridge, Georgia, United States, 2001.
175. Bucci, D.Z. and Tavares, L.B.B., PHB packaging for the storage of food products. *Polym. Test.*, 24, 564, 2005.
176. Lee, S.H., Oh, D.H., Ahn, W.S., Lee, Y., Choi, J.I., Lee, S.Y., Production of poly(3-hydroxybutyrate-co-3hydroxyhexanoate) by high-cell-density cultivation of *Aeromonas hydrophila*. *Biotechnol. Bioeng.*, 67, 240, 2000.
177. Philip, S., Keshavarz, T., Roy, I., Polyhydroxyalkanoates: biodegradable polymers with a range of applications. *J. Chem. Technol. Biotechnol.*, 82, 233, 2007.
178. Paik, Y.H., Simon, E.S., Swift, G., A review of synthetic approaches to biodegradable polymeric carboxylic acids for detergent applications. *Adv. Chem. Ser.*, 248, 79, 1996.
179. Williams, S.F. and Peoples, O.P., Making plastics green. *Chem. Br.*, 33, 29, 1997.

7

Microbial Synthesis of Polyhydroxyalkanoates (PHAs) and Their Applications

N.N.N. Anitha and Rajesh K. Srivastava*

Department of Biotechnology, GIT, Gandhi Institute of Technology and Management (GITAM), Deemed to be University, Visakhapatnam, AP, India

Abstract

The current period is facing the hazardous effects of synthetic plastics or polymers, reported as non-degradable with the big quantity of its accumulation in our healthy environments. Synthetic plastics are used for packaging of food items, medicine, water, cloths pharmaceutical, and also carrying vegetables or fruits or others application. To minimize or control this issue or problem, we need to utilize the biodegradable plastics or bio-plastic for various applications in our modern life. For the last two decades, microbial-derived polyhydroxyalkanoates (PHAs) gained more attention for various industrial applications. An effective microbial system (i.e., bacterial strain) is required for the maximum quantity of PHAs production at optimized cultural conditions via D-optimal statistical design. PHA-producing bacterial strain are isolated from various soil and spring samples and characterized using morphological, biochemical, and 16S rDNA sequencing method. Crude glycerin concentration and carbon to nitrogen ratio (C:N) in the culture medium for PHB growth were optimized by Response Surface Methodology (RSM). Accumulation of PHAs is found in mixed microb ial culture (MMC), and active-biomass yield coefficient (Y), observed PHA yield coefficient (Y), biomass PHAs content (X), and volumetric productivity (Pr) are four indicators influenced by culture media. Polyhydroxybutyrate (PHB) is reported in a combined form with good plastic nature with the natural nature of terpene and D-limonene organic compound. It has exhibited the dual objective of increasing PLA crystalline nature and also obtained flexible films for food packaging application tasks.

**Corresponding author*: rajeshksrivastava73@yahoo.co.in

Bibhuti Bhusan Mishra, Suraja Kumar Nayak, Swati Mohapatra, and Deviprasad Samantaray (eds.) Environmental and Agricultural Microbiology: Applications for Sustainability, (151–182) © 2021 Scrivener Publishing LLC

The functional properties of PHB are examined through colorimetric variables, oxygen permeability, water-resistance nature evaluation, thermal stability, crystalline response, mechanical, and nano-material characteristics. FTIR spectra have revealed for their characteristics bands equivalent in PLA and PHB, provided the information for their relative molecular interaction. PyGC-MS demonstrated the D-limonene characteristic peaks along with PLA and PHB thermal degradation product profiles. In this chapter, we emphasize recent development in the various types of PHB synthesis processes along with their characterization and applications.

Keywords: Microbial system, bacteria, biopolymer, polyhydroxyalkanoates, crystalline, polyhydroxybutyrate, culture media

Abbreviations

DHm	Melting enthalpy
(P3HB4HB)	Poly(3-hydroxybutyrate-co-4-hydroxybutyrate)
3H2MV	3-hydroxy-2-methylvalerate
3HB	3-hydroxybutyrate
AFD	Active feeding and discharge
BTEX	Benzene, toluene, ethylbenzene and p-xylene
C:N	Carbon to nitrogen
CI	Compression ignition
CDW	Cell dry weight
COD	Chemical oxygen demand
Da	Dalton
FTIR	Fourier-transform infrared spectroscopy
GC/MS	Gaschromatography/mass spectroscopy
HDPE	High density polyethylene
IRR	Internal rate of return
K_2HPO_4	Di-potassium hydrogen phosphate
LIM	D-limonene
MBC	Mixed bacterial community
mcl-PHA	Medium-chain length polyhydroxyalkanoate
MMC	Mixed microbial cultures
NA	nonanoic acid
NaClO	Sodium hypochlorite
NGIB	Next generation industrial biotechnology
OMW	Oil mill wastewater
OPTS	Oil palm trunk sap
P(3-HB)	poly(3-hydroxybutyrate)
P(3HBco-3HV)	Hydrooxybutyrate-co-3-hydroxyvalerate

P(3HB-co-3HV)	poly(3-hydroxybutyrate-co-3-hydroxyvalerate)
P(3HO)	(3-hydroxyoctanoate)
PBA	Poly-3-hydroxybutyrate
PC	Polycarbonate
Pco	P. corrugate
PE	Polyphenylene ether
PHA (PHA(MC))	Mixed culture of PHA
PHAs	Polyhydroxyalkanoates
PHB	Poly (hydroxybutyrate)
PHBV	Hydroxybutyrate-co-3-hydroxyvalerate)
Pme	*P. mediterranea*
POM	Polyoxymethylene
Pr	Productivity
PVC	Polyvinylchloride
rDNA	Recombinant deoxyribonucleic acid
RSM	Response Surface Methodology
SBR	Sequencing batch reactors
SPO	crude sludge palm oil
TG	Thermogravimeter
Tm	Melting temperature
UDA	10-undecenoic acid; v/v Volume by volume
VFA	Volatile fatty acid
WAS	Waste activated sludge
X	Biomass
Y	yield

7.1 Introduction

Biopolymer produced by several microbial strains is the polymeric biomolecules that are made up of several units of monomers with covalently bonded to each other to form larger structures. One of the best examples of a biopolymer is PHAs that synthesize as an adaptive feature of effective and specific microbial strains under stressed conditions. It is produced on the pivotal stage of many bacteria and Achaea to reserve carbon and energy in the cells. It is intracellular polyester, regarded as an essential biological macromolecule, beneficial for the biomaterial industry, which possesses many properties such as biodegradability and eco-friendly nature. Biodegradation of PHAs is reported in soil and aquatic sources and their samples have been detected after proceeding of anaerobic digestion towards the biggest amounts or quantity of its products that can help its application in bioremediation with reduction of conventional nature of plastic concentration. By the solvent extraction method, PHAs were extracted from marine algal species. It was found that for all these origin PHAs, these were thermally stable at 260°C [1].

We found the increased limitation of fossil resources utilized in chemical plastics development due to non-biodegradable nature. Increased resource efficiency and more persistence to microbial degradation are also reported for chemical plastics. Now, people are demanding eco-friendly and sustainable processes for bio-based plastic development in the current market. In this regard, bioplastic is seemed to be very interesting because of minimized carbon footprint profiles during their production and processing. Next, persistent and biomass-derived plastics can be a very promising replacement for traditional fossil-based plastics which are used in many applications with durability and longevity requirements [2].

There are reports for the non-biodegradable nature of conventional or synthetic plastics and bioplastics like PHA or PHB is obtained from renewable resources via the utilization of suitable and efficient microbial strain. There are reports on bioplastics degradation within periods in all the seasons and around 57% of total bioplastic or among all the bioplastics that are slightly bio-based plastics such as bio polyethylene, bioterepthalate, or bio polyamides, used in the current market and followed by fully biomasses derived fully biodegradable polyethylene. Most of the biopolymers have been discovered that shown the best thermoplastic properties from the last few decades and first biopolymers polythioesters is also discussed as bioplastics [3]. This non-biodegradable nature of biopolymer was biosynthesized by bacterial strains and its example is polythioesters that were found to nonbiodegradable by microbial strains that are a contrast to all others biopolymers and it was represented as novel nonbiodegradable plastics materials [2, 4].

PHAs are the most favorable alternative bioplastics for conventional non-biodegradable plastic due to its biodegradability. PHAs coating materials can enhance the hydrophobicity and transparency due to bacterial cellulose-based nanopores. The intensive nature of PHAs pulls them towards the manufacture of nanoparticles and nanocomposites. The combination of PHAs with nanoparticles enforced in various areas like drug delivery, antibacterial agents and bioengineering, and their establishment in various fields such as medicines, catalysis, biosensors, and adsorbents [2, 3].

Development of a bioreactor-based method of high cell density can help to generate medium-chain length PHAs by utilization of cheap substrate like cooking oil waste. Thus, nine Halomonas strains were experimentally tested to conceal the PHAs production from cooking or fried oil as waste organic sources. *Halomonas hydrothermalis* is seemed to be a fascinating halophilic strain for the production of PHAs from the lipid substrate. Fourier transform infrared (FTIR) spectroscopy analysis is done for the detection of molecular or atomic mass or weights of various bioplastic materials that shown its crystalline nature and monomers of its structure or concentration that are utilized the correct information on bioplastics structures or its functional groups. There are different extraction methods, used to get in pure form with the evaluation of its efficacy. Spirulina species LEB-18 is a species of microalgae that is known for PHAs biosynthesis. This information on the various origins of bioplastics can be obtained from various extraction methods. Different PHAs is reported to accumulate in different quantity or capacity, in much microbial strain including bacterial strain and their synthesized bioplastics components. But, the initial uses of sodium hydrochloride in the extraction process can be increased accumulation of PHB. Additionally, the use of ethanol at the end stage of the process can increase the purity of polymers. This has increased the ability to manufacture long-chain polymers from monomers such as 11-hydrohexadecanoate (in more quantity), with hydrohexadecanoate as well as hydroheptanoate in different bacterial strains [4].

Food and industrial wastewater are reported to contain high organic carbon sources that can be retrieved for the production of PHAs. The MMC subjected fermentation in a sequential batch reactor under an anaerobic environment and dynamic feeding rate, where it demonstrated PHAs accumulation up to 72% in batch and 65% in continuous feeding mode per dry cell weight (DCW). Despite having lower PHAs accumulation in continuous feeding strategy, it shows four times more biomass growth. Hence, the continuous strategy is a suitable investigating tool for PHAs production [5].

Several samples collected from the soil as well as spring season are reported that have reported producing PHAs with accumulation in soil bacterial cell (30% of total) and is conceived to its identity as the bacterial strains with its developing PHAs. This information has been confirmed by morphological or biochemical tests for strains and also 16rDNA sequencing methods that are the most efficient methods for identifying the PHA-producing capabilities in different microbial strains. Pseudomonas species are reported as a PHA producer bacterial strain that is found in Antarctic soil. In the detection or analysis method, maximum PHA biosynthesis is attained in the exponential phase and a lesser amount of it in the stationary phase of the bacterial growth reported. Renewable and agro-industrial byproduct can be used as an effective and fermentative carbon source for PHAs synthesis. Lignocellulosic waste can be used for PHAs using wild type *Bacillus megaterium* strain Ti3 and intrinsic or innate hydrolytic enzymes [6].

PHAs is a microbial biopolymer that acts as a counteract to the problem of microplastics, obtained by the usage of microbeads. These microbeads are used in cosmetics which are non-biodegradable. Since some of PHAs are reported to utilize in the cosmetic sector due to its biodegradability. Further, biodegradable microbeads are reported to develop via using the double emulsion solvent evaporation technique or approach that is used in the cosmetic sector. Eco-friendly nature bioprocess such as fermentation is used with pretreated wood waste matter that can promote a greener material. Sewage waste having a high carbon source is used for the synthesis of PHA by adding mixed microbial consortium that can replace traditional petroleum-based polymers waste [7].

Favorable nature of microbial cells or strain is required to minimize the intricacy or complex nature of PHA development that can improve the resistance to contamination and are targeted particularly for extremophile nature such as *Halomonas* species are known as salt-tolerant proteobacteria (can grow at 5% to 25% sodium chloride concentration). This strain can help in the production of PHA and can be successfully controlled and classified as next-generation industrial biotechnology (NGIB) era. Various natures of PHAs can be produced from this method. Polyhydroxyalkanoate (PHA) can be produced from crude glycerol substrate as a raw carbon source by using a new bacterial strain, i.e., *Bhurkholderia glumae* M13 that is isolated from an Atlantic rain forest ecosystem. This strain is expressed as an adaptive bacterium necessary for the synthesis of PHA from biofuels [8]. This chapter will discuss PHB production from various bacteria through different downstream processing along with their multiple applications.

7.2 Conventional Plastics and Its Issues in Utility

In the recent era, synthetic polymers are derived from petroleum oil and there are many examples of plastics such as nylon, polyethylene, polyester, teon, and epoxy compounds. However, bio-based polymers are synthesized and obtained from natural sources that include silk, wool, DNA, cellulose, or protein [9]. Further, pectin polymers in most of the fruits are reported with its monomers chains that form a network from a segment of pectin chains. They can join together with the forming of 3D-network and crystalline structures via attaching to water, sugars, or other organic materials. Gel formation is reported in these polymers that can be caused by physical or chemical changes in polymers with a decrease solubility of the pectin, and it also favors small localized crystal formation. There are shown as some important factors (i.e., temperature) that influence the gel formation capacity in pectin. This pectin material can also use bioplastic material synthesis [10]. After cooled from hot solution, containing pectin can decrease the movement of pectin and induced the get networks. This property of pectin can apply as food packaging material in food products like jellies or jams with a sufficient amount of sugars in pectin mixture at gel state [10, 11].

Plastics are an important commodity used extensively at commercial levels as the greatest innovations of soft nature. Plastics are extensively used in the market due to lightweight, cheap, flexible, and reusable property. Globally, annual plastics production has increased by up to 10% from 1950 (1.5 million metric tonnes) and till 2018 (359 million metric tonnes), while Asia was established as the largest polymer consumer by accounting 36.5% of global consumption. Plastics are used as a major segment of packaging materials and continue to be utilized in packaging tasks with 35% of global demand [12].

Synthesis of petroleum-based fuel is also reported that uses the catalytic pyrolysis process for the degradation of waste plastics material or stocks. In this approach, a polymeric material is degraded by the heating condition at high temperature and anoxic conditions in the presence of catalysts. This process leads to the synthesis of byproducts such as waste oil, toxic gases causing environmental pollution. In this process, produced oil samples are gone for parametric study, based on oil yields, electicivity of oils, fuel properties or reaction temperatures, and also an optimal catalyst performance and reaction conditions [13].

The gas chromatography/mass spectroscopy (GC/MS) technique was applied for selected and optimized oil samples that were detected for their chemical compositions or functional groups. Further, performance analysis was done for the other selected oil samples via the use of a compression ignition (CI) engine tool. Finally, most of the polythene bags are used for carrying materials from various shops that are reported as sources of plastic wastes in our earth and it has created the big challenge for our worlds for their degradation as well as the reduction in concentration, reported from various sources of water or soil bodies.

For degradation of plastics, it needs to convert into usable products (fuels) that can be achieved by using various catalysts such as silica, alumina, Y-zeolite, various carbonate, alone zeolite, or their combination. Normally, in pyrolysis reaction, it is carried out for polymers to catalyst ratio (10:1) and temperature at the 400°C–550°C range. Further, the immobile or inert nature of the atmosphere for the pyrolysis process is required (such as the use of N_2 (nitrogen) gas as the carrier gas) [14].

Microplastics pollution is reported to be a big challenge in the ocean due to its detrimental effects on aquatic animals and plants found in the depth of the ocean. Further, ingested microplastics concentration are found in the hindgut of amphipods (Lysianassoidea, a crustacea population), and these studies were done for samples from six deep ocean trenches or wells across the Pacific Rim city, Japan, Izu-Bonin, Mariana, Kermadec, as well as Peru-Chile Mariana and these have shown the depths from 7,000 to 10,890 m in the ground [15, 16]. The presence of microplastics contaminants in this animal is reported to vary in concentrations from the deepest trenches of oceans and around 72% Individual Ocean had been examined that had contained at least one microplastic particles in that animal gut. The numbers of microplastic particles were reported to ingest per individual animal in all the trenches can be found from 1 to 8 [16]. Sub-samples of microfibers and fragments had been analyzed by using the FTIR technique, and these analyses were recorded for collected plastic and synthetic materials from their gut. Nylon, polyethylene, polyamides, and polyvinyl alcohol are recorded as inorganic filler materials and semi-synthetic (rayon or lyocell products) with certain natural fibers (ramie). The deepest record of microplastic ingestion has indicated by anthropogenic activity as waste debris and is available for the organism at some extensive locations in the Earth's or ocean's site [16, 17]. The author will discuss bioplastic production and its application.

7.2.1 Synthetic Plastic and Its Accumulation or Degradation Impacts

The authors have discussed synthetic plastic materials, utilized for packaging tasks of different food or other products, and now, it has collected as plastic waste in our environment and these are reported to contain various types of plastic such as 50% of higher density nature polyethylene (HDPE), 30% of polyethylene terephthalate and 20% as other plastic components. The various techniques are well-known for evaluating the effect of different catalysts on the decomposition of waste plastics and these are screened by high-pressure thermogravimeter (TG) and GC/MS tools. The catalyst used with its access for the conversion of plastic waste and that are superacids, conventional cracking catalysts, hydrocracking catalysts, and zeolite catalysts [18, 19]. Common wettable agents such as sodium-containing lignin sulfonate, aerosol OT, tannic acid, and saponin were used for the segregation of four important plastics like polyacetal (POM), polyvinylchloride (PVC), polyphenylene ether (PEE), and polycarbonate (PC), and these are reported to obtain from their synthetic mixtures. Exploding synthetic plastics releases toxic chemicals into the environment that will harm peoples, plants, and animals, and also recycling of plastics is very costly. To reduce this impact, bioplastics came into existence [20, 21].

7.3 Bioplastics

Now, our researcher groups are found to focus on various approaches of biological processes, parameters, or microbial systems that are utilized for bioplastic production with the promotion of greenery evolution that can minimize the impact of the accumulation of synthetic polymers in soil or water bodies. This synthetic plastic degradation is reported to our environment. In this regard, we need to promote the synthesis of bioplastics such as PHBs or other bioplastics. These bioplastics can minimize the negative impact of plastic usage and its degradation or accumulation. In this regard, the bio-based economy of the country or world plays a crucial role in making avail of bioresources to produce many bioproducts like paper, pulp, chemicals, biofuels, and bioplastics. An advantage of utilizing lignocellulosic or non-food crops as the raw carbon source for the production of bioplastics. This bioplastic eventually shows competent with plastics and replacement of synthetic plastics which is beneficial to the environment [21, 22]. There are three bioplastic forms.

- Starch material can generate the bioplastics, and in this regard, corn starch derived sugar can be used to get the simple nature of bioplastics.
- Cellulose material can also generate various types of bioplastics that can be found at a cheap price. This can help to generate esters components and also derivatives of cellulose.
- Protein-derived bioplastics are reported that can be generated using various sources protein such as wheat gluten, and milk casein protein. Further, PHAs as bioplastic belong to cellulose-based bioplastics or its derived polyester.

7.3.1 Polyhydroxyalkanoates

PHAs are reported as biomaterial or biopolymer and that can be biosynthesized with its extraction from fermentation broth. This fermentation is utilized for PHAs production from mixed or pure bacterial culture strains (MBC) of *Pseudomonas* species and these strains have shown the best PHA, production with good storing capacity in sequence batch reactors (SBRs) during fermentation that fed with unnatural or synthetic nature of effluents [i.e., from fermented oil mill wastewater (OMW)]. PHAs are reported in three categories such as small chain (3 to 4 carbon atoms with hard, crystalline or brittle nature or high M.P. values), medium (6 to 14 carbon atoms), or large numbers (more than 14 carbon atoms) of a carbon chain with soft and elastomers or low M.P., and these are reported for various types of PHAs. Another bioplastic such poly-β-hydroxybutyrate (PHB) is also reported to produce from mixed bacterial culture. This production approach for bioplastic has shown the highest yield (74%) and a high rate of purity (100%) by applying NH4-Laurate reagent with the best extraction operating parameters such as concentration, temperature, and contact periods. PHA extracted from MMC culture was found to go for NaClO pretreatment at 85°C for 1 h of contact time and lauric acid or lauric acid to biomass ratio (2:1) for 3 h contact time and structure of PHA in Figure 7.1 is shown [7, 23].

The feast or famine technique is used to select the culture of PHA-producing and -accumulating strain from mixed microbial strain and eco-physiology of microbial communities is used in selection processes. Dynamic feast or famine regime was utilized for leading the repetitive cycles with moderate change monitoring tasks in substrate availability in an aerobic consumption (i.e., glucose) during the cultivation of yeast species (*Saccharomyces cerevisiae*). In this regard, three well-defined Feast or Famine technique systems with well optimal and defined culture conditions

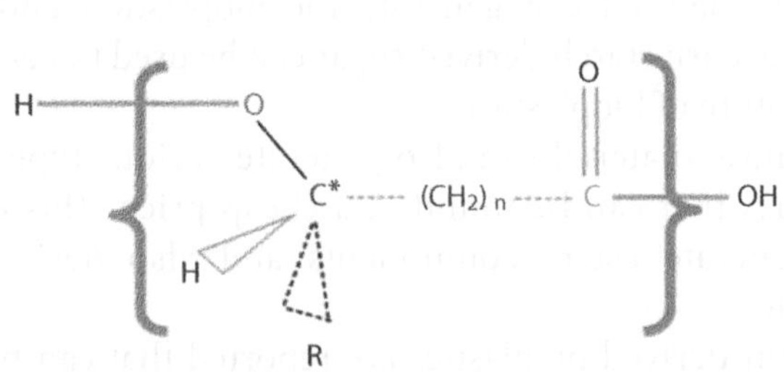

Figure 7.1 General structure of Polyhydroxyalkanoate (PHA) shown in this figure and R indicates the side chain of groups or monomers and n is the number of methyl groups in monomer's backbone. The asterisk spots indicate the chiral center of most PHA building block.

were used to synchronize the variation at a higher level of behavior of mixed culture (MC) and it is shown the more complication in microbial communities succession (enrichment or elimination of non-top competitors) [24, 25]. PHA-accumulating function genes (*pha*C) in bacterial cells have been quantified with the best performance by the biggest competitors that can be established the PHA biosynthesis with the consecutive position for high rate or rapid turnover. It was taken the consideration of specific physiological properties for PHA-producing microbial system processes, and in this regard, *Thauera* strain OUT 7 was reported to responsible for affecting or fluctuating with a threat to the reliability or robustness of the Famine or Feast tool system. The competitiveness of other PHA producer strain is reported as *Paracoccus* strain OTU1 [26].

Deterministic processes are reported to dominate for the entire FF system that resultant in the predictable microbial community succession profiles in the acclimatization phase as well as maintenance of the stable PHA-accumulating function or operations system in the maturation phase. The predation of bacterial phages with unreliable temporal dynamics of the top competitors can be provoked by neutral processes [27]. Production of PHA is reported from fermentation that utilized the thermal-hydrolyzed sludge and it has applied for mixed cultures (MMCs) with enriched species of *Brachymonas denitrificans* (61%) under aerobic feast or famine regime with the capability of denitrification and accumulation of PHA and it can be shown in Figure 7.2 [28].

The benefits of the PHA-accumulating denitrifier strains were demonstrated by an aerobic feast or anoxic regime system and the combination of culture selection with the denitrification process was implemented. Outcomes can be shown for culture enriched with PHA-accumulating

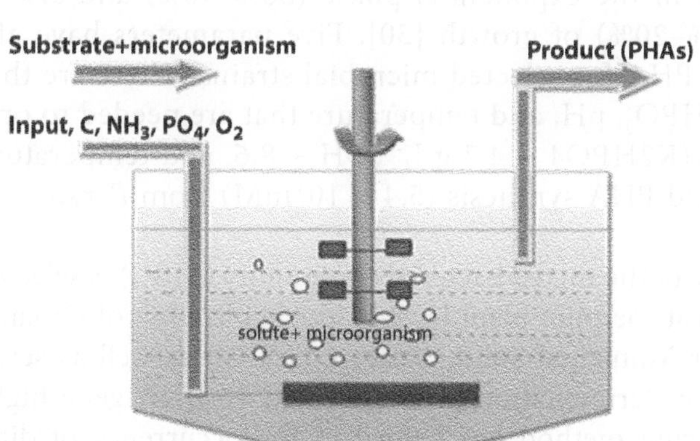

Figure 7.2 Diagram of fermentation process for PHA production via microbial strains.

capability with its yielding of VFA (0.47 g COD per gram COD) with 98% nitrate removal under denitrification processes. The aerobic-feast/famine procedure can achieve the maximum PHA quantity to the complete or whole aerobic feast or famine regime system (50 wt% vs. 47 wt%) and it reduced the aeration input energy by 79% in the culture selection stage. There is a deposition of nitrate or nitrous oxides compounds during the PHA synthesis, due to denitrification with the integration of the wastewater treatment processes [29].

The hazardous effects of synthetic polymers can be minimized by encouraging the investigators to find out biodegradable polymer or bioplastics and microbial-derived PHA bioplastics can achieve good attention from the last few decades. For this effort, isolation and characterization of bacterial cells or species is necessary that produced PHA as well as reported with the enhancement of optimal culture conditions for PHA synthesis. In this regard, various PHA-producing bacteria strain was identified with the application of the reported protocol, and 16rDNA gene sequencing techniques were applied for analysis morphological or biochemical properties of bacterial species, and they used D-optimal statistical design for optimal production media compositions for PHA biosynthesis [24].

The screening phase using Nile blue comprising the microbial culture obtained from the plate approach, and these are reported for morphological and biochemical analysis that can help in the prediction of bioplastic production and these advanced studies were done by 16rDNA gene analysis. This approach was introduced for the quantification of efficient PHA producers (such as *P. psuedoalcaligenes* strain Te). This bacterial strain has shown for the maximum quantity of PHA production at optimum process conditions. PHA biosynthesis from this strain is reported in the exponential phase (80%–90%) and also stationary phase (10%–20%) of growth [30]. Five parameters have affected the efficacy of PHA for selected microbial strains. There are three factors such as K_2HPO_4, pH, and temperature that are needed to optimize the conditions (K2HPO4 ~ 4.7 g L^{-1}, pH ~ 8.6, and temperature ~ 25°C) for increased PHA synthesis (5.4×10^3 mM) from *P. psuedoalcaligenes* [24, 30].

Dynamic of the mixed bacterial community (MBC) is selected for PHA production at the pilot plant level by the application of Organic Fraction of urban or Municipal Solid Waste (OFMSW) as well as sewage sludge (SS) waste in fermentation processes, and 16rRNA gene high throughput sequencing method has helped in the occurrence of different species of PHA-accumulating bacteria in open operating conditions with real substrates and without temperature control [31]. The volatile fatty

Table 7.1 The various types of microbial strain used for efficient mode production of PHAs or PHB.

Microorganism	Efficiency production of PHAs/PHB	Reference
E. coli	In laboratory conditions, *E. coli* produced polybeta hydroxybutyrate (95% from generated cell dry weight)	[50]
Alcaligenes eutrophus	*Alcaligenes species* synthesized copolymers containing 3-hydroxybutyrate and 3-hydroxy valerate. This species contained the genes encoding in the single operon that can produces enzymes beta-ketothiolase, NADPH-dependent acetoacetyl-CoA reductase and poly (beta-hydroxybutyric acid) synthase (PHB synthase) for the synthesis of three-step PHB-biosynthetic pathway.	[51] [52]
Metabolically engineered *E. coli*	Overexpress the phosphotransacetylase or acetate kinase enzymes and induced the operon for P3HB (1.27 g L^{-1}) at minimal medium supplemented with 10 g L^{-1} yeast extract and 5 g L^{-1} of acetate in shake flask experiment	[39, 40]
Fungus	Different systematic and economical related 159 fungal strains are used for the degradation of PHAs into monomers in microorganisms. It produces 0.1% of BIOPOL in each media of PHB.	[53]
Psuedomonas putida S12	Conversion of crude sludge (SPO) as expensive renewable raw material to PHA that highest yield (41%) of elastomeric medium-chain length (mcl)-PHA from SPO substrate	[46]
Pseudomonas mosselli TO_7	Yielded high content of mcl-PHA with 48% cell dry weight in 48 hours produces maximum content of PHAs, i.e., 13.16 mg PHAs $L^{-1} h^{-1}$.	[43]

(*Continued*)

Table 7.1 The various types of microbial strain used for efficient mode production of PHAs or PHB. (*Continued*)

Microorganism	Efficiency production of PHAs/PHB	Reference
Arabidopsis thaliana	Addition of PHA enzymes into the plant through suitable microbes and produced saturated and unsaturated 3-hydroxyalkanoic acids (6% to 16% monomers 3-hydroxyoctanoic and 3-hydroxyoctenoic acid). GC/MS illustrated the transgenic plants that produced 4 mg per g dry weight of mcl-PHA.	[54]
E. coli and *Klebsiella* strains	Initiation for the synthesis of 3-hydroxyburyrate in *E. coli* and *Klebsiella* strains reported due to introduced cloned genes (from *Alcaligenes eutrophus*). PHB polymer (1×10^6 Da) and (2×10^6 Da) produced in *K. aerogenes* and *E. coli*, respectively. *K. aerogenes* utilized sugar cane molasses with production of PHB (1 g L^{-1}). *K. oxytoca* fadR strain can integrate 3-hydroxybutyrate into poly-(3-hydroxybutyrate-co-3-hydroxyvalerate).	[55]

acid (VFA) can change in feed and temperature condition that affects the dynamic of the PHA-accumulating bacteria with operating conditions and higher PHA contents is associated with MBC groups that comprised the *Hydrogenophaga* species at high-temperature condition during operation of PHA biosynthesis. This group is related to heterogeneous PHA-accumulating MBC with association with high *Pha*C synthase genes in biodiversity with confirmation of occurrence of functional redundancy [31, 32]. There are some microbial strains shown that are involved in PHA biosynthesis and can be also seen in Table 7.1.

7.3.1.1 *Microorganisms in the Production of PHAs*

7.3.1.1.1 *Rhodospirillum rubrum*

This bacterium is known for the synthesis of industrial relevant copolymers like poly (hydroxybutyrate-co-3-hydroxyvalerate) with 56% moles of 3HV contents through syngas fermentation. The capability of *Rhodospirillus rubrum* is found as simple biological vehicles for the conversion of simple carbohydrate precursor into desirable or value-added bio-based products including PHA. In this regard, *Rhodospirillium rubrum* was genetically engineered for expression of individual or all six PHA biosynthetic genes (*pha* C1, *pha* A, *pha*B, *pha*C2, *pha* C3, or *pha*J) and researcher have studied nine overexpressing genes for evaluation of their effects on PHA contents with the effect of microbial growth pattern or profiles and it can be seen in Figure 7.3 [21].

Various experiments from researcher have been done on these strains (PHAs producer) and their aims were shown at genetically mode evaluating of these genes with their role of each PHA polymerase enzymes and it was apparently identified their influence on the productivity of PHA and also to identified the relevant genes in PHA biosynthetic operon that can control the PHA productivity and other factors to support the PHA productivity. It has been found that overexpressed of each PHA polymerase gene has indicated by *pha*C1 and A or B gene that are found significant contributors for the productivity of PHA but phaC3 has shown the little impact on PHA productivity. So, it can be concluded that overexpression of individual genes or in a mixture of the three PHA biosynthetic genes are situated in *pha* operon with information of *pha*B gene (is the crucial component for PHA productivity) impact. Further, from the equivalent or analogous experiment, it has shown that the *pha*J are not reported to effect

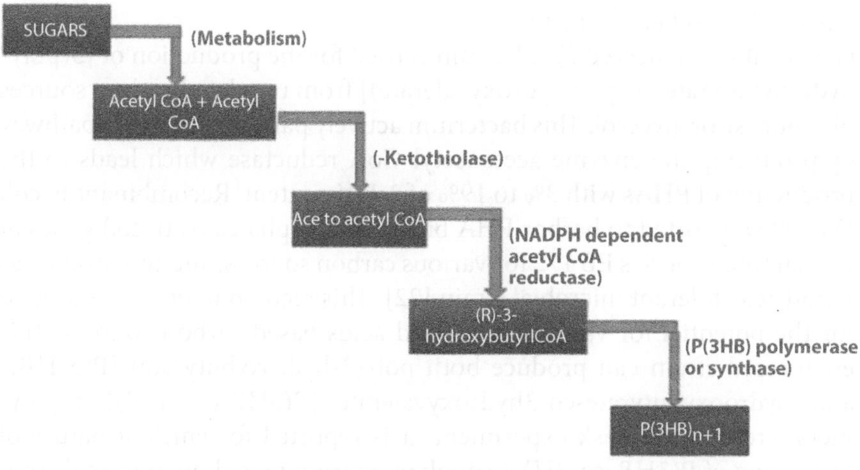

Figure 7.3 Poly(3-hydroxybutyrate~ PHB) biosynthesis shown in R. eutropha for its carbohydrate pathway, utilized for PHB inside cell.

on the PHA productivity and bioengineered *R. rubrum* strain has achieved the higher PHA production rate (till 30% of dry cell mass or weight) and it has shown 2.5-fold larger biomass growth than control strain and it is shown in Table 7.1 [33].

Syngas [a gaseous organic mixture of H_2, CO_2, carbon monoxide (CO), methane, and nitrogen in various ratios] has been reported to obtain or produce from effective feedstock (biomass, plastic, coal, or municipal wastes), and this fuel can be used to produce bioplastic such as poly ([R]-3-hydrobutyrate) (PHB) from *R. rubrum* strain in microbial fermentation. For this strain, growth phases was studied with the use of gas-tight serum vials and it has found that syngas (with composition 40% of CO, 40% of H_2, 10% of CO_2, and 10% of N_2 v/v) where N_2 was diluted to 60% concentration and these are reported the four-fold higher biomass with the sample grown on 100% syngas and it has shown the growth inhibitory effects [34]. The best syngas combination was used for C-C, N-, C, and P-substrates that are found to restrict the fed-batch fermentation mode in the bioreactor with a continuous supply of syngas and acetate substrates. It was reported that C and P-substrates have reduced the PHB productivity with five times greater than C-substrate limited microbial growth, reaching a maximum content (30%w/w). It has been found that growth and production of PHB are ended due to N-substrate as the second nutrient in production media and is reported as a growth-limiting substrate. It is found that a minimum supply of 0.2 g $CO.g^{-1}$ biomass per hour can be shown a guarantee to protect the cellular maintenance energy [35].

7.3.1.1.2 *Escherichia coli*

Genetically, engineered *E. coli* strain is used for the production of [3(polyhydroxybutyrate-co-polyhydroxyvalerate)] from unrelated carbon sources like glucose or glycerol. This bacterium actively participates PHAs pathway by producing the enzyme acetoacetyl-CoA reductase which leads to the production of PHAs with 3% to 19% of 3-HV content. Recombinant E. coli JM 109 is reported to harbor PHA biosynthesis (pha C, A, B, co,) genes of Comamonas species EB 172 for various carbon sources, and this strain was found acid-tolerant microbial strain [22]. This recombinant was examined for the potential for various sugar and acids based carbon sources. This engineered strain can produce both poly(3-hydroxybutyrate) [P(3-HB)] and hydrooxybutyrate-co-3hydroxyvalerate [P(3HB-co-3HV)] copolymers. From shake flask experiment, it is reported for efficient nature of producing of P(3HB-co-3HV) copolymers from mixed organic acid solution or media [36].

But larger productivity of P(3HB-co-3HV) is acquired from glucose compared to mixed organic acids. It has found the PHA accumulation similar to different carbon sources. Nitrogen supplement in production media is established to enhance the DCW but has negatively impacted 3HV in the production co-polymer.

Optimum 3HV monomer quantity (3%) with C/N ratio (42:1) was achieved by utilizing a mixture of organic acids as a carbon source. The productivity and yield of PHA in 2-L bioreactor was found the 0.16-g PHA/L h and 0.41-g PHA/g substrates, respectively, by using the C/N ratio and 20 g L^{-1} of glucose concentration and 0.5 g L^{-1} of ammonium sulfate, respectively. Recombinant strain had produced the polymers with the mol. wt. 8.5×10^5 and 1.4×106 Da [37, 38].

Metabolically engineered *E. coli* is reported to biosynthesize the poly(3-hydroxybutyrate-co-4-hydroxybutyrate) (P3HB4HB), poly-3-hydroxybutyrate (PBA), and poly (3-hydroxybutyrate-co-3-hydroxyvalerate (PHBV) using acetate as the primary caron substrates. It has been reported that overexpressed of phosphotransacetylase or acetate kinase pathway can be an effective strategy for the assimilation of acetates and the development of biopolymers. This recombinant strain is shown to overexpress the phosphotransacetylase or acetate kinase enzymes as well as induced the operon system P3HB (1.27 g L^{-1}) at a minimum concentration of medium accompanied with 10 g L^{-1} yeast extract and 5 g L^{-1} of acetate substrates in a shake flask experiment [39, 40].

Also, improvement in the synthesis of P3 HB4HV content is reported with the introduction of 4-hydroxybutyrate dehydrogenase, succinate semialdehyde dehydrogenase, and CoA transferase enzymes that can lead to the accumulation of P3 HB4HV polymers (titer ~ 1.7 g L^{-1}) with 4-hydroxybutyrate monomers contents (5.6 % molar content) with 4-hydroxybutyrate monomer content, and 1 g L^{-1} of α-ketoglutarate or citrate substrates addition into the production medium can enhance the titer of P3HB4HV (1.99 and 2.15 g L^{-1}, respectively). Acetate or propionate was supplied for PHBV production, and propionyl-CoA transferase was overexpressed to produce the 3-hydroxyvalerate precursor. It can produce the PBHV (0.33 g L^{-1}) and a 3-hydroxyvalerate monomers contents (6.6 mol %) [41, 42].

Further, overexpressed propionate permease has improved PHBV titer (1.1 g L^{-1}) and 3-hydroxyvalerate monomers content (10.4 mol %). The use or application of acetate substrate as a source of carbon in bacterial fermentation can reduce food consumption and agro-renewable bioresources for biorefineries. This metabolic engineered strategy will increase the production of PHA using acetate (abundant and cheap or cost-effective feedstock for the production of chemicals, materials, or biofuels) [40, 42].

7.3.1.1.3 *Pseudomonas* Species

Using biodiesel-extracted crude glycerol as a source of carbon, *Pseudomonas mosselli* TO7 delivered or yielded the high concentration of mcl-PHA with 48% cell dry cell mass or weight in 48 hours that produces maximum amounts or content of PHAs, i.e., 13.16 mh PHAs L^{-1} h^{-1}. These microorganisms are environmentally friendly used for the production of PHAs to replace plastics. Hence, they are called bioplastics and it can be shown in Figure 7.4 [43].

P. corrugata (Pco) and *P. mediterranea* (Pme) are reported to biosynthesize the medium-chain of PHA (mcl-PHA) in elastomers nature and also extracellular substances or products and these are based on relevant or unrelated carbon sources, fermentation process or any additives as effective parameters. These parameters have affected the yield and compositions

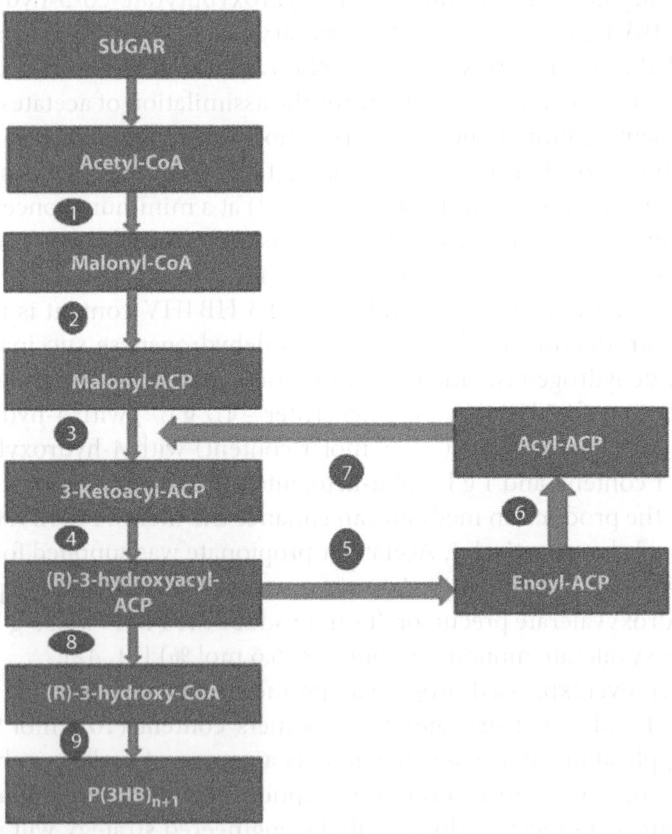

Figure 7.4 PHB biosynthesis shown in *Pseudomonas* species that represented the de-novo fatty acid synthesis for PHB inside cell.

of PHAs and are also dependent on microbial strain, carbon sources or fermentation condition, or additives concentration or types. Selected Pco microbial strain has shown its capability to produce the indeterminant or amorphous and sticky nature of mcl-PHA, but microbial strains of Pme can generate the high level and filmable PHA, from biodiesel or glycerol as a raw carbon substrates and it is found to distinct from a conventional microbial system produced mcl-PHAs. It is considered that accepting of producing these blends with polylactide acids [44]. It is necessary to improve the yield with reduction of production cost via using integrated processes and it also needed to develop the recovering of intracellular mcl-PHA that has biosynthesized the extracellular bioactive compounds. Further, transcriptional regulation strategies can help in improving production of PHA, by recognizing the metabolic potential of Pco and Pme strain, and these can provide the information of biosynthetic genes and their regulation for developing the cost-effective PHA production and can be seen in Figure 7.5 [44, 45].

PHAs are biodegradable plastics produced by bacteria and it has shown broad applications but prohibited by high cost of production. So, the reduction of cost of PHA biosynthesis can be done by conversion of crude sludge (SPO) as expensive renewable raw material to PHA via strain *Pseudomonas putida* S12. This microbial strain has provided the maximum yield (41%) of elastomeric medium-chain length (mcl)-PHA from SPO substances. This polymers character was analyzed by GC/MS or gel permeation chromatography or differential scanning calorimetric. This approach is found to reduce the cost of PHA production cost with widespread application in many sectors [46].

PHA biosynthesis capability with the growth of strain P. mosselii TO7 strain is found from wastewater of a vegetable processing location and it was studied by phenotypic and phylogenetic analysis of the 16S-rRNA gene. These microbial strains can utilize the palm kernel and soybean oils for the production of cell dry weight (CDW-up to 50%) and mcl nature of PKA that contained the high or more quantity of poly (3-hydroxy octannoate, P3HO) contents. Further, P3HO polymer can be enhanced to 45% DCW while growing in octanoate components by using a single-step microbial culture method. PHA monomers have been detected by 13C nuclear magnetic resonance spectroscopy. Average molecular mass and PHA polydispersity index are the 218 and 2.21, respectively [46].

P. mosselii TO7 microbial strain has produced PHA from palm kernel oil and it has shown two melting temperature (Tm) value of 37.2°C and 55.7°C with melting enthalpy (DHM) value of 51.1 and 26.6 jg^{-1}, respectively. Inhibition analyses using acrylic and 2-bromooctanoic acids have shown

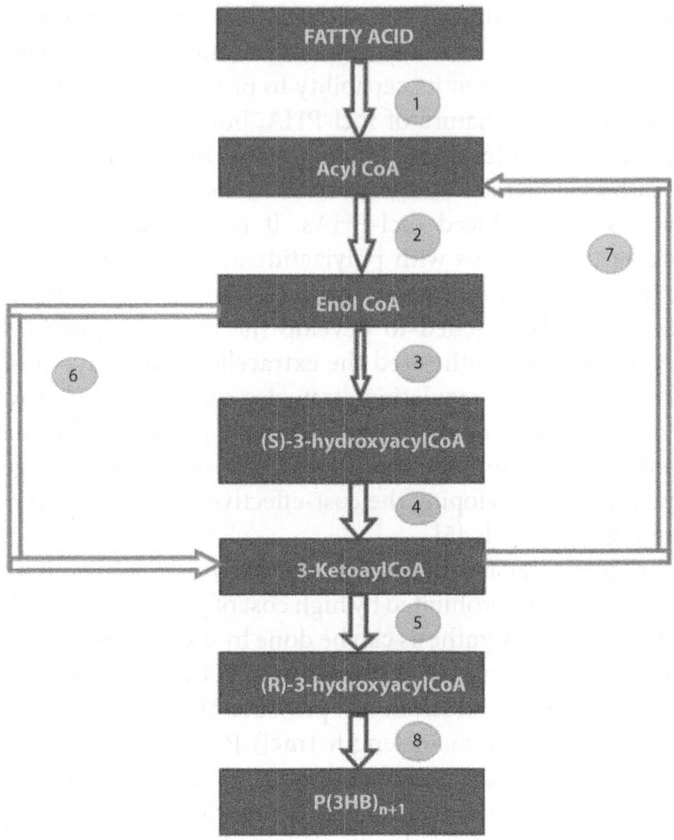

Figure 7.5 PHB in *R. rubrum* that utilized fatty acid biosynthesis pathway for fatty acid oxidation.

that β-oxidation can be identified as primary pathways for octanoic acid biosynthesis of mcl-PHA. Stain *P. putida* GPp104 PHA is reported to harbor PHA synthase genes of strain *P. mosselii* (*pha*C1pm and *pha*C2pm) and it has been heterologous expressed with the demonstration of *pha*C1pm as main PHA biosynthesis enzymes by using 3-hydroxyoctanoyl-CoA as its major substrates [47].

7.3.1.1.4 Stain *A. quitalea* Species USM4

Class 1 model of *pha*C (PHA synthase) enzyme from *Aquitalea* sp. USM4 expresses three-branch structure where two of them are responsible for the entrance of substrate and exit of product and third is extended in the class 2 model of *pha*C in *Pseudomonas aeruginosa* which place a role for the production PHAs [48].

7.3.1.1.5 Strain *Natrinema altunense* RM-10

Strain *Natrinema altunense* RM-10 pile-up 61.02 g/L of cell dry mass with PHA for 72 days continuous batch culture and yield 0.210 g L^{-1} h^{-1}. This bacterium belongs to domain Archaea and class strain *Halobacteria* which decreases the cost of producing the PHA. Transmission electron microscopy can able for the visibility of PHA granules within the archaeal cells. Extremely halophilic archaeon are used for the small-scale production of PHAs [49].

7.4 Fermentation for PHAs Production

Pseudomonas putida CA-3 has been reported to pile up the PHA production by the addition of polystyrene pyrolysis oil (as a source of carbon) as well as energy maintenance under nitrogen-limited growth conditions in continuously stirred tank reactors (CSTRs). The increase in the feeding concentration of nitrogen (1 mg N/L/h) has led to an increase in the DCW with the accumulation of PHA content in the fermentation broth. Further, an increase in feeding concentration of N2 (1.5 mg N/L/h) can lead to an increase in the percentage of production of DCW with PHA accumulation and can be seen in Figure 7.6. However, a further increase in nitrogen feeding concentration (1.7 to 2 mg N/L/h) can lead to a significant reduction

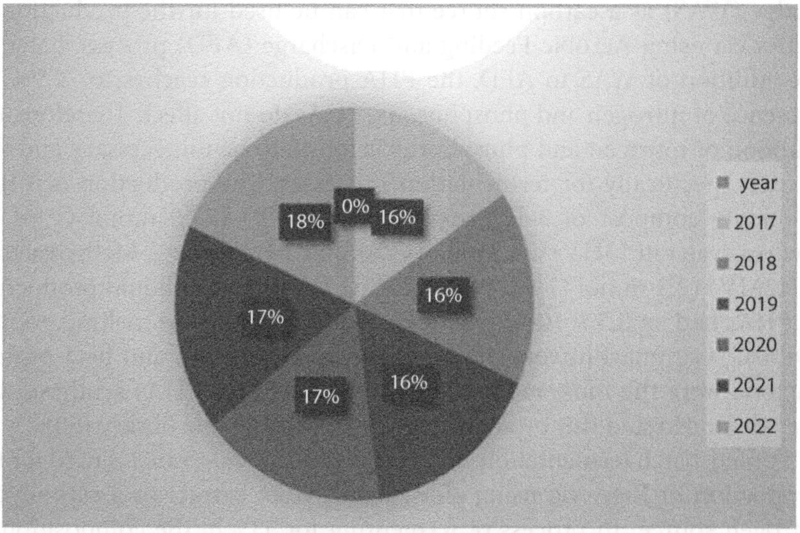

Figure 7.6 Estimated production capacity of bioplastics worldwide per annum.

in the % of PHA biosynthesis. The highest yield obtained is 0.28g PHA per gram styrene polymer has been provided with 1.5 mg/L/h of nitrogen feed rate [56].

Benzene, toluene ethylbenzene, and xylene (BTEX) compounds are used as a common source of carbon in *P. putida* F1 for accumulating PHA of medium-chain length (mcl). A synthetic mixture containing BTEX compounds along with styrene was added to a given defined mixture of P.putida F1 mt-2 and CA-3 in the shake flask as well as bioreactors experiments The PHA contained is around to 24% and also a round of the DCW in shake flasks or bioreactors. A 5-L fermenter resulted in the usage of BTEXS compounds around 59.6 g and produce 6 g of mcl-PHA [57].

A three-stage process was established for the production of PHAs even from sugar cane molasses: (1) acidic fermentation by molasses; (2) selection of accumulating cultures of PHAs; (3) accumulation of PHA batch using abundant sludge and fermented molasses. Storage yields approximately 0.37 to 0.50 Cm.mol. PHA/C.mmol VFA in selected acetate culture. The association of a few organic load capacities with a high concentration of ammonia generates the yield around 0.6 Cm.mol. PHA/C.m. VFA with good storage capacity is similar to acetate-selected culture [58]. In the last 30 years, PHAs are biodegradable plastic produced by a submerged fermentation process by utilizing prokaryotic organisms. At present, solid-state fermentation and production of PHAs from transgenic plants have been proposed for better results [59].

Alkaline fermentation liquid is reported to utilize for waste activated sludge (WAS) as a carbon source that can be used for the production of PHAs via using Aerobic Feeding and Discharge (AFD) process. Later, in the addition of WAS to AFD, the PHA production reaches to 72.9%. In presence of nitrogen and phosphorous, PHAs do not affect. Therefore, the addition of nitrogen and phosphorus is found to be unnecessary and this is used specifically for fermentation processes. The production of PHAs is mainly compost of 3-Hydroxybutyrate (3HB) (73.5 mmol C %), 3-Hydroxyvalerate (3HV) (24.3 mmol C %), and 3-Hydroxy-2-Methylvalerate (3H2MV) (2.2 mmol C %). WAS alkaline fermentation liquid production of PHAs had an 8.5 × 10(5) Da molecular weight and the melting point is 101.4°C. Gamma-Proteobacteria, alpha-proteobacteria, and beta proteobacteria were the most moderate microbial cells for PHAs synthesis and they demonstrated this by analyzing the 16S-RNA gene library [60].

For fed-batch fermentation, strain *Haloferax mediterranei* is used for the production of PHA via using glucose and yeast extract as a carbon and nitrogen source. In process of fermenting for 117 h, the composition of strain H. mediterranei and PHA get to 85.6 g L^{-1} and 48.6%, respectively.

Through spectroscopy analysis, it reveals that copolymers of PHAs are produced. It is poly (3-hydroxybutyrate-co-hydroxyvalerate) [P(3HB-co-3HV)] [61]. By utilizing acetone/chloroform, it was identified that the produced PHAs are two concentration all different co-polymers (P1 and P2). One of the co-polymer P(3HB-co-3HV) P1: 93.4% by wt) has 10.7 mol% of 3-HV contents in chain structure and has a molecular mass of nearly 570 kg/mol. Another copolymer (P2: 6.6% by wt) has a significantly higher the 3-HV content and it also contains 12.3% but its mol. mass is low (78.2 kg/mol) [61, 62].

The optimum production of PHAs from industrial wastes like ice cream residues is reported by strain *Ralstonia eutropha* and other bacteria through novel statistical experimental design. A frequentative stream of experimental design was performed for optimal conditions in the order of factorial design, the path of steepest ascent, and full factorial augmented with axial design (rotational central composite design) [63, 64]. The optimum production of lipid (15 mg/ml) and % lipid (88%) and advance examination to validate the optimum conditions for PHAs production form ice cream residues [56.68% ice cream in water or 56.68 ml of ice cream (v/v), 1 ml of mineral salts solution, 5.03 ml of buffer, 100 ml of seed culture, 100 ml of trace element solution and 213.8 h of fermentation time] [65].

Fed-batch fermentation of strain *Pseudomonas putida* KT2440 was carried out to produce unsaturated mcl-PHAs with productivity of 0.63–1.09 g PHA L^{-1} h^{-1} with final PHA content ranging from 42.6% to 55.8% at carbon-limited single stage. A mixture of non-anionic acid (NA) and 10-undecenoic acid (UDA) to monitor growth rate was exponentially fed. In the fermentation process, the molar fraction of PHAs monomer production is constant throughout the process, specifying that the end product is homogenous rather than a combination of different copolymers [66].

7.5 Downstream Process for PHAs

The downstream process of PHAs is reported as the process of recovery and purification of biosynthetic products like PHAs or other pharmaceuticals. These can also be applied to extract natural or synthetic metabolites from natural resources like plants, animals, and fermentation broth after proper treatment as well as disposal of waste. There are two strategies for recovering PHAs from the medium of the post-fermentation process such as dissolving biomass with strong oxidants to separate PHAs granules from the mixture and dissolving biomass with suitable solvents for the purification of PHAs. Economical and ecological purification methods are necessary for industrial biosynthesis of PHAs [62]. Through the homogenization process, PHA-producing cells are broken and PHA granules are cleansed and recovered as latexes. However, cell breakdown can lead to the release of a large amount of DNA which can increase the viscosity of the medium. To minimize the viscosity, the medium is exposed to high temperature or enlarge supplementing with hypochlorite or treated with a commercially available nuclease enzyme [63].

The applications of three different charcoals were compared to remove impurities in the product. They are charcoal activated, powder, and pure forms. The biomass in the freeze-dried state is extracted and purified by using ethyl acetate as a solvent for extraction and activated charcoal for the purification before the extraction process. The solvent ratio to the biomass is 15:1 and even the extraction process occurred for an hour [64]. Based on treating an industrial waste on the MC of PHA and biogas production, the lifetime assessment and financial evaluation are undertaken. To quantify financial feasibility and environmental effect on the production, the Internal Rate of Return (IRR) and CO_2 emission were used. PHA (MC) is also favored for producing biogas for the treatment of the specified industrial effluent [65].

The maximum PHB content obtained by using glycerol as a substrate, fed-batch, and pilot-scale fermenters, strain *Zobellella dentrificans* MW1, a newly isolated bacterium as well as the content is recovered through a sample organic sample extraction process while downstream processing. The next self-flotation of cell debris after the collection of PHB with chloroform allows the cells to be easily isolated from the PHB-solvent solution. The optimum or maximum purity reached with a polymer purity of 98.3% after 72 hours of collection process with chloroform at 30°C [68].

To recover co-polymers of PHA from strain *Ralstonia eutropha* biomass, a specialized process was designed through non-halogenated solvent methyl ethyl ketone, methyl isobutyl ketone, or ethyl acetate and butyl acetate. All PHAs extraction process was obtained both from dry cell mass as well as wet cells with varying quantities of 2 to 3 ml and by using a solvent ration of 2% (w/v). Ethylethanoate exhibit high levels of recovery and also product purity quantity (up to 00%) when used dry cell as a precursor source. When wet was used, the methyl isobutyl ketone that showed the beneficial effect of solvent for PHA extraction with recovery (up to 84%) and also purity level (99%) [69].

Halomonas species SK 5 isolated from hypersaline microbiological mats was able to synthesize both copolymers and homopolymers of PHA from various utilized carbon sources or substrates. Mainly oil palm sap (OPTS) and seawater as carbon sources and cultivation media enable a significant amount of accumulation of P(3HB). In the existence of alkali or detergent, a general form of downstream processing relying on the osmotic lysis for both dry and wet form biomass evolved an approximate polymer recovery (90%–100%) with the purity of up to 90%. The range of recovered polymers with average mol.wt (M (w)) was in the range of $1–2 \times 10^6$ Da [70].

7.6 Conclusions

Biopolymer is produced by several microbial strains and polymeric biomolecules like PHAs that are made up of several monomers via covalently. Synthetic plastics or polymers are non-degradable nature and also found in big quantities that are accumulated in soils or waste sources. Synthetic plastics are used for packaging the food items, medicine, water, cloths pharmaceutical, and also carrying vegetables or fruit or other applications. Microbial-derived PHA is bioplastics and gained more attention due to provide alternative options to synthetic plastics and can be used for various industrial applications. The bacterial strain is applied for gaining the maximum quantity of PHAs production at optimized cultural conditions via D-optimal statistical design. PHA-producing bacterial strain can be collected and screened from various soil (including Nile blue soil) and other soil samples via application of morphological, biochemical, and 16S-rDNA gene sequencing methods. Crude glycerin as well as the carbon to nitrogen (C:N) ratio in the culture medium were used for PHB production at optimal production media via applying of RSM tool. PHA accumulation is also found in MMC. The development of a bioreactor-based method of high cell density can be helped to generate medium-chain length PHA (mcl-PHA) by utilization of cheap substrate like cooking oil waste. PHAs accumulation is found up to 72% in batch and 65% in continuous feeding mode per DCW. Despite having lower PHAs accumulation in continuous feeding strategy, it shows four times more biomass growth. PHB is combined with natural terpene or D-limonene (LIM) and plasticized flexible film required for food packaging application. It is beneficial for the biomaterial industry and possesses many properties such as biodegradable and eco-friendly. Biodegradation of PHAs is also reported in soil and aquatic environments due to the proceeding of anaerobic digestion via microbial strain systems. By the solvent extraction method, PHAs were extracted from marine algal species. PHAs are reported to thermally stable at 260°C. Most of the biopolymers have been discovered with thermoplastic properties. The combination of PHAs with nanoparticles can be enforced in various areas like drug delivery or antibacterial agents, and bioengineering with their establishment in various fields such as medicines, catalysis, biosensors, and adsorbents is also reported.

References

1. Kumar, V., Kumar, S., Singh, D., Microbial polyhydroxyalkanoates from extreme niches: Bio prospection status, opportunities and challenges. *Int. J. Biol. Macromol.*, 147, 1255, 2020.
2. Andreeßen, C. and Steinbüchel, A., Recent developments in non-biodegradable biopolymers: Precursors, production processes, and future perspectives. *Appl. Microbiol. Biotechnol.*, 103, 143, 2019.
3. Allison, E.H. and Bassett, H.R., Climate change in the oceans: human impacts and responses. *Science*, 350, 778, 2015.
4. Arutchelvi, J., Sudhakar, M., Arkatkar, A., Doble, M., Bhaduri, S., Uppara, P.V., Biodegradation of polyethylene and polypropylene. *Indian J. Biotechnol.*, 7, 9, 2008.
5. Vu, D.H., Akesson, D., Taherzadeh, M.J., Ferreira, J.A., Recycling strategies for polyhydroxyalkanoate-based waste materials: An overview. *Bioresour. Technol.*, 298, 122393, 2020.
6. Roja, K., Sudhakar, D.R., Anto, S., Mathimani, T., Extraction and characterization of polyhydroxyalkanoates from marine green alga and cyanobacteria. *Biocatalys. Agricult. Biotechnol.*, 22, 101358, 2019.
7. Mannina, G., Presti, D., Montiel-Jarillo, G., Carrera, J., Suárez-Ojeda, M.E., Recovery of polyhydroxyalkanoates (PHAs) from wastewater: A review. *Bioresour. Technol.*, 297, 122478, 2020.
8. Urbina, L., Eceiza, A., Gabilondo, N., Corcuera, M.A., Retegi, A., Valorization of apple waste for active packaging: multicomponent polyhydroxyalkanoate coated nanopapers with improved hydrophobicity and antioxidant capacity. *Food Packag. Shelf lif.*, 21, 100356, 2019.
9. Tarrahi, R., Fathi, Z., Seydibeyoglu, M.O., Doustkhah, E., Khataee, A., Polyhydroxyalkanoates (PHA): From Production to Nanoarchitecture. *Int. J. Biol. Macromol.*, 146, 596, 2020.
10. Rinaudo, M., Chitin and Chitosan Properties and Application. *Polymer Sci.*, 31, 603, 2006.
11. Arikan, E.B. and Ozsoy, H.D., A Review: Investigation of Bioplastic. *J. Civil Eng.*, 9, 188, 2015.
12. Bharti, S.N. and Swetha, G., Need for bioplastics and role of biopolymer PHB: A short review. *J. Pet. Environ. Biotechnol.*, 7, 272, 2016.
13. Panda, A.K., Singh, R.K., Mishra, D.K., Thermolysis of waste plastics to liquid fuel: a suitable method for plastic waste management and manufacture of value added products: a world prospective. *Renew. Sustain. Ener. Rev.*, 14, 233, 2010.

14. Cleetus, C., Thomas, S., Varghese, S., Synthesis of Petroleum-Based Fuel from Waste Plastics and Performance Analysis in a CI Engine. *J. Ener.*, 2013, 608797, 2013.
15. Gulab, H., Jan, M.R., Shah, J., Manos, G., Plastic catalytic pyrolysis to fuels as tertiary polymer recycling method: effect of process conditions. *J. Environ. Sci. Health. A Tox. Hazard. Subst. Environ. Eng.*, 45, 908, 2010.
16. Jamieson, A.J., Brooks, L.S.R., Reid, W.D.K., Piertney, S.B., Narayanaswamy, B.E., Linley, T.D., Microplastics and synthetic particles ingested by deep-sea amphipods in six of the deepest marine ecosystems on Earth. *R. Soc. Open Sci.*, 6, 180667, 2019.
17. Schlining, K., von Thun, S., Kuhnz, L., Schlining, B., Lundsten, L., Stout, N.J., Chaney, L., Connor, J., Debris in the deep: using a 22-year video annotation database to survey marine litter in Monterey Canyon, central California, USA. *Deep Sea Res. (Part I)*, 79, 96, 2013.
18. Courtene-Jones, W., Quinn, B., Gary, S.F., Mogg, A.O.M., Narayanaswamy, B.E., Microplastic pollution identified in deep-sea water and ingested by benthic invertebrates in the Rockall Trough, North Atlantic Ocean. *Environ. Pollut.*, 231, 271, 2017.
19. Ruiz, C., Kenny, S.T., Narancic, T., Babu, R., Connor, K.O., Conversion of Waste Cooking Oil Into Medium Chain Polyhydroxyalkanoates in a High Cell Density Fermentation. *J. Biotechnol.*, 306, 9, 2019.
20. Pernicova, I., Kucera, D., Nebesarova, J., Kalina, M., Novackova, I., Koller, M., Obruca, S., Production of polyhydroxyalkanoates on waste frying oil employing selected *Halomonas* strains. *Bioresour. Technol.*, 292, 122028, 2019.
21. Costa, S.S., Miranda, A.L., Assis, D.D.J., Souza, C.O., Morais, M.G., Costa, J.A.V., Druzian, J.I., Efficacy of *Spirulina* sp. polyhydroxyalkanoates extraction methods and influence on polymer properties and composition. *Algal Res.*, 33, 231, 2018.
22. Bhalerao, A., Banerjee, R., Nogueira, R., Continuous cultivation strategy for yeast industrial wastewater-based polyhydroxyalkanoate production. *J. Biosci. Bioeng.*, 129, 595, 2019.
23. Huang, L., Chen, Z., Wen, Q., Zhao, L., Lee, D.J., Yang, L., Wang, Y., Insights into Feast-Famine polyhydroxyalkanoate (PHA) producer selection: Microbial community succession, relationships with system function and underlying driving forces. *Water Res.*, 131, 67, 2018.
24. Zihayat, B., Shakibaie, M., Shahrbabak, S.S., Doostmohammadi, M., Ameri, A., Adeli-Sardou, M., Forootanfar, H., Medium optimization for polyhydroxyalkanoate production by *Pseudomonas pseudoalcaligenes* strain Te using D-optimal design. *Biocatalys. Agricult. Biotechnol.*, 18, 101001, 2019.
25. Israni, N. and Shivakumar, S., Polyhydroxyalkanoate (PHA) biosynthesis from directly valorized ragi husk and sesame oil cake by *Bacillus megaterium* strain Ti3: Statistical optimization and characterization. *Int. J. Biol. Macromol.*, 148, 20, 2020.

26. Govindasamy, S., Syafiq, I.M., Amirul, A.A.A., Amin, R.M., Bhubalan, K., Dataset on controlled production of polyhydroxyalkanoate-based microbead using double emulsion solvent evaporation technique. *Data in Brief.*, 23, 103675, 2019.
27. Li, D., Yin, F., Ma, X., Towards biodegradable polyhydroxyalkanoate production from wood waste: Using volatile fatty acids as conversion medium. *Bioresour. Technol.*, 299, 122629, 2020.
28. Tu, W., Zhang, D., Wang, H., Lin, Z., Polyhydroxyalkanoates (PHA) production from fermented thermal-hydrolyzed sludge by PHA-storing denitrifiers integrating PHA accumulation with nitrate removal. *Bioresour. Technol.*, 292, 121895, 2019.
29. Chen, G.Q., Chen, X.Y., Wu, F.Q., Chen, J.C., Polyhydroxyalkanoates (PHA) toward cost competitiveness and functionality. *Adv. Indust. Eng. Polymer Res.*, 3, 1, 2020.
30. Paula, F.C.D., Kakazu, S., Bilia, C., Paula, C.B.C.D., Almeida, A.F.D., Gomez, J.G.C., Contiero, J., *Burkholderia glumae* MA13: A newly isolated bacterial strain suitable for polyhydroxyalkanoate production from crude glycerol. *Biocatalys. Agricult. Biotechnol.*, 20, 101268, 2019.
31. Crognale, S., Tonanzi, B., Valentino, F., Majone, M., Rossetti, S., Microbiome dynamics and phaC synthase genes selected in a pilot plant producing polyhydroxyalkanoate from the organic fraction of urban waste. *Sci. Total Environ.*, 689, 765, 2019.
32. Luo, M. and Curtis, C.W., Effect of reaction parameters and catalyst type onwaste plastics liquefaction and coprocessing with coal. *Fuel Process. Technol.*, 49, 177, 1996.
33. Jin, H. and Nikolau, B.J., Evaluating PHA Productivity of Bioengineered *Rhodosprillum rubrum*. *PLoS ONE*, 9, e96621, 2014.
34. Karmann, S., Panke, S., Zinn, M., Fed-Batch Cultivations of *Rhodospirillum rubrum* Under Multiple Nutrient-Limited Growth Conditions on Syngas as a Novel Option to Produce Poly(3-Hydroxybutyrate) (PHB). *Front Bioeng. Biotechnol.*, 7, 59, 2019.
35. Do, Y.S., Smeenk, J., Broer, K.M., Kisting, C.J., Brown, R., Heindel, T.J., Growth of *Rhodospirillum rubrum* on synthesis gas: conversion of CO to H and poly-β-hydroxyalkanoate. *Biotechnol. Bioeng.*, 97, 279, 2007.
36. Mumtaz, T., Mohammadi, M., Phang, L.Y., Ando, Y., Raha, A.R., Hassan, K.A., Zakaria, M.R., Polyhydroxyalkanoate Synthesis by Recombinant *Escherichia coli* JM109 Expressing PHA Biosynthesis Genes from Comamonas sp. EB172. *J. Microb. Biochem. Technol.*, 4, 103, 2012.
37. Chen, J., Li, W., Zhang, Z.Z., Tan., T.W., Li, Z.J., Metabolic engineering of *Escherichia coli* for the synthesis of polyhydroxyalkanoates using acetate as a main carbon source. *Microb. Cell Fact.*, 17, 102, 2018.
38. Liu, K. and Meuzelaar, H.L.C., Catalytic reactions in waste plastics, HDPE and coal studied by high-pressure thermogravimetry with on-line GC/MS. *Fuel Process. Technol.*, 49, 1, 1996.

39. Shibata, J., Matsumoto, S., Yamamoto, H., Kusaka, E., Pradip, Flotation separation of plastics using selective depressants. *Int. J. Miner. Process.*, 48, 127, 1996.
40. Brodin, M., Vallejos, M., Opedal, M.T., Area, M.C., Carrasco, G.C., Lignocellulosics as sustainable resources for production of bioplastics. *J. Cleaner Product.*, 162, 646, 2017.
41. Heinrich, D., Raberg, M., Steinbuchel, A., Synthesis of poly(3-hydroxybutyrate-co-3-hydroxyvalerate) From Unrelated Carbon Sources in Engineered *Rhodospirillum rubrum*. *FEMS Microbiol. Lett.*, 362, 38, 2015.
42. Srirangan, K., Liu, X., Tran, T.T., Charles, T.C., Young, M.M., Chou, C.P., Engineering of *Escherichia coli* for Direct and Modulated Biosynthesis of poly(3-hydroxybutyrate-co-3-hydroxyvalerate) Copolymer Using Unrelated Carbon Sources. *Sci. Rep.*, 6, 36470, 2016.
43. Liu, M.H., Chen, Y.J., Lee, C.J., Characterization of Medium-Chain-Length Polyhydroxyalkanoate Biosynthesis by *Pseudomonas Mosselii* TO7 Using Crude Glycerol. *Biosci. Biotechnol. Biochem.*, 82, 532, 2018.
44. Licciardello, G., Catara, A.F., Catara, V., Production of Polyhydroxyalkanoates and Extracellular Products Using *Pseudomonas corrugata* and *P. Mediterranea*: A Review. *Bioengineering (Basel)*, 6, 105, 2019.
45. Prieto, A., Escapa, I.F., Martínez, V., Dinjaski, N., Herencias, C., de la Peña, F., Tarazona, N., Revelles, O., A holistic view of polyhydroxyalkanoate metabolism in *Pseudomonas putida*. *Environ. Microbiol.*, 18, 341, 2016.
46. Kang, D.K., Lee, C.R., Lee, S.H., Bae, J.H., Park, Y.K., Rhee, Y.H., Sung, B.H., Sohn, J.H., Production of Polyhydroxyalkanoates from Sludge Palm Oil Using *Pseudomonas putida* S12. *J. Microbiol. Biotechnol.*, 27, 990, 2017.
47. Chen, Y.J., Huang, Y.C., Lee, C.Y., Production and characterization of medium-chain-length polyhydroxyalkanoates by *Pseudomonas mosselii TO7*. *J. Biosci. Bioeng.*, 118, 45, 2014.
48. Teh, A.H., Chiam, N.C., Furusawa, G., Sudesh, K., Modelling of Polyhydroxyalkanoate Synthase From *Aquitalea* Sp. USM4 Suggests a Novel Mechanism for Polymer Elongation. *Int. J. Biol. Macromol.*, 119, 438, 2018.
49. Mahansaria, R., Dhara, A., Saha, A., Haldar, S., Mukherjee, J., Production enhancement and characterization of the polyhydroxyalkanoate produced by *Natrinema ajinwuensis* (as synonym)- *Natrinema altunense* strain RM-G10. *Int. J. Biol. Macromol.*, 107, 1480, 2018.
50. Fidler, S. and Dennis, D., Polyhydroxyalkanoate Production in Recombinant *Escherichia coli*. *FEMS Microbiol Lett.*, 9, 231, 1992.
51. Anderson, A.J., Haywood, G.W., Dawes, E.A., Biosynthesis and Composition of Bacterial Poly(hydroxyalkanoates). *Int. J. Biol. Macromol.*, 12, 102, 1990.
52. Steinbuchel, A. and Schlegel, H.G., Physiology and Molecular Genetics of Poly(beta-Hydroxy-Alkanoic Acid) Synthesis in *Alcaligenes eutrophus*. *Mol. Microbiol.*, 5, 535, 1991.

53. Matavulj, M. and Molitoris, H.P., Fungal Degradation of Polyhydroxyalkanoates and a Semiquantitative Assay for Screening Their Degradation by Terrestrial Fungi. *FEMS Microbiol. Rev.*, 9, 323, 1992.
54. Mittendorf, V., Robertson, E.J., Leech, R.M., Kruger, Steinbuchel, A., Poirier, Y., Synthesis of Medium-Chain-Length Polyhydroxyalkanoates in *Arabidopsis thaliana* Using Intermediates of Peroxisomal Fatty Acid Beta-Oxidation. *PNAS*, 95, 13397, 1998.
55. Zhang, H., Obias, V., Gonyer, K., Dennis, D., Production of Polyhydroxyalkanoates in Sucrose-Utilizing Recombinant *Escherichia coli* and *Klebsiella Strains. Appl. Environ. Microbiol.*, 60, 1198, 1994.
56. Goff, M., Ward, P.G., O'Connor, K.E., Improvement of the Conversion of Polystyrene to Polyhydroxyalkanoate Through the Manipulation of the Microbial Aspect of the Process: A Nitrogen Feeding Strategy for Bacterial Cells in a Stirred Tank Reactor. *J. Biotechnol.*, 132, 283, 2007.
57. Nikodinovic, J., Kenny, S.T., Babu, R.P., Woods, T., Blau, W.J., O'Connor, K.E., The Conversion of BTEX Compounds by Single and Defined Mixed Cultures to Medium-Chain-Length Polyhydroxyalkanoate. *Appl. Microbiol. Biotechnol.*, 80, 665, 2008.
58. Albuquerque, M.G.E., Eiroa, M., Torres, C., Nunes, B.R., Reis, M.A.M., Strategies for the Development of a Side Stream Process for Polyhydroxyalkanoate (PHA) Production From Sugar Cane Molasses. *J. Biotechnol.*, 130, 411, 2007.
59. Castilho, L.R., Mitchell, D.A., Freire, D.M.G., Production of Polyhydroxyalkanoates (PHAs) From Waste Materials and By-Products by Submerged and Solid-State Fermentation. *Bioresour. Technol.*, 100, 5996, 2009.
60. Jiang, Y., Chen, Y., Zheng, X., Efficient Polyhydroxyalkanoates Production From a Waste-Activated Sludge Alkaline Fermentation Liquid by Activated Sludge Submitted to the Aerobic Feeding and Discharge Process. *Environ. Sci. Technol.*, 43, 7734, 2009.
61. Don, T.M., Chen, C.W., Chan, T.H., Preparation and Characterization of Poly(hydroxyalkanoate) From the Fermentation of *Haloferax mediterranei. J. Biomater. Sci. Polym. Ed.*, 17, 1425, 2006.
62. Kosseva, M.R. and Rusbandi, E., Trends in the Biomanufacture of Polyhydroxyalkanoates With Focus on Downstream Processing. *Int. J. Biol. Macromol.*, 107, 762, 2018.
63. Gamero, J.E.R., Favaro, L., Pizzocchero, V., Lomolino Basaglia, G.M., Casella, S., Nuclease Expression in Efficient Polyhydroxyalkanoates-Producing Bacteria Could Yield Cost Reduction During Downstream Processing. *Bioresour. Technol.*, 261, 176, 2018.
64. Wampfler, B., Ramsauer, T., Kehl, K., Zinn, M., Meyer, L.T., Application of Activated Charcoal in the Downstream Processing of Bacterial Olefinic poly(3-hydroxyalkanoates). *Chimia (Aarau)*, 64, 784, 2010.

65. Lee, K.M. and Gilmore, D.F., Modelling and Optimization of Biopolymer (Polyhydroxyalkanoates) Production From Ice Cream Residue by Novel Statistical Experimental Design. *Appl. Biochem. Biotechnol.*, 133, 113, 2006.
66. Sun, Z., Ramsay, J.A., Guay, M., Ramsay, B.A., Fed-batch Production of Unsaturated Medium-Chain-Length Polyhydroxyalkanoates With Controlled Composition by Pseudomonas Putida KT2440. *Appl. Microbiol. Biotechnol.*, 82, 657, 2009.
67. Gurieff, N. and Lant, P., Comparative Life Cycle Assessment and Financial Analysis of Mixed Culture Polyhydroxyalkanoate Production. *Bioresour. Technol.*, 98, 3393, 2007.
68. Ibrahim, M.H.A. and Steinbuchel, A., Poly(3-hydroxybutyrate) Production From Glycerol by Zobellella Denitrificans MW1 via High-Cell-Density Fed-Batch Fermentation and Simplified Solvent Extraction. *Appl. Environ. Microbiol.*, 75, 6222, 2009.
69. Riedel, S.L., Brigham, C.J., Budde, C.F., Bader, J., Rha, C., Stahl, U., Sinskey, A.J., Recovery of poly(3-hydroxybutyrate-co-3-hydroxyhexanoate) From *Ralstonia eutropha* Cultures With Non-Halogenated Solvents. *Biotechnol. Bioeng.*, 110, 461, 2013.
70. Rathi, D.N., Amir, H.G., Abed, R.M., Kosugi, A., Arai, T., Sulaiman, O., Hashim, R., Sudesh, K., Polyhydroxyalkanoate Biosynthesis and Simplified Polymer Recovery by a Novel Moderately Halophilic Bacterium Isolated From Hypersaline Microbial Mats. *J. Appl. Microbiol.*, 114, 384, 2013.

8

Polyhydroxyalkanoates for Sustainable Smart Packaging of Fruits

S. Pati[1], S. Mohapatra[2], S. Maity[3], A. Dash[1] and D. P. Samantaray[1]*

[1]*Department of Microbiology, Odisha University of Agriculture & Technology, Bhubaneswar-3, Odisha*
[2]*Department of Microbial Technology, Amity University Utter Pradesh, Noida, India*
[3]*University Innovation Cluster Biotechnology, University of Rajasthan, Rajasthan, India*

Abstract
In recent years, changes in consumer preference for safe and fresh fruits has emerged smart and active biodegradable polymers technology which drives the development of a new generation of smart packaging systems in food industries. The principal function of smart packaging is protection and preservation of fruits from external contamination such as deterioration and excess ripening. In the smart packaging these polymers have been successfully applied as carrier to entrap micronutrients, antioxidant and fruit quality indicators that monitor the condition of packaged fruits and provide information regarding the quality of the packaged fruits during transportation and storage. This review focuses on techniques for the preparation and characterization of smart polymers films and microparticles, as well as its potential applications in fruits packaging.

Keywords: Biodegradable, smart packaging, deterioration, micronutrients, quality indicator

8.1 Introduction

Globally, health has an unprecedented importance; consequently, humans are exhibiting inflated concerns in nutrition, fitness, and beauty. Thus, food is the prime desire of life. However, fruits have a mystical power to maintain

**Corresponding author*: dpsamantaray@yahoo.com

Bibhuti Bhusan Mishra, Suraja Kumar Nayak, Swati Mohapatra, and Deviprasad Samantaray (eds.) Environmental and Agricultural Microbiology: Applications for Sustainability, (183–196) © 2021 Scrivener Publishing LLC

good health by providing physiological benefits. Fruits, being the fountain of youth, refer to live tissues with high moisture content (60%–95%) and are rich in adequate amount of proteins, phytochemicals, vitamins, minerals, fibers, and antioxidants [1]. In addition, antioxidants like, phenolic flavonoids, lycopene, carotenoids, glucosinolates, and ascorbic acid of fruits stimulates several health benefits and alleviates disorder/diseases [2, 3]. Mostly, fruits are used fresh or a non-thermal preservation method is applied before consumption. These minimal processing reduces shelf life (1–3 days) of fruits during storage under refrigeration and humidity [4–6]. After harvesting and processing, respiration, transpiration, and enzymatic activity of fruits result in spoilage by pathogenic microbes [7, 8], increased oxidation, loss of tissue texture, and enzymatic browning [7].

Thus, traditional packaging came to the lime light for protection of fruits from deteriorative effects, convenience of time and use, and containment of fruits of various size and shape [9]. However, traditional packaging used the non-renewable petroleum-based plastic as feed stock, which is in the verge of extinction and leading to environmental pollution [10]. In order to overcome this snag, an innovative packaging idea came into the mind of academia such as smart packaging using biopolymer. Biopolymers are the renewable sources with a sustainable disposal option and degraded to water and CO_2 by microbial action [11]. Polyhydroxyalkanoates (PHAs) is the most significant biopolymer, is accumulated in the cytoplasm of bacterial cell as energy and carbon storage granules. Biocompatibility, water-vapor barrier and biodegradability properties of PHAs makes it a suitable candidate for fruits packaging [12, 13]. Nevertheless, use of PHAs is restricted because of its costs of production, low thermal stability, and brittleness [14–16].

Hence, smart packaging using biopolymer emerged as a prominent option that can be applied to the food industry. However, use of PHAs as a packaging material is still in the stage of infancy [17, 13]. In smart packaging, biopolymers are used for development of active films and microencapsulation that packed with bioactive compounds, antimicrobial, antioxidant, and chemical sensor to indicate food spoilage [18–22]. Smart packaging of fresh fruits using an edible biopolymer modifies internal atmosphere of fruits, causing moderate oxygen and permeability, delayed senecscene [23] and low water vapor permeability to prevent desiccation & maintain fruit firmness [24]. Additionally, the chemical sensor attached to the packaging system indicates spoilage of fruits, while biopolymer provide mechanical protection [22]. The present review focuses on the different smart packaging strategies for fresh fruits with special reference to PHAs.

8.2 Physiological Changes of Fresh Fruits During Ripening and Minimal Processing

Fruits with high vitamins, dietary fiber, and minerals plays an imperative role in the human diet. Despite of being the second highest producer of fruits, India still faces 20%–50% of post harvesting losses [25]. Ever since harvesting, these living entities exposed to physiological changes, out of which some are desirable while others are undesirable from consumer standpoint. Though physiological changes are not stationary, they can be decelerated by taking proper measures during handling, transportation, and packaging. Thus, fresh fruits must be maintained in an excellent condition during post harvestation and packaging to be of top quality for consumers.

The major physiological changes such as excessive ripening, skin breaks, bruises, injuries, and other mechanical damages up during post harvesting condition of fruits can modify its firmness, color, taste, starch content, flavors, and organic acids [26, 27]. Hence, it is highly essential to understand that the packaging of fruits is directly proportional to envisioned end use. Each fruit has an optimum ripening stage at which natural organic acids and starch are broken down into simpler compounds, leading to an alkaline condition [28] and produce aromatic volatile compounds & phenolics [29, 30]. As a result, a preferred level of sugars, firmness, flavors, and color is maintained in the fruits. Nevertheless, overripening leads to excessive tissue softening, loss of pigments, and sugar reduction, which provides favorable condition for growth of spoilage causing as well as pathogenic microbes [31].

Processing of fruits degrades appearance, textural quality, and freshness by increasing the rate of respiration, transpiration, and ethylene production. The rate of these changes depends on the types and degree of processing [32, 33]. Examples are browning reactions in apples and deterioration as well as loss of firmness of watermelon within 3 days of cutting due to ethylene production [34, 35]. Thus, storage, packaging, and handling play significant role in keeping fruits at the optimum maturation stage. Packaging suppresses respiration and transpiration consequently and reduces the metabolic activity. It also prevents food contact with atmospheric oxygen and accelerates deterioration reactions [36]. Moreover, smart packaging is designed in such a way that it not only prevents microbial deterioration but also indicates minor changes in fruit quality. Additionally, smart packaging using plastics can store fruits for 3–4 weeks at 10°C–15°C [35]. Traditionally, low cost synthetic polymers with good mechanical properties were used for fruit packaging. However, ecological problems created by synthetic plastics shifted the attention of food industries to use biocompatible PHAs for fruit packaging [7]. Therefore, this chapter briefly described the use of PHAs in sustainable smart technology for fruit packaging.

8.3 Smart Packaging

Globalization and dynamism led to development of new food wrapping strategies to maintain the quality and safety of food as per the need of consumer. The prime focus of these packaging technologies is to prevent physical, chemical, or biological damage of food, thereby delivering preserved, fresh, tasty, and appropriate food products with extended shelf-life and quality [37]. Basing on fundamental assets of package, packaging can be comprised into passive, active, and smart packaging [38]. Passive packaging is an ordinary, traditional one, where a protective, innate shield preserves the food. However, traditional packaging being a source of huge waste causes environmental pollutions as well as increase product complexity [39]. Thus, it is the need of the hour to promote global economy and minimize carbon foot print, which paved the way for active and smart packaging [40]. In active packaging subsidiary constituents such as, oxygen scavengers, moisture, ethylene, antioxidants, and antimicrobial agents are included intentionally to augment the performance of package system [41]. Here, the package, environment, and product interrelate positively to achieve the goal [9]. However, in active packaging accidental breakage can cause release of constituents into the food which ultimately adversely affect the food quality [42].

Thus, by keeping an aerial view on the above snag, researchers developed smart packaging, which also has contribution toward improvement of Hazard Analysis and Critical Control Points (HACCP) and Quality Analysis and Critical Control Points (QACCP) systems for onsite detection of unsafe food, identification of potential health hazards, and establishment of strategies to reduce hazards. Additionally, it improves the food quality by identifying the process that affect the quality attributes of food [43]. Smart packaging also known as intelligent packaging refers to packaging, which comprises an external/internal indicator to deliver information about pack integrity, tamper evidence, food safety, and food quality [41]. Smart packaging devices include sensors, indicators, and radiofrequency identification (RFID) systems [43, 44]. Sensor like biosensor, gas, chemical, or electronic offers continuous signal to detect and measure any physical or chemical changes in product to which the device retorts. Biosensor quantifies and transmits info regarding any biological reactions or pathogens present in the food whereas, gas sensor detects any gaseous reactions occurring in the package. The gaseous sensor includes CO_2, O_2, water vapors, metal oxide, organic conducting polymer, or ethanol sensor. Additionally, chemical sensor selectively provides details of undesirable chemicals or gas present in the package through surface adsorption.

Smart packaging can also perceive unwanted changes in package using printed electronic or electronic nose. Printed electronic merged and form bond with pre-polymeric mixture to measure the change while, electronic nose designed to mimic human olfactory in the packaging for identification and classification of each aroma in the odor by giving a unique response [9, 44, 45]. Apart from sensors, use of indicators in smart packaging is a well-known technology to indicate presence or absence of unknown matter in packaging and to quantify concentration in terms of any distinguishable change [46]. Inclusion of freshness indicator in the packaging provides visual information regarding microbial or chemical changes within the product by reacting with the respective metabolites. Integrity indicator ensures food veracity throughout the supply chain. Time temperature indicator (TTI) is a simple, cost affordable device that monitors and commutatively determines the physical, chemical, and biological food spoilage basing on time and temperature history. It is categorized into partial history, full history, and critical temperature indicator [9, 47]. RFID is an automatic, electronic information-based technology that gives real-time, accurate info to the user. RFID is more advanced than any other system for successfully tracing labor saving costs, supply chain management, quality, and safety of food [37].

Whether it is about application of sensor or RFID or indicators, polymer has a potential role in smart packaging. At first, low cost, thermal stability, good O_2 and CO_2 barrier, and efficient mechanical performance make the way for food corporations to exploit synthetic polymers for fruit packaging [48]. ToxinGuard® (Toxin Alert, Canada) developed an antibody printed polymer packaging system to detect targeted pathogens like *Salmonella* sp., *E. coli*, *Campylobacter* sp., and *Listeria* sp. [49]. Many researches have been conducted in favor of use of polymer in smart packaging [50, 51]. However, nowadays, by keeping a view on functional requirement of biodegradability and environmental sustainability, researchers are shifting their attention to replace synthetic polymers by biopolymers. Hence, this chapter focuses on different packaging strategies of fruits using the most common biopolymer like PHAs.

8.4 Biodegradable Polymers for Fruit Packaging

Though synthetic polymer is well-recognized in packaging industry, its disposal, recycling costs, and harmful impact on human health led to development of biodegradable polymers for smart fruit packaging [52, 53]. Moreover, in smart packaging, the word "bio-based material" assigned to renewable, cytocompatible biopolymer. Basing on source of origin these are categorized into plant-based, bio-based, and microbial biopolymer [7]. Traditionally, plant-based biopolymers like chitosan, starch dominated packaging market but high-water vapor permeability, reduced biodegradability after polymerization and low thermal stability restricted their thriving commercialization [7, 54, 55]. Besides, polyethylene (PE), polyvinylchloride (PVC), polypropylene (PP), polystyrene (PS), polyethylene terephthalate (PET), low density polyethylene, high density polyethylene, and polyamide (PA) are the common polymers used in food packaging [53, 56]. Among them, PHAs is the most preferred biopolymer for fruit packaging due to significant properties such as renewability, biodegradability, low H_2O permeability, visible and UV resistant, low CO_2, O_2, and water solubility [56, 57].

PHAs are the fascinating group of biopolymers synthesized by a wide array of Gram-positive and Gram-negative bacteria as carbon and energy storage granule [58]. These biomaterials imitate the attribute of synthetic polymer and recyclable to CO_2 and H_2O in the natural condition [59]. In consequence, this thermoplastic, water insoluble, non-toxic, and moldable [60] biopolymer has wide range of applications including packaging materials, medical implants, drug delivery carriers, nutritional supplements, drugs, and fine chemicals [61]. The inherent hydrophobicity, biodegradability, and enormous property range have branded the microbial PHAs as promising competitors of synthetic plastics in the packaging market. Furthermore, PHAs is a tempting option for packaging due to the following properties:

i. PHAs with thermal properties *Tg* from −52°C to 4°C, *Tm* >177°C and *Td* from 227°C to 256°C along with high plasticity, melt extrusion, and thermoforming is an appropriate option for packaging.
ii. Comparatively high-water vapor permeability of PHAs than synthetic polymer also encourages PHAs to be used in packaging system.
iii. Biodegradable under natural condition is another positive aspect for PHAs.

iv. PHAs also have a dimensional stability in wet environment resulting in making itself a potential candidate for fruits packaging [62, 63].

Apart from this, in smart packaging incorporation of several active ingredients, such as anti-browning, anti-microbial, nutraceuticals, texture enhancer, sensors, TTI, and RFID, freshness indicator into PHAs boost the safety, nutritional, and sensory attributes of fruits [9, 23]. Many reports are available in support of edible smart packaging of fruits using polymer but in this matter, PHAs is still in research phase. Moreover, there is no report of smart packaging of fruits using PHAs. Thus, more research is highly indispensable in this sector.

8.5 Legal Aspects of Smart Packaging

As per article 3 of EC/1935/2004 in smart packaging, neither carrier nor smart device should transfer its constituents to food. Articles 4(d) and 11 of EC No 450/2009 also spell out that smart ingredients should be labelled properly as non-edible to avoid the accidental consumption so that it will not mislead the consumer. The smart packages should undergo risk and safety assessments laid by the Council of the European Communities and the Scientific Committee (Council Directive 89/107/EEC) which includes scientific and technical criteria prior to their authorization. This authorization recognized the smart ingredients as food additives or food ingredients after which it can be released to the market [9, 35].

8.6 Pros and Cons of Smart Packaging Using PHAs

Smart packaging has a prime role in supply chain and is advantageous to the consumer, thus, this chapter addresses the basic characteristics and alluring advantages and disadvantages of smart packaging (Figure 8.1). Biodegradable, cytocompatible PHAs used in smart packaging have a wide range of production source along with less energy dependency. In this system consistency in quality observation reduces time and overall analysis cost of fruit packaging. Most importantly, smart packaging in combination with PHAs reduces the waste production during processing. Regardless of the comparisons of PHAs with synthetic polymer, its immense use in packaging industry has been limited due to drawbacks of cost, high glass

Traditional fruit packaging	Smart fruit packaging
Advantage: • Easy handling. • Physical protection. • Reduces contamination. • Cost-effective. Disadvantage: • Generation of waste. • Product complexity. • Uncontrolled internal environment. • No indication of quality.	Advantage: • Less time consumption & waste generation. • Controlled internal environment. • Extended shelf life. • Maintain & trace quality. • Personalization of product. • Edible packaging. Disadvantage: • Expensive & complex processing. • Increase unsold fruits.

Figure 8.1 Traditional vs. smart fruit packaging.

transition temperature, low resistance to thermal degradation, brittleness, and low thermal stability in the molten state [11, 17]. In addition to this printing of biosensor in smart packaging requires polymer processing step as some polymers are antibiotic sensitive. Besides, the indicators and sensor can make the consumer deprive of buying food by depicting the food quality which leads to increase in unsold fruits [64]. The smart device must have compatibility with the food type because not every smart device can be used for any food. Only the package used in smart packaging is biodegradable, the waste generated during entire supply chain sometimes counter strike the goal of reducing the food wastage [43, 65]. Demand of smart packaging is growing day by day and it is not appropriate to completely rely on this for fruit quality. Observing only one or two aspects cannot provide a complete statement about a system. Therefore, a bird's eye view on the entire smart packaging and higher-level research on increasing flexibility of PHAs is highly essential for exploitation of more advantages of this system.

8.7 Conclusion

Food stuffs especially fruits have specific requirements in terms of packaging. Thus, in the present scenario different smart packaging strategies are developing to meet the requirements of food supply chain in food industry. This review focuses on the prospective applications of biodegradable and edible PHAs coating in combination with smart ingredients for storage of minimally processed fruits. Moreover, these smart packaging technologies can extend the shelf life and improve the quality and safety of foods. Thus, despite of many limitations, research on these smart technologies are still going on for further improvements to provide benefits and convenience to consumer.

References

1. Liu, R.H., Health benefits of fruit and vegetables are from additive and synergistic combinations of phytochemicals. *Am. J. Clin. Nutr.*, 78, 517–520, 2003. https://doi.org/10.1093/ajcn/78.3.517S.
2. Kaur, C. and Kapoor, H.C., Antioxidants in fruits and vegetables- the millennium's health. *Int. J. Food Sci. Technol.*, 36, 703–725, 2001.
3. Rico, D., Martin-Diana, A.B., Barat, J.M., Barry-Ryan, C., Extending and measuring the quality of fresh-cut fruit and vegetables: A review. *Trends Food Sci. Technol.*, 18, 373, 2007.
4. Allende, A., Tomas-Barberan, F.A., Gil, M.I., Minimal processing for healthy traditional foods. *Trends Food Sci. Technol.*, 17, 513–519, 2006.
5. Abadias, M., Usall, J., Anguera, M., Solsona, C., Viñas, I., Microbiological quality of fresh, minimally-processed fruit and vegetables and sprouts from retail establishments. *Int. J. Food Microbiol.*, 123, 121–129, 2008. https://doi.org/10.1016/j.ijfoodmicro.2007.12.013.
6. Galindo, F.G., Herppich, W., Gekas, V., Sjöholm, I., Factors affecting quality and post-harvest properties of vegetables: Integration of water relations and metabolism. *Crit. Rev. Food Sci. Nutr.*, 44, 139–154, 2004. https://doi.org/10.1080/10408690490424649.
7. Galgano, F., Condelli, N., Favati, F., Dibianco, V., Perretti, G., Caruso, M.C., Biodegradable packaging and edible coating for fresh-cut fruits and vegetables. *Italian J. Food Sci.*, 27, 1–20, 2015.
8. Nguyen-the, C. and Carlin, F., The microbiology of minimally processed fresh fruits and vegetables. *Crit. Rev. Food Sci. Nutr.*, 34, 371, 1994.
9. Biji, K.B., Ravishankar, C.N., Mohan, C.O., Gopal, T.K.S., Smart packaging systems for food applications: A review. *J. Food Sci. Technol.*, 52, 6125–6135, 2015. https://doi.org/10.1007/s13197-015-1766-7.
10. Kumar, P., Green plastic: A new plastic for packaging. *Int. J. Eng. Sci. Res. Technol.*, 5, 168–174, Sonia2016.
11. Arrieta, M.P., Samper, M.D., Aldas, M., López, J., On the use of PLA-PHB blends for sustainable food packaging applications. *Materials*, 10, 1008, 1–26, 2017. http://dx.doi.org/10.3390/ma10091008.

12. Plackett, D. and Siro, I., Polyhydroxyalkanoates (PHAs) for food packaging, in: *Multifunctional and nano-reinforced polymers for food packaging*, J.M. Lagaron (Ed.), pp. 498–526, Woodhead Publishing Ltd, Cambridge, United Kingdom, 2011.
13. Pawar, P.A. and Purwar, A.H., Biodegradable polymers in food packaging. *Am. J. Eng. Res.*, 2, 151–164, 2013.
14. Ohashi, E., Drumond, W.S., Zane, N.P., Barros, P.W.D.F., Lachtermacher, M.G., Wiebeck, H., Wang, S.H., `Biodegradable poly(3-hydroxybutyrate) nanocomposite. *Macromol. Symp.*, 279, 138–144 2009.
15. Gregorova, A., Wimmer, R., Hrabalova, M., Koller, M., Ters, T., Mundigler, N., Effect of surface modification of beech wood flour on mechanical and thermal properties of poly(3-hydroxybutyrate)/wood flour composites. *Holzforschung*, 63, 565–570, 2009.
16. Erceg, M., KovacÏicÂ, T., Sanja, P., Isothermal degradation of poly(3-hydroxybutyrate)/organically modified montmorillonite nanocomposites. *Polym. Comp.*, 31, 272–278, 2010.
17. Gabor, D. and Tita, O., Biopolymers used in food packaging: A review. *Acta Universitatis Cibiniensis Series E: food Technol.*, 16, 3–19, 2012.
18. Bhattarai, J.K., Sharma, A., Fujikawa, K., Demchenko, A.V., Stine, K.J., Electrochemical synthesis of nanostructured gold film for the study of carbohydrate-lectin interactions using localized surface plasmon resonance spectroscopy. *Carbohydrate Res.*, 405, 55–65, 2014. https://doi.org/10.1016/j.carres.2014.08.019.
19. Rukchon, C., Nopwinyuwong, A., Trevanich, S., Jinkarn, T., Suppakul, P., Development of a food spoilage indicator for monitoring freshness of skinless chicken breast. *Talanta*, 130, 547–554, 2014. https://doi.org/10.1016/j.talanta.2014.07.048.
20. Yoshida, C.M.P., Maciel, V.B.V., Mendonça, M.E.D., Franco, T.T., Chitosan biobased and intelligent films: Monitoring pH variations. *LWT- Food Sci. Technol.*, 55, 83–89, 2014.
21. Silva-Pereira, M.C., Teixeira, J.A., Pereira-Júnior, V.A., Stefani, R., Chitosan/corn starch blend films with extract from *Brassica oleraceae* (red cabbage) as a visual indicator of fish deterioration. *LWT- Food Sci. Technol.*, 61, 258–262, 2015.
22. Pereira, V.A., Arruda., I.N.Q., Stefani, R., Active chitosan/PVA films with anthocyanins from *Brassica oleraceae* (Red Cabbage) as time-temperature indicators for application in intelligent food packaging. *Food Hydrocolloids*, 43, 180–188, 2015.
23. Rojas-Grau, M.A., Oms-Oliu, G., Soliva-Fortuny, R., Martin-Belloso, O., The use of packaging techniques to maintain freshness in fresh-cut fruits and vegetables: A review. *Int. J. Food Sci. Technol.*, 44, 875–889, 2009.
24. Del-Valle, V., Hernandez-Munoz, P., Guarda, A., Galotto, M.J., Development of a cactus-mucilage edible coating (*Opuntia ficus indica*) and its application

to extend strawberry (*Fragaria ananassa*) shelf-life. *Food Chem.*, 91, 751–756, 2005.
25. Verma, L.R. and Joshi, V.K., *Postharvest technology of fruits and vegetables-Handling processing fermentation and waste management*, pp. 77–94, Indus Publishing, Delhi, 2000.
26. Gundewadi, G., Reddy, V.R., Bhimappa, B.B., Physiological and biochemical basis of fruit development and ripening: A review. *J. Hill Agri.*, 9, 7–21, 2018. http://dx.doi.org/10.5958/2230-7338.2018.00003.4.
27. Watada, A.E., Abe, K., Yamuchi, N., Physiological activities of partially processed fruits and vegetables. *Food Technol.*, 44, 116–122, 1990.
28. Galindo, F.G., Sjöholm, I., Rasmusson, A.G., Widell, S., Kaack, K., Plant stress physiology: Opportunities and challenges for the food industry. *Crit. Rev. Food Sci. Nutr.*, 46, 749–763, 2007. https://doi.org/10.1080/10408390601062211.
29. Chaïb, J., Devaux, M., Grotte, M., Robini, K., Causse, M., Lahaye, M., Marty, I., Physiological relationships among physical, sensory and morphological attributes of textures in tomato fruits. *J. Exp. Bot.*, 58, 1915–1925, 2007. https://doi.org/10.1093/jxb/erm046.
30. Coolong, T.W., Randle, W.M., Wicker, L., Structural and chemical differences in the cell wall regions in relation to scale firmness of their onion (*Allium cepa* L.) selections at harvest and during storage. *J. Sci. Food Agri.*, 88, 1277–1286, 2008.
31. Poole, N.D. and Gray, K., Quality in citrus fruit: to degreen or not degreen. *British Food J.*, 104, 492–505, 2002.
32. Iqbal, T., Rodrigues, F.A.S., Mahajan, P.V., Kerry, J.P., Gil, L., Manso, M.C., Cunha, L.M., Effect of minimally processing conditions on respiration rate of carrots. *J. Food Sci.*, 73, E396–402, 2008. https://doi.org/10.1111/j.1750-3841.2008.00923.x.
33. Park, W.P., Cho, S.H., Lee, D.S., Effect of minimally processing operations on the quality of garlic, green onion, soybean sprouts and watercress. *J. Sci. Food Agri.*, 77, 282–286, 1998.
34. Toivonen, P.M.A. and Brummell, D.A., Biochemical base of appearance and texture changes in fresh-cut fruit and vegetables. *Postharvest Biol. Technol.*, 48, 1–14, 2008.
35. Mehyar, G.F. and Han, J.H., Active packaging for fresh-cut fruits and vegetables, in: *Modified atmosphere packaging for fresh-cut fruits and vegetables*, A.L. Brody, H. Zhuang, J.H. Han (Eds.), pp. 267–284, Blackwell Publishing Ltd, Hoboken, New Jersey, United States, 2010.
36. Abbas, M.F. and Ibrahim, M.A., The role of ethylene in the regulation of fruit ripening in the Hillawi date palm (*Phoenix dactylifera* L). *J. Sci. Food Agri.*, 72, 306–308, 1996.
37. Dobrucka, R. and Cierpiszewski, R., Active and intelligent packaging food-research and development: A review. *Pol. J. Food Nutr. Sci.*, 64, 7–15, 2014. http://dx.doi.org/10.2478/v10222-012-0091-3.

38. Brockgreitens, J. and Abbas, A., Responsive food packaging: Recent progress and technological prospects. *Compr. Rev. Food Sci. Food Saf.*, 5, 115, 2016. http://dx.doi.org/10.1111/1541-4337.12174.
39. Schaefer, D. and Cheung, W.M., Smart packaging: Opportunities and challenges. 51st CIRP Conference on Manufacturing Systems. *Procedia CIRP*, 72, 1022–1027, 2018.
40. Gontard, N., Panorama des emballages alimentaire actif, in: *Les Emballages Actifs, Paris,* N. Gontard (Ed.), France: Tech & Doc Editions, Paris, France, 2000.
41. Robertson, G.L., Active and intelligent packaging, in: *Food packaging: principles and practice*, CRC Press, Boca Raton, FL, 2006.
42. Yildirim, S., Rocker, B., Pettersen, M.K., Nilsen-Nygaard, J., Ayhan, Z., Rutkaite, R., Radusin, T., Suminska, P., Marcos, B., Coma, V., Active packaging applications for food. *Compr. Rev. Food Sci. Food Saf.*, 17, 165–199, 2018. http://dx.doi.org/10.1111/1541-4337.12322.
43. Vanderroost, M., Ragaert, P., Devlieghere, F., Meulenaer, B.D., Intelligent food packaging: The next generation. *Trends Food Sci. Technol.*, 39, 47–62, 2014.
44. Kerry, J.P., Ogrady, M.N., Hogan, S.A., Past, current and potential utilization of active and intelligent packaging systems for meat and muscle-based products: A review. *Meat Sci.*, 74, 113–130, 2006.
45. Yam, K.L., Takhistov, P.T., Miltz, J., Intelligent packaging: Concepts and applications. *J. Food Sci.*, 70, 1–10, 2005.
46. Hogan, S.A. and Kerry, J.P., Smart packaging of meat and poultry products, in: *Smart packaging technologies for fast moving consumer goods*, J. Kerry and P. Butler (Eds.), pp. 33–59, John Wiley & Sons Ltd, West Sussex, England, 2008.
47. Kuswandi, B., Maryska, C., Jayus, A.A., Heng, L.Y., Real time on package freshness indicator for guavas packaging. *Food Measure*, 7, 29–39, 2013.
48. Siracusa, V., Rocculi, P., Romani, S., Dalla-Rosa, M., Biodegradable polymers for food packaging: A review. *Trends Food Sci. Technol.*, 19, 634, 2008.
49. Bodenhamer, W.T., Jackowski, G., Davies, E., *Toxin Alert: Surface binding of an immunoglobulin to a flexible polymer using a water-soluble varnish matrix*, p. 66992973, United States Patent, United States, 2004.
50. Cha, D.S. and Chinnan, M.S., Biopolymer-based antimicrobial packaging: A review. *Crit. Rev. Food Sci. Nutr.*, 44, 223–237, 2004.
51. Dolmaci, N., Çetes, A.F., Yaşar, A., An amperometric biosensor for fish freshness detection from xanthine oxidase immobilized in polypyrrole-polyvinyl-sulphonate film. *Artif. Cells Nanomed. Biotechnol.*, 40, 275–279, 2012.
52. Moncayo, D., Buitrago, G., Algecira, N., The surface properties of biopolymer-coated fruit: A review. *Ingeniería E Investigación*, 33, 11–16, 2013.

53. Bratovčić, A., Odobašić, A., Ćatić, S., Šestan, I., Application of polymer nanocomposite materials in food packaging. *Croatian J. Food Sci. Technol.*, 7, 86–94, 2015.
54. Caner, C., Vergano, P.J., Wiles, J.L., Chitosan film mechanical and permeation properties as affected by acid, plasticizer and storage. *J. Food Sci.*, 68, 1049, 1998.
55. Claus, J.W., Biobased packaging materials for the food industry, http://www.biodeg.net/fichiers/Book%20on%20biopolymers%20(Eng).pdf.
56. Khosravi-Darania, K. and Bucci, D.Z., Application of poly(hydroxyalkanoate) in food packaging: Improvements by nanotechnology. *Chem. Biochem. Eng. Quart.*, 29, 275–285, 2015.
57. Bucci, D.Z., Tavares, L.B.B., Sell, I., Biodegradation and physical evaluation of PHB packaging. *Polym. Test.*, 26, 908–915, 2007. http://dx.doi.org/10.1016/j.polymertesting.2007.06.013.
58. Sudesh, K., Abe, H., Doi, Y., Synthesis, structure and properties of polyhydroxyalkanoates: Biological polyesters. *Prog. Polym. Sci.*, 25, 1503–1555, 2000.
59. Khanna, S. and Srivastava, A.K., Recent advances in microbial polyhydroxyalkanoates. *Proc. Biochem.*, 40, 607–619, 2005.
60. Mayeli, N., Motamedi, H., Heidarizadeh, F., Production of poly-hydroxybutyrate by *Bacillus axaraqunsis* BIPC01 using petrochemical wastewater as carbon source. *Braz. Arch. Biol. Technol.*, 58, 643–650, 2015.
61. Maity, S., Das, S., Samantaray, D.P., Effect of vitamin on accumulation of PHB by *Zobellella* species under submerged fermentation process. *Int. J. Cur. Microbiol. Appl. Sci.*, 6, 1310–1315, 2017.
62. Israni, N. and Shivakumar, S., Polyhydroxyalkanoates in packaging, in: *Biotechnological applications of polyhydroxyalkanoates*, V.C. Kalia (Ed.), pp. 363–388, Springer Nature Singapore Pvt Ltd, The Gateway West, Singapore, 2019, https://doi.org/10.1007/978-981-13-3759-8_14.
63. Laycock, B., Halley, P., Pratt, S., Werker, A., Lanta, P., The chemo mechanical properties of microbial polyhydroxyalkanoates. *Prog. Polym. Sci.*, 38, 536–583, 2013. https://doi.org/10.1016/j.progpolymsci.2012.06.003.
64. Dainelli, D., Gontrad, N., Spyropoulos, D., Zondervan-van den Beuken, B., Tobback, P., Active and intelligent food packaging: Legal aspects and safety concerns. *Trends Food Sci. Technol.*, 19, 103–112, 2008.
65. Müller, P. and Schmid, M., Intelligent packaging in the food sector: A brief overview. *Foods*, 8, 1–12, 2019. https://doi.org/10.3390/foods8010016.

9

Biosurfactants Production and Their Commercial Importance

Saishree Rath and Rajesh K. Srivastava*

Department of Biotechnology, GIT, Gandhi Institute of Technology and Management (GITAM), Deemed to be University, Visakhapatnam, AP, India

Abstract

Surfactants are an amphiphilic molecule with exhibiting extensive industrial applications, produced by chemical techniques from ancient to current periods. Biological operations can feasible and sustainable processes for the production of biosurfactant via using favorable microorganisms, alternative options to chemical surfactant. Strain *Serratia marcescens* shown its capacity for biosurfactant production via uses of various sources of carbon and nitrogen substrates, salinity, pH, temperature, and agitation speed. Conditions for achieving the desired quantity (glycerol, ammonium sulfate and peptone) of biosurfactant with surface tension properties are discussed via applying central composite design (CCD), a statistical optimization modeling tool. Various microbial strains showed good and promising potentiality for a biosurfactant production, used in any environment set-up of biological agent processes. Use of hydrolyzate of sugarcane bagasse hemicellulosic component as a carbon source for the production of biosurfactants reported. Culture media screening with xylose, detoxified, non-detoxified sugarcane bagasse hydrolyzate are used commercially with potential microbial strains for emulsifying and tensoactive. A production media contained hemicellulosic hydrolyzate or yeast extract with the best workability of these products in biorefineries, now reported for sustainable production of biosurfactant. In this chapter, the author discusses different types of biosurfactants, its production processes with their application for commercial tasks.

**Corresponding author*: rajeshksrivastava73@yahoo.co.in

Bibhuti Bhusan Mishra, Suraja Kumar Nayak, Swati Mohapatra, and Deviprasad Samantaray (eds.) Environmental and Agricultural Microbiology: Applications for Sustainability, (197–218) © 2021 Scrivener Publishing LLC

Keywords: Biosurfactant, microbial strain, bioreactor, optimal conditions, properties, production, emulsifier

Abbreviations

BCL	*Bacillus* identification card
Bp	Base pair
$C_{18}H_{36}O_3$	3-Hydroxy-16-ethylheptadecanoic acid
Ca	Calcium
CCD	Central composite design
CMC	Critical Micelle Concentration
dTDP	Deoxythymidine diphosphate
EC 50	Half maximal effective concentration
FTIR	Fourier-transform infrared
GFP	Green fluorescent protein
HAA	3-[3-hydroxyalkanoyloxy) alkanoic acid
HLB	Hydrophilic lipophilic balance
HPLC	High-performance liquid chromatography
HWCO	Heavy weight crude oil
LC-MS	Liquid chromatography mass chromatography
MEOR	Microbial Enhanced Oil Recovery
Mg	Magnesium
NaCl	Sodium chloride
NMR	Nuclear magnetic resonance spectroscopy
PAH	Polycyclic aromatic hydrocarbons
Rha-C10-C10	Monorhamnolipid Rhamnosyl-β-hydroxydecanoyl-β-hydroxydecanoate
Rha-Rha-C10-C10	Dirhamnolipid (rhamnosyl-rhamnosyl-β-hydroxydecanoyl-β-hydroxydecanoate
RNA	Ribonucleicacid
rRNA	RibosomalRNA
TLC	Thin layer chromatography
ULO	Used lubricant oil

9.1 Introduction

Biosurfactants are found in the amphiphilic nature of compounds and that are produced by many species plants and microorganisms strains, having the property of eco-friendly and competent nature to biological processes. These compounds are mainly produced by many strains of bacteria, yeast, or fungi. This compound has exhibited hydrophilic and hydrophobic moieties with a reduction of surface tension property for liquid. Biosurfactant is reported to show the main characteristics such as limiting surface and interfacial tension [1]. These compounds can use the same mechanism as chemical surfactants but it shows more promising results compared to chemical surfactants especially in fermentation processes (to reduce the bubble formation). Surfactants are reported as an active surface compound that reduced the surface tension between two phases. In general, surfactants are used for the separation of oil particles. This compound has shown the high commercial and industrial importance and is used in areas of petroleum, pharmaceuticals, detergents, paints, cosmetics, and water treatment [1, 2].

Conditions such as the carbon substrate, ion concentration, pH, temperature and salinity, bacterial strains, and crop conditions, are found as significant factors for regulating of the biosurfactant biosynthesis with physicochemical properties. Currently, the production of biosurfactants can be found to be more expensive, due to the involvement of various processes. To reduce the cost of production of biosurfactants, adaptation of techniques and methods is necessary with advanced equipment applications with lowcost media, renewable carbon feedstocks, and a knowledgeable process optimization approach are used [3]. Biosurfactants have exhibited superior surfactant abilities when compared to synthetic or artificial surfactants. The main characteristic feature of biosurfactants is to show physiochemical properties along with biological activities. Biosurfactants have been reported for its important roles in many industries such as pharmaceutical and environmental bioremediation tasks such as detergence (improving detergent quality), foaming and wetting or emulsifying agent, or stabilization, lubricating, dispersion, and solubilizing agent of hydrophobic compounds with reducing adverse effects on biological processes and components of the environment [3, 4].

In the current period, the biosurfactants are use as laundry detergent (called as potential substitutes of chemical surfactant). These compounds have attained more popularity in recent years and have been considered as an alternative for remediation technology. Biosurfactants are also known for showing anti-adhesive and anti-biofilm activities. Various strains of Bacillus have also shown properties like antimicrobial activities and natural antioxidants [5]. Biosurfactants have been found various types based on their structure and microbial source. The key categories of biosurfactant are glycolipids, lipopeptides, phospholipids, fatty acids, or neutral lipids nature as polymeric compounds, and particulate matters. Biosurfactants can also be categorized as anionic, cationic and nonionic, and amphoteric nature, depending on the ionizing state in aqueous solution [6]. Biosurfactants are considered to be eco-friendly and now have been considered for various industrial and agricultural processes. They also are known to exhibit antimicrobial and anti-adhesive properties against various types of microbial growth and infections. Due to these properties, it can use as eco-friendly and eco-accommodating materials for bioremediation technology [6, 7].

Surfactants are capable to lower the surface tension due to the presence of hydrophilic and hydrophobic groups between liquid phases like oil/water. They are used mostly for cleaning purposes like emulsification. Alkylbenzene sulfonates (detergents), (fatty acid) soaps, lauryl sulfate (foaming agent), di-alkyl sulfosuccinate (wetting agent), and lignosulfonates (dispersants) are the few known categories, widely used as surfactants. Uses of chemical surfactant with their non-biodegradable nature are found that has caused some problems such as use in difficulties, the authorization of ecological conservation and adverse environmental change for the next generations and we need to use the biomaterial such as biosurfactants at industrial application [7]. The author will discuss the various types of surfactant chemicals, microbial sources for biosurfactant production, and also their application in various sectors.

9.2 Chemical Surfactant Compounds

Chemical approach synthesized surfactants can exhibit the surface tension properties in decreasing nature between any two liquids or gasses or liquid or solid. It can function as a detergent, wetting agents, and emulsifier, or anti-foaming agent or dispersant compounds. It has also been created as the self-assembled molecular clusters in a solution known as micelles (between the water and the oil phase). It can absorb to the interfaces between a solution and various phases (forming gases or solids) [8].

The chemical surfactant can exhibit the different properties, due to variation in its chemical structure (this can be reported due presence of different functional groups in chemical structures of surfactant). These surfactants can exhibit the different degree or extent of affinity within the same surfactant molecule. Normally, the surfactant can also contain both groups such as an alkyl chain and a 6-22 carbon chain (known as hydrophobic groups). This group can not show any affinity to water molecules while another group is reported as lipophilic groups that often surfactant used in the water system. The functional group of surfactant molecules is reported as hydrophilic and shown more affinity to water molecules. In surfactant compounds, there are two opposing functions that are called amphiphilic structures [9].

Normally surfactant compounds are categorized into the ionic or non-ionic surfactant. Ionic surfactants are further divided into three types such as anionic, cationic, or amphoteric surfactant compounds. Anionic surfactants are those compounds that can dissociate in aqueous solution via changing the hydrophilic groups into anions. But in aqueous solution, the cationic surfactant may dissociate into a cation. In the case of amphoteric surfactants, dissociation of it may create cations or anions depending on pH conditions. In aqueous solution, nonionic surfactants do not dissociate into ions [10].

And two types of biosurfactant compounds are found with hydrophilic or hydrophobic groups. Some, common hydrophilic biosurfactant is reported to contain the carboxylate (-Coo-), sulfate ($-OSO_3-$), sulfonate (SO_3^-), carboxybetaine ($-NR_2CH_2COO^-$), sulfobetaine (-N(CH3) $C3H6SO3^-$), or quaternary ammonium ($-R_4N^+$) groups and some of these compounds are reported to contain an anionic surfactant. A soap molecule is reported to consist of a hydrocarbon chain (its lipophilic functional groups) and also lipid affinity group (lipophilic). Next, carboxylate anion and hydrophilic groups showed more water affinity. The carboxylate anion group can form a structure in an aqueous solution, with counterions parts such as Na^+, K^+, or Mg^{2+} ions. Non-ionic surfactant hydrophilic groups are usually found to contain the polyoxyethylene group along with the glyceryl groups or sorbitol groups [11].

Due to different hydrophilic groups in non-ionic surfactant, it can make more applications depending on need. It has been reported that surfactant compounds can spontaneously construct a variety of self-organized structures in solvents. These properties can depend on the chemical structures as well as interactions manner with the solvents. Key tools have been given for an interpretation of surfactant properties. Accurate information on the structure character of surfactant is salvation and the fundamental concept of phase behavior can be found in surfactant molecules. Surfactant compounds can exhibit phase behavior and physicochemical properties in the different functional systems [12].

Due to the wider use of detergent and other cleaning agent, these have created several challenges such as residual surfactants that are found to discharge into the sewage system or directly reached into surface water. Most surfactant compounds are reported to disperse into various compartments of the environment such as soil, water, or sediment systems. These compounds exhibited toxic effects on different aquatic organisms [13].

Chemical surfactant compounds can exhibit a more toxic effect on the environment, due to their concentration exceed to the critical level. Croatia, a country of Southeast Europe, has reported presence of chemical surfactant in the environment, below the national boundary or limit. Most chemical surfactant compounds are found to be poor or least in biodegradable nature, and their amounts or concentrations in wastewater treatment plants can be greatly affected to reduce after secondary treatment approaches. Chemicals surfactant is shown as the highest concern, to release in untreated wastewater or wastewater that can also be present after primary treatment of wastewater. With their massive quantity in the environment, its discharge can be polluted and these can cause serious effects on the ecosystem. Future studies and advanced treatment plants can reduce the highly toxic and non-biodegradable surfactant compounds and biosurfactants can replace the chemical surfactant, via its synthesizing at a commercial scale that can reduce its production costs [11, 14].

9.2.1 Biosurfactant Compounds

Biosurfactant molecules are found as biological and surface-active agents that shown the ability to reduce the interfacial tension between any two solutions (i.e., liquid, solid, and gas). These biological compounds can allow going for mixing and dispersing properties in water or other liquids as it has shown similar properties like a chemical surfactant. It has exhibited as an amphiphilic nature that meant consisting of a hydrophilic and a hydrophobic group or moiety and in a heterogeneous system, it can interact with the phase boundary [15]. It also consists of non-polar tails (because it contains a hydrocarbon chain) and also contains the polar head due to the presence of different or varied functional groups such as carbohydrates, amino acids, or groups of phosphates) [16].

As observed in chemical counterpart, surfactant properties, and characteristics, biosurfactant has also shown multiples applications in households, industries, and agricultural sectors, shown in Figure 9.1. Biosurfactant can also be used in cleaning applications and formulations and it promoted the solubilization, emulsification, or dispersion of other molecules in the chemicals industry or cosmetics industry as well as detergents, food, textiles, and pharmaceutical industries. These biomolecules can not interfere with every product and human life [17]. Its additional applications are shown to exhibit the antitumor, antiviral, or antimicrobial activity or properties. Further, immunological properties or inducer of cell differentiation is shown by biosurfactant. These properties of biosurfactant can show the potential applications in multiples fields including biomedical sciences [18].

Biosurfactant is found to be involved in plant protection with its general use as formulating or dispersing aid tasks. Some of the biosurfactant compounds are rhamnolipids that exhibit high and specific antimicrobial activity against the Phytophthora zoospores (under the group of most important phytopathogenic fungi) (Figure 9.2). In current periods also, most of the people are using the chemical surfactant (which is the petroleum-based origin) and is synthesized by chemical approach or routes. These compounds are found often environmentally toxic and their use can lead to environmental problems (in washing applications) and is inevitably responsible for ending or damaging to our environment after its uses [19, 20]. There are a lot of problems from chemical surfactant uses

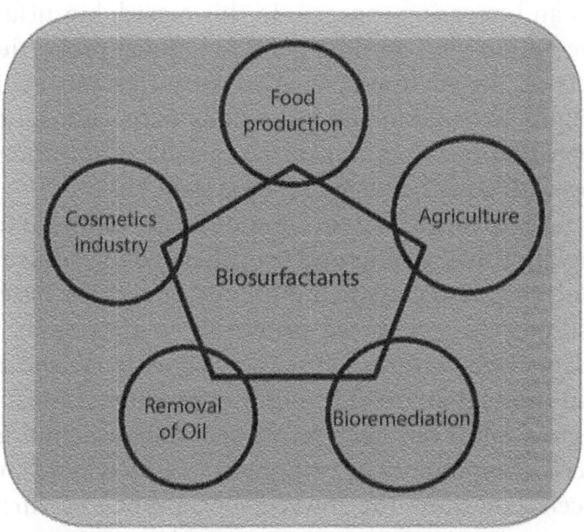

Figure 9.1 Biosurfactants applications in various sectors.

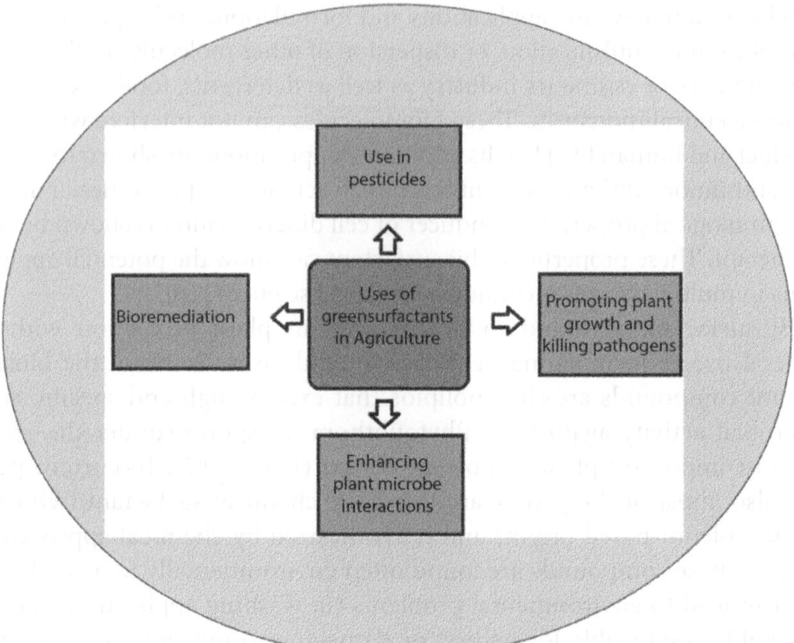

Figure 9.2 Biosurfactants in agricultural field for various pollutant removal.

and it has shown the eco-toxicity, bioaccumulation, or least biodegradation issues as an increasing concern. In this regard, biosurfactant can be an alternative option with availing good functional properties and lower adverse effects on the environment. Also, biosurfactants exhibited excellent compatibility to the skin and other body parts and can be produced from renewable resources (e.g., sugars or vegetable oils) through fermentation processes with an effective microbial system [20].

And structures of biosurfactant compound are reported to determine by producing microbial strains and is influenced by culture conditions. This biomolecule is also classified into four categories, based on chemical compositions. It is reported primarily as glycolipids nature, and oligopeptides or lipopeptides nature as well as fatty acid phospholipids or neutral lipids that are also found in polymeric biosurfactants as shown in Figure 9.3. It has been found that biosurfactant is composed of external cell components of carbohydrates, fatty acids, or peptides. The whole-cell can show the lowering properties of surface tension [21]. There are some useful properties of biosurfactant, discussed in the below sections.

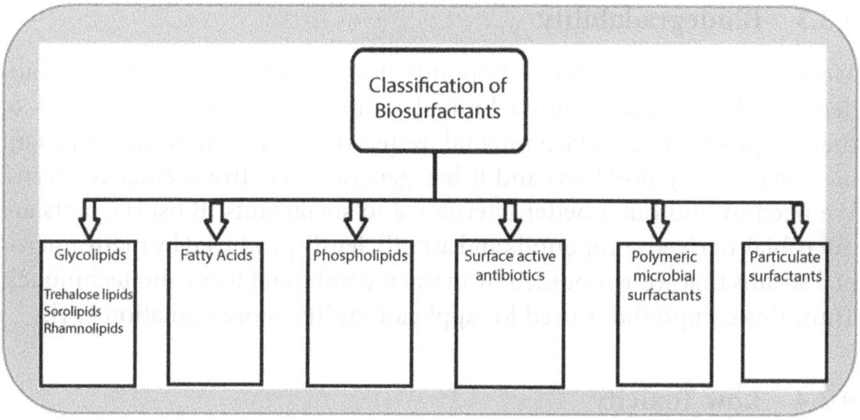

Figure 9.3 Various types of biosurfactants produced by microbial sources.

9.3 Properties of Biosurfactant Compound

9.3.1 Activities of Surface and Interface Location

Normally, most surfactants have been found to reduce surface tension and interface tension of various types of solutions (especially heterogeneous nature). This compound capacity has exhibited these activities as reported in different surfactant compounds. It can reduce the various extent of water surface tension (up to 25 m Nm^{-1}) and also reported to reduce the water or hexadecane interfacial tension (less than 1 m Nm^{-1}). Biosurfactants are found to be more effective than chemical surfactants. Further, it can exhibit the low Critical Micelle Concentration (CMC) that helps them lower the surface tension furthermore capacities [22].

9.3.2 Temperature and pH Tolerance

Biosurfactants, and their surface activity, have been reported to exhibit the more resistant nature toward physical factors such as temperature and pH. Biosurfactants are produced by many microbial strains including Arthrobacter protophormiae and B. licheniformis and are found more thermostable (up to 30°C–50°C) as well as pH stable (~2–10). This property has gained more the interest of its production from using of extremophiles nature of microbial species. These biosurfactants (produced from extremophiles) can be considered of great commercial value due to their high or commercial uses in many industries and sectors [23].

9.3.3 Biodegradability

Most of the biosurfactant compounds have shown a higher rate of biodegradability nature within a limited period and there are no issues of their deposition in environmental water or soils. Chemical surfactant has raised many problems and it has generated environmental concerns. We need to find out a better alternative to surfactants. Biosurfactants are microbial-derived compounds and are efficiently produced by many microbial strains that were modified by many recombinant tools and techniques. Thus, this compound is used for applications like bioremediation [24].

9.3.4 Low Toxicity

We have found that synthetic surfactant usages are highly toxic to aquatic animals and can be more harmful to the environment or its compounds. In this regard, biosurfactants are found to be low or non-toxic products in chemical nature. This property can encourage our industries to use in many sectors. Biosurfactants is mostly applied in the field of pharmaceuticals, cosmetic, and food industries. Sophorolipids from C. bombicola is used in the food industry. Biosurfactants produced from P. aeruginosa are found to be non-toxic and non-mutagenic and can be applied in different industries including detergent [25].

9.3.5 Emulsion Forming and Breaking

Biosurfactants can be used as emulsifiers. Emulsions can be termed as a fine dispersion of minute droplets of one liquid in and other liquid where the two liquids are soluble. Additives like biosurfactants can stabilize emulsions further which can be maintained up to a few years. Liposan produced from Candida lipolytica is used to emulsify edible oils. Liposans are also used in the cosmetics and food industries [26, 27].

9.4 Production of Biosurfactant by Microbial Fermentation

Microbial surfactants or biosurfactant have shown their capability to reduce the superficial and interfacial tension in the water solution or mixture of hydrocarbon. This compound has shown lower toxicity, higher biodegradability, and also environmental compatibility with a specific activity in extreme conditions. It has shown many advantages compared to chemical surfactants. Biosurfactant is reported to biosynthesize by batch cultivations of Bacillus atrophaeus ATCC 9372, B. subtilis or B. subtilis W strain. These bacterial strains have grown in culture media containing glucose, digested casein or soy flour, and salts (NaCl or potassium dibasic phosphate) and is shown in Table 9.1 [28].

Bacterial strains (including B. subtilis) were genetically engineered with GFP expression to control the microbial viability. The control of microbial viability can be achieved by potentially using casein and glucose in culture media. This favorable media is reported for biosurfactant production via using Bacillus atrophaeus ATCC 9372 (635 mg L^{-1}), B. subtilis or B. subtilis W strain (GFPuv). These bacterial cultures were maintained at 150 rpm, temperature ~ 35°C for 24 h in batch fermentation [29]. Further, maintaining of the glucose (14 g L^{-1}) and hydrolyzed casein (10 g L^{-1}) concentration in culture media can be found to improve 17 fold productions in only exponential cell growth. Glucose (18 g L^{-1}) concentration is maintained in the culture media for biosurfactant production during all growth stages [28, 30].

Table 9.1 Various types of surfactant structures from various microbial sources [31].

Structure of surfactant	Source
Trehalose dimycolates	*Mycobacterium* sp.
Trehalose dicorynmycolates	*Nocardia* sp., *Rhodococcus* sp., *Arthrobacter* sp., *Corynebacterium* sp.
Rhamnolipids	*Pseudomonas* sp.
Sorolipids, Amino lipids	*Torulopsis* sp.
Lipopeptidases	*Bacillus* sp., *Streptomyces* sp. *Corynebacterium* sp., *Mycobacterium* sp.
Ornithine lipid	*Pseudomonas* sp., *Thlobacillus* sp. *Agrobacterium* sp. *Gluconobacter* sp.
Phosholipids	*Conida* sp., *Corynebacterium* sp. *Micrococcus* sp., *Thiobacillus* sp.
Fatty acids/Natural lipids	*Acinetobacter* sp., *Pseudomonas* sp. *Micrococcus* sp., *Mycococcus* sp., *Canidida* sp., *Penicillium* sp., *Aspergillus* sp.

1-Pentanonacontene (C95H190) is reported a fatty alkene, and 3-Hydroxy-16-methylheptadecanoic acid ($C_{18}H_{36}O_3$) is also reported to fatty acids. These are reported to produce and isolate from marine microbial strains Bacillus species SGD-AC-13 and it is a novel thermostable strain that is known as a biosurfactant overproducer. This microbial strain needed to maintain the conditions of the optimal parameters and these can help to the production of biosurfactants using a culture medium consisted of 1% yeast extract in tap water for 24 h in shake flasks and 12 h in a bioreactor system. Using the 16S-rRNA gene sequence (1,515 bp) and BCL cards (bioMerieux) tools and this strain has been identified as *Bacillus* genera strains [31].

Crude type of biosurfactant can show reduction capability of surface tension of distilled water (31.3 m Nm^{-1}) with its critical micelle concentration (CMC value ~0.3 mg ml). *Bacillus* species in fermentation broth needs to supernatant cell-free portion and shows the excellent emulsification and oil displacement activity or properties with its good stability up to the temperature (of 160°C) and pH (~6–12) and also high salt concentration (NaCl ~ 50 g L^{-1}), (Table 9.2). These biosurfactants were characterized by different analytical tools such as FTIR, TLC, LC-MS, as well as NMR spectroscopy techniques [32].

Further, these crude biosurfactant compounds are found to reduce the contact angle of a distilled water droplet (117 to 52.6) or 2% solution of pesticide (78.8 to 73.4). It has been found that 750 μg/ml of crude biosurfactant can recover 35% used lubricant oil (ULO) and 12% heavyweight of crude oil (HWCO) from contaminated sand. As a biosurfactant, this

Table 9.2 Biosurfactants exhibit various functions and application [38].

Functions	Elastomers	Textiles	Building	Food	Leathers	Paper or petroleum
Emulsification	Yes	Yes	Yes	---	Yes	Yes
Demulsification	Yes	---	Yes	---	Yes	Yes
Wetting and Penetrating	Yes	Yes	---	Yes	Yes	Yes
Solubility	Yes	Yes	Yes	---	Yes	Yes
Air entrapment	Yes	Yes	Yes	---	Yes	Yes
Detergent ability	Yes	Yes	Yes	---	Yes	Yes
Defoaming ability	Yes	---	Yes	---	---	---
Antistatic inhibition to Corrosion	Yes	---	Yes	Yes	Yes	Yes

thermostable fatty acid alkene is found structurally different from other reported biosurfactants with potential applications in agriculture, oil recovery, and bioremediation processes [32, 33].

In other reports, the *Bacillus safensis* J2 strain has been reported to produce biosurfactants under submerged cultivation conditions of a fermentation process using agro-industrial waste (i.e., bagasse) is the main source of carbon. Biosurfactant produced from this strain was characterized for its properties and stability under varying stresses. From the media, biosurfactant was extracted and tested for oil recovering and restoration of diesel from contaminated soils. It has been found that sugarcane bagasse (for this biosurfactant utilized concentration ~15 g L^{-1}) is served as an efficient substrate for cost-effective biosynthesis of biosurfactant (0.92 g L^{-1}) for this particular strain [33, 34].

These microbially produced biosurfactant has shown the stability property at various temperatures and pH with efficient emulsifier activity at varying salt concentrations. It has been demonstrated with good oil recovery capacity (up to 46.5 percent) from trapped column sand-packed oil and acute earthworm toxicity tests. Further, this surfactant is applied with E5 soil treatment (100g soil, 10g bagasse, 250ml distilled water, 10ml diesel oil and J2 strain, and 25ml crude biosurfactant) and is found to restore and detoxify the diesel contaminated soil effectively [34, 35].

9.4.1 Factors Influencing the Production of Biosurfactants

Many factors can influence the biosurfactant production, yield, or productivity. These factors can either boost or lower biosurfactant productivity in fermentation processes. The factors include environmental conditions as well as sources of biosurfactant from microbial growth conditions [22, 36].

9.4.1.1 Environmental Conditions

Various environmental factors like salinity, pH, and temperature can found to alter the production of biosurfactants and it showed effects on their cellular growth and activity. Few microorganisms are found to grow *in situ* like Microbial Enhanced Oil Recovery (MEOR) and survived in low oxygen levels, high temperatures, salinity, and pressures. Salt concentrations can affect the production of biosurfactants depending on the microbial strain and cellular activity. Few strains of Pseudomonas (i.e. MEOR 171 and 172) are found that are not affected by higher Ca and Mg concentrations and also temperatures and pH values [24].

9.4.1.2 Carbon Substrates

Various types of carbon substrates are used for the production of biosurfactants. Carbon sources and culture age are found to affect the yield and productivity of rhamnolipids (Table 9.3). Even in presence of hydrophobic substrates (such as corn oil and also long-chain alcohol), microbial fermentation is reported in rich, in unsaturated and saturated fats that induced the

Table 9.3 Different types of biosurfactants based on functional groups and their sources of microbial strains [39].

Class of surfactants	Microorganisms
Trehalose lipids	*Arthbacter paraffineus*, *Corynbacterium* sp., *Mycobacterium* sp., *Rhodococcus erythropolis*
Rhamnolipids	*Pseudomonas aeruginosa*, *Pseudomonas* sp.
Sophorose lipids	*Candida apicola*, *C. bombicola*, *C. lipolytica*, *C. bogoriensis*
Glusoce-, fructose-, saccharose lipids	*Arthobacter* sp., *Corynebacterium* sp., *R. erthropolis*
Cellobiose lipids	*Ustilago maydis*
Polyol lipids	*Rhodotorula glutinous*, *R. graminis*
Diglycosyl diglycerides	*Lactobacillus fermentii*
Lipopolysaccharides	*Acinetobacter calcoaceticus* (RAG1), *Pseudomonas* sp., *C. lipolytica*
Lipopeptides	*Arthrobacter* sp., *Bacillus pumilus*, *B. licheniformis*
Surfactin	*B. subtilis*
Viscosin	*P. fluorescens*
Ornithine, lysine peptides	*Thiobacillus thiooxidants*, *Sterptomyces siamensis* *Gluconobacter cerinus*
Phospholipids	*Acinetobacter* sp.
Sulfonylipids	*T. thiooxidans*, *Corynebacterium alkanolyticum*
Fatty acids (corynomycolic acids, spiculisporic acids, etc.)	*Capnocytophaga* sp., *Penicillium spiculisporum*, *Corynebacterium lepus*, *Arthrobacter paraffineus*, *Talaromyces trachyspermus*, *Nocardia erythropolis*

maximum quantity of biosurfactants production [27]. Strain P. aeruginosa can use the various carbon sources (such as C11 and C12 alkanes, citrate, fructose, glycerol, olive oil, glucose, and mannitol) and these are reported to produce for rhamnolipids. However, the nature of the carbon source can affect the type, quantity, and quality of the biosurfactant that is produced. Biosurfactants production is reported by using much microbial strain which uses diesel and crude oil as a carbon source. Other good sources include glucose, sucrose, and glycerol [28].

9.4.1.3 Estimation of Biosurfactants Activity

We can estimate and study the effect in levels of surface and interfacial tensions, hydrophilic-lipophilic balance (HLB), as well as stabilization/ destabilization of emulsions after adding a biosurfactant. A device (called tensiometer) is used to measure the surface tensions of various emulsions. The surface tension of the distilled water has been measured (i.e., 72 $mN.m^{-1}$) whereas the addition of biosurfactant can reduce the surface tension (i.e., up to 28$mN.m^{-1}$) in various solutions depending of nature or its concentration. It also helps us to determine CMC is the minimum concentration of any surfactant compounds, needed to form micelles [27].

9.5 Advantages, Microorganisms Involved, and Applications of Biosurfactants

9.5.1 Advantages of Using Biosurfactants

The recent growth in the industrial and commercial use of biosurfactants is reported due to their useful properties as shown below sections.

9.5.1.1 Easy Raw Materials for Biosurfactant Biosynthesis

Biosurfactants can be biosynthesized from various types of cheap raw materials that are present on our planet in abundant quantities with easy availability. The source of carbon may come from hydrocarbons, carbohydrates, or lipids that are used separately or in synergy [1].

9.5.1.2 Low Toxic Levels for Environment

Most biosurfactants compounds can show lower toxicity than the chemical surfactants that are derived from chemicals via using petroleum products.

It reported exhibiting the higher EC 50 values for biosurfactants than synthetic or chemical dispersants [2].

9.5.1.3 *Best Operation With Surface and Interface Activity*

A good biosurfactant compound lowers the water surface tension (from 75 m Nm^{-1} or J m^{-1} to 35 m Nm^{-1} or J m^{-1}). In the water and hexadecane mixture, interfacial tension may lower the values of surface tension (from 40 to 1 m Nm^{-1}). Any biosurfactant can exhibit the lowering of the water surface tension (up to 25 m Nm^{-1}) and also lower the water/hexadecane interface tension (up to < 1 m Nm^{-1}) [6].

9.5.1.4 *Good Biodegradability*

Biological processes mediated surfactants are found to easily degrade by many microorganisms present in soils [15].

9.5.1.5 *Physical Variables*

Many biosurfactants compounds are reported to be unaffected by environmental factors (such as tolerances of temperature, pH, and ionic strength ranges). Lichenysin is reported to be produced by the Bacillus licheniformis strain and it has not been affected by temperature ranges (up to 50°C), pH ranges (4-5-9.0), and concentrations of NaCl (50 g L^{-1}) or Ca (25 g. L^{-1}) compounds [4, 5].

9.5.2 Microbial Sources

Surfactants produced from biological sources are known as biosurfactants. These sources may include microorganisms of various types like bacteria, yeasts, or fungi; the various substrates used for the production of biosurfactants include sugars, oils, alkanes, and wastes [5, 7]. In general, most of the biosurfactants are produced using hydrocarbon substrates. Biosurfactants can also be produced from carbohydrates to make the biosurfactants very soluble. The typical growth of biosurfactants is during the late logarithmic and/or stationary phases [25].

Studies on the effectiveness of microbially produced biosurfactants on bioremediation as well as oil recovery have come into view in recent years. Microorganisms in presence of an insoluble carbon source diffuse into the cell by producing biosurfactants. Some bacteria as well as yeasts produce ionic surfactants that are used in the emulsification of CxHy substances in the growth media. Some microorganisms produce surfactants that are non-ionic or lipopolysaccharides in their cell wall by modifying their cell wall structure [35]. Some examples also include the Rhodococcus erythropolis and different species of Mycobacterium. It is known that Fe species of fungi produce the biosurfactant compound at cheap raw materials. Examples are some other microbial cells or species that include *C. bombicola, C. lipolytica, C. ishiwadae, C. bastisae,* as well as *Aspergillus ustus* and *Trichosporan asahii.* These microbial strains are reported to produce the different types of biosurfactant, and fungal strains produced biosurfactant that is sophorolipids (glycolipids) [16, 19, 20]. These are few useful microorganisms used to produce biosurfactants at commercial levels.

9.5.3 Production of Biosurfactants

Due to their biodegradable nature, biosurfactants are extremely useful and efficient and can be used for large-scale production. Rhamnolipids are mostly studied and produced for industrial purposes among the various biosurfactants [5, 17].

9.5.3.1 Production of Rhamnolipids

P. aeruginosa produces more than 27 types of rhamnolipids which can vary in chain length size and to the extent of saturation. The two main types of rhamnolipids produced in liquid crops are monorhamnolipid (rhamnosyl-β-hydroxydecanoyl-β-hydroxydecanoate~Rha-C10-C10) and also the rhamnolipid (rhamnosyl-rhamnosyl-β-hydroxydecanoyl-β-hydroxydecanoate~Rha-Rha-C10-C10).

Two different metabolic pathways are reported for biosurfactants synthesis. These are known as the dTDP-L-rhamnose biosynthesis and *Rhl*A diversion of the β-hydroxydecanoyl-ACP intermediate from the FASII cycle and are reported to synthesize the fatty acid dimer moiety of rhamnolipids and free alkanoic acid (HAA) 3-[3-hydroxyalkanoyloxy] [20].

Rhamnosyltransferases *Rhl*B and *Rhl*C can catalyze the transfer of dTDP-L-rhamnose to either HAA or to a monorhamnolipid that was previously generated. The rule is the protein used for converting two β-hydroxyacyl-ACP molecules into an HAA. Free HAA shows surface tension properties and has been used in the promotion of swarming motility. Recently it was discovered that *Rhl*A diverts the intermediate β-hydroxydecanoyl-ACP from the FASII cycle. This diversion provides a substrate for *Rhl*AB enzyme which produces the rhamnolipid molecule hydrocarbon chain. This is a clinically important step yet to be properly studied [8, 25].

9.5.3.2 *Regulation of Rhamnolipids Synthesis*

The regulation of the synthesis is essential for the large production of rhamnolipids. It is highly controlled by gene regulations that interact with a variety of environmental and physiological signals and is highly capable of combining various signals to create unique and specific responses. It has been seen that the key to regulation is the proper functioning of the *rml*B-DAC and rhlAB operons [8, 24].

Other factors favoring high levels of production include high carbon-to-nitrogen ratios, nitrogen exhaustion, stress conditions, and high cell densities. Various environmental stress conditions can influence the production of rhamnolipids. Nutrient deprivation and nitrogen exhaustion can easily accomplish an increase in rhamnolipid production, even in a QS-independent manner. Various regulatory factors can deal with specific gene regulation patterns and few include sigma factors of RNA polymerase RpoS (σS or σ 38) and RpoN (σ54) [8, 37].

9.5.3.3 *Commercial Use of Biosurfactants*

9.5.3.3.1 Food Industries

Biosurfactants are used to stabilize the aggregation of fat globules in aerated systems, to improve the texture and product shelf life. It is also used in baking for consistency control, slowing down and solubilizing flavor oils agents in fat and oil cooking. Biosurfactants are also used to enhance the properties of creamed butter and frozen confectionery products [18, 35].

9.5.3.3.2 Bioremediation

Bioremediation includes increasing the availability of natural materials to enhance the biodegradation of components in a contaminated environment. Biosurfactants are used when the growth of bacterial culture is slow or when degradation is difficult. These compounds are in general used to stimulate bioremediation technique [24, 34].

9.5.3.3.3 Removal of Oil and Petroleum

Biosurfactants have been experimentally used to along with hydrocarbons to enhance water solubility and oil displacing properties from various solid compounds. Biosurfactants are also known to increase microbial uptake [26, 37].

9.5.3.3.4 Agricultural Use of Biosurfactants

Biosurfactants are used as mobilizing agents to increase the solubility of bio-hazardous chemicals like PAH. Biosurfactants are also known to increase the ability of microbes to absorb pollutants that cover soil particles. Biosurfactant can be used in heavy soils for hydrophilization to obtain good wettability with even soil fertility distribution. Biosurfactants prevent certain fertilizers from clogging and increase the spread and penetration of toxicants in pesticides [16, 38, 39].

Agricultural uses of biosurfactants include:

- Improving soil quality
- Plant pathogen elimination
- Encouraging plant-microbe interaction [36]

9.6 Conclusions

Chemical surfactants can be categorized into the ionic or non-ionic surfactant. Ionic surfactant is also found in anionic, cationic, or amphoteric. Normally, most surfactant compounds are found to reduce the surface tension and the interfacial tension of various types of solutions. Due to the presence of hydrophilic and hydrophobic groups, between liquid phases like oil/water, surfactants can lower the surface tension of any liquid. Biosurfactant is used for cleaning purposes like emulsification. Various types of chemical surfactants have exhibited different properties and characteristics and these are alkylbenzene sulfonates (detergents), (fatty acid) soaps, lauryl sulfate (foaming agent), di-alkyl sulfosuccinate (wetting agent), lignosulfonates (dispersants) which are few commonly used surfactants. But this chemical surfactant has raised many environmental problems, so we need to utilize the biosurfactants. These are also found to amphiphilic compounds and also shown the multiples applications in households, industries, and agricultural sectors. Glycolipids, oligopeptides and lipopeptides, fatty acid phospholipids, or neutral lipids, and polymeric biosurfactants are reported as various categories. Many conditions such as the nature of the carbon substrate, medium ion concentration, pH, temperature, salinity, bacterial strains, and conditions of culture, are reported as important factors, controlling the production and physicochemical properties. Biosurfactants can help in the building of carbohydrates, fatty acids or peptides as external components of the cells. Biosurfactants are mostly applied in the field of pharmaceuticals, cosmetics, and food industries. Sophorolipids from C. bombicola is used in the food industry. Biosurfactants produced from P. aeruginosa are found to be non-toxic and non-mutagenic and can be applied in different industries including detergent.

References

1. Jahan, R., Bodratti, A.M., Tsianou, M., Alexandridis, P., Biosurfactants, natural alternatives to synthetic surfactants: Physicochemical properties and applications. *Adv. Colloid Interface Sci.*, 275, 102061, 2020.
2. Jimoh, A.A. and Lin, J., Biosurfactant: A new frontier for greener technology and environmental sustainability. *Ecotoxicol. Environ. Saf.*, 184, 109607, 2019.
3. Akbari, S., Abdurahman, N.H., Yunus, R.M., Fayaz, F., Alara, R.O., Biosurfactants—a new frontier for social and environmental safety: a mini review. *Biotechnol. Res. Innov.*, 2, 81, 2018.
4. Vijayakumar, S. and Saravanan, V., Biosurfactants-Types, Sources and Applications. *Res. J. Microbiol.*, 10, 181, 2015.
5. Fakruddin, M., Biosurfactant: Production and Application. *J. Pet. Environ. Biotechnol.*, 3, 1000124, 2012.
6. Santos, A.P.P., Silva, M.D.S., Costa, E.V.L., Rufino, R.D., Santos, V.A., Ramos, C.S., Sarubbo, L.A., Porto, A.L.F., Production and characterization of a biosurfactant produced by *Streptomyces* sp. DPUA 1559 isolated from lichens of the Amazon region. *Braz. J. Med. Biolog. Res.*, 51, e6657, 2017.
7. Mulligan, C.N. and Gibbs, B.F., Types, Production and Applications of biosurfactants. *Proc. Indian Natn. Sci. Acad.*, B70, 31, 2004.
8. Kambo, N. and Upadhyay, S.K., Antagonism in (conventional anionic-gemini anionic) mixed micelle catalyzed oxidation of D-fructose by alkaline chloramines. *T. Internat. J. Chem. Kinet.*, 41, 123, 2009.
9. Mubofu, E.B. and Engberts, J.B.F.N., Surfactant-assisted specific-acid catalysis of Diels-Alder reactions in aqueous media. *J. Physic. Organic. Chem.*, 20, 764, 2007.
10. Larson, R.A., Jafvert, C.T., Bosca, F.K., Marley, A., Miller, P.L., Effects of surfactans on reduction and photolysis (>290 nm) of nitroaromatic compound. *Enviro. Sci. Technol.*, 34, 505, 2000.
11. Ivanković, T. and Hrenović, J., Surfactants in the environment. *Arh. Hig. Rada. Toksikol.*, 61, 95, 2010.
12. Sakamoto, K., Lochhead, R.Y., Maibach, H.I., Yamashita, Y. (Eds.), *Cosmetic Science and Technology, Theoretical Principles and Applications*, Elsevier, USA, 2017.

13. Romanelli, M.F., Moraes, M.C.F., Villavicencio, A.L.C.H., Borrely, S.I., Evaluation of toxicity reduction of sodium dodecyl sulfate submitted to electron beam radiation. *Radiat. Phys. Chem.*, 71, 411, 2004.
14. Rosen, M.J., Li, F., Morrall, S.W., Versteeg, D.J., The relationship between the interfacial properties of surfactants and their toxicity to aquatic organisms. *Environ. Sci. Technol.*, 35, 954, 2001.
15. Banat, I.M., Makkar, R.S., Cameotra, S.S., Potential commercial applications of microbial surfactants. *Appl. Microbiol. Biotechnol.*, 53, 495, 2000.
16. Desai, J.D. and Banat, I.M., Microbial production of surfactants and their commercial potential. *Microbiol. Molecul. Rev.*, 61, 47, 1997.
17. Bodour, A.A., Guerrero-Barajas, C., Jiorle, B.V., Malcomson, M.E., Paull, A.K., Somogyi, A., Structure and characterization of flavolipids, a novel class of biosurfactants produced by *Flavobacterium* sp. strain MTN11. *Appl. Environ. Microbiol.*, 70, 114, 2004.
18. Nitschke, M. and Costa, S.G.V.A.O., Biosurfactants in food industry. *Trends Food Sci. Technol.*, 18, 252, 2007.
19. Cameotra, S.S., Makkar, R.S., Kaur, J., Mehta, S.K., Synthesis of biosurfactants and their advantages to microorganisms and mankind. *Adv. Exp. Med. Biol.*, 672, 261, 2010.
20. Neto, D.C., Meira, J.A., Tiburtius, E., Zamora, P.P., Bugay, C., Mitchell, D.A., Krieger, N., Production of rhamnolipids in solid-state cultivation: characterization, downstream processing and application in the cleaning of contaminated soils. *Biotechnol. J.*, 4, 748, 2009.
21. Sachdev, D.P. and Cameotra, S.S., Biosurfactants in agriculture. *Appl. Microbiol. Biotechnol.*, 97, 1005, 2013.
22. Banat, I.M., Satpute, S.K., Cameotra, S.S., Patil, R., Nyayanit, N.V., Cost effective technologies and renewable substrates for biosurfactants' production. *Front. Microbiol.*, 5, 697, 2014.
23. Rebello., S., Anoopkumar, A.N., Raveendran, S., Parameswaran, B., Pandey, A., Aneesh., E.M., Comparative, life-cycle analysis of synthetic detergents and biosurfactants—an Overview, in: *Refining Biomass Residues for Sustainable Energy and Bioproducts, Technology, Advances, Life Cycle Assessment, and Economic*, R.P. Kumar, E. Gnansounou, J.K. Raman, G. Baskar (Eds.), pp. 511–521, Academic Press, USA, 2019.
24. Yalaoui-Guellal, D., Fella-Temzi, S., Djafri-Dib, S., BrahmibIbrahim, F., Banat, I.M., Madani, K., Biodegradation potential of crude petroleum by hydrocarbonoclastic bacteria isolated from Soummam wadi sediment and chemical-biological proprieties of their biosurfactants. *J. Petrol. Sci. Eng.*, 184, 106554, 2020.
25. Singh, A.K. and Sharma, P., Disinfectant-like activity of lipopeptide biosurfactant produced by *Bacillus tequilensis* strain SDS21. *Colloid, Surfaces B: Biointerf.*, 185, 110514, 2019.
26. Purwasena, I.A., Astuti, D.I., Syukron, M., Amaniyah, M., Sugai, Y., Stability test of biosurfactant produced by *B. licheniformis* DS1 using

experimental design and its application for MEOR. *J. Petrol. Sci. Eng.*, 183, 106383, 2019.

27. Giri, S.S., Ryu, E.C., Sukumaran, V., Park, S.C., Antioxidant, antibacterial, and anti-adhesive activities of biosurfactants isolated from Bacillus strains. *Microb. Pathogen.*, 132, 66, 2019.
28. Mohammadipour, M., Mousivand, M., Salehi, Jouzani, G., Abbasalizadeh, S., Molecular and biochemical characterization of Iranian surfactin-producing *Bacillus subtilis* isolates and evaluation of their biocontrol potential against *Aspergillus flavus* and *Colletotrichum gloeosporioides*. *Can. J. Microbiol.*, 55, 395, 2009.
29. Morikawa, M., Beneficial biofilm formation by industrial bacteria Bacillus subtilis and related species. *J. Biosci. Bioeng.*, 101, 1, 2006.
30. Mukherjee, A.K., Potential application of cyclic lipopeptide biosurfactants produced by Bacillus subtilis in laundry detergent formulations. *Lett. Appl. Microbiol.*, 45, 330, 20072007.
31. Ram, H., Sahu, A.K., Said, M.S., Banpurkar, A.G., Gajbhiye, J.M., Dastage, S.G., A novel fatty alkene from marine bacteria: A thermo stable biosurfactant and its applications. *J. Hazard. Mater.*, 380, 120868, 2019.
32. Hisham, N.H.M.B., Ibrahim, M.F., Ramli, N., Abd-Aziz, S., Production of Biosurfactant Produced from Used Cooking Oil by *Bacillus* sp. HIP3 for Heavy Metals Removal. *Molec.*, 24, 26, 2019.
33. De Oliveira, J.G. and Garcia-Cruz, C.H., Properties of a biosurfactant produced by Bacillus pumilus using vinasse and waste frying oil as alternative carbon sources. *Braz. Arch. Biol. Technol.*, 56, 155, 2013.
34. Das, A.J. and Kumar, R., Production of biosurfactant from agro-industrial waste by Bacillus safensis J2 and exploring its oil recovery efficiency and role in restoration of diesel contaminated soil. *Environ. Technol. Innovat.*, 16, 100450, 2019.
35. Sharma, R., Singh, J., Verma, N., Production, characterization and environmental applications of biosurfactants from Bacillus amyloliquefaciens and Bacillus subtilis. *Biocataly. Agric. Biotechnol.*, 16, 132, 2018.
36. Otzen, D.E., Biosurfactants and surfactants interacting with membranes and proteins: Same but different? *Biochim. Biophys. Acta Biomembr.*, 1859, 639, 2016.
37. Karanth, N.G.K., Deo, P.G., Veenanadig, N.K., Microbial production of biosurfactants and their Importance. *Curr. Sci.*, 77, 116, 1999.
38. Sachdev, D.P. and Cameotral, S.S., Biosurfactants in agriculture. *Appl. Microbiol. Biotechnol.*, 97, 1005, 2013.
39. Banat, I.M., Franzetti, A., Gandolfi, I., Bestetti, G., Martinotti, M.G., Fracchia, L., Smyth, T.J., Marchant, R., Microbial biosurfactants production, applications and future potential. *Appl. Microbiol. Biotechnol.*, 87, 427, 2010.

Part 2

MICROBES IN SUSTAINABLE AGRICULTURE AND BIOTECHNOLOGICAL APPLICATIONS

10

Functional Soil Microbes: An Approach Toward Sustainable Horticulture

C. Sarathambal*, R. Dinesh and V. Srinivasan

ICAR-Indian Institute of Spices Research, Kozhikode, India

Abstract

Plant growth–promoting rhizobacteria (PGPR) directly encourage plant growth through acquisition of nutrients, production of regulatory chemicals, and influencing plant hormone levels, while indirectly they suppress pathogens that inhibit plant growth and development. Numerous studies have irrevocably proven the multi-trait beneficial characteristics of such bacteria, and an array of PGPR is now being deployed in crop production, thereby decreasing the global dependence on exogenous chemical inputs that have continued to threaten the fragile agro-ecosystems. A further complete understanding of the function and diversity of PGPR, with special emphasis on those that have non-symbiotic and symbiotic associations with horticultural crops especially fruits, vegetables, and spices will be dealt in this chapter.

Keywords: Plant growth–promoting rhizobacteria, rhizosphere, phytohormone production, mineral solubilization, biocontrol, sustainable horticulture

10.1 Introduction

Among the soil bacterial communities, the free-living bacteria that colonize the plant root region and have the capacity to promote plant growth directly or indirectly are referred to as plant growth–promoting rhizobacteria (PGPR). The bacteria have a great affinity to the rhizosphere, considered a significant soil environment with intense plant–microbe communications. Prospecting root-associated microbiomes for natural

**Corresponding author*: saratha6@gmail.com

Bibhuti Bhusan Mishra, Suraja Kumar Nayak, Swati Mohapatra, and Deviprasad Samantaray (eds.) Environmental and Agricultural Microbiology: Applications for Sustainability, (221–242) © 2021 Scrivener Publishing LLC

products that boost agricultural production without damaging the environment is a current trend [1]. Many studies have shown that crop plants are interacting with diverse rhizosphere microorganisms such as both free-living and symbiotic microbes. Inoculation with these microbes has resulted in enhanced growth and yield of crops, due mainly to increased nutrient uptake and the disease resistance. However, due to reckless use of chemical pesticides and fertilizers, the importance of PGPR has been seldom recognized in intensive crop production systems, which has led to severe erosion in biodiversity as well as its qualities. Nevertheless, better awareness in various nations, including India, is gradually leading to improved utilization of such beneficial microbes, leading to the sustainable agriculture. In this chapter, the role and mechanisms of plant growth promotion involving PGPR and their importance in horticulture crops are discussed.

10.2 Rhizosphere Microbial Diversity

Microbial communities are a critical component of ecosystems that play crucial roles in decomposition of organic matter and bio-geochemical nutrient transformations like N_2 fixation, nutrient solubilization, and nutrient mobilization in soil. In a soil profile, the top 20- to 40-cm horizon mostly constitutes the rooting zone that encompasses nutrient cycling and biological activity. This region with intense microbial activity is also called the rhizosphere wherein intricate interactions occur among soil, plants, and microbial communities. Consequently, this region is rich in substrates like C compounds, amino acids, sugars, and organic acids. All plants secrete distinctive compounds into the rhizosphere, which decides the microbial population of that rhizosphere. In the battle for such unique compounds, bacterial strains either secrete antibiotic molecules and other compounds that remove competition or produce chemicals that enhance root density and root volume. This, in turn, enhances the plant's capacity to draw more nutrients from the rhizosphere.

Conversely, plants encourage or suppress microbial communities and their functions through the secretion of an array of chemicals. The plant root exudates are organic compounds that are water-soluble and consist of mainly amino acids, carbohydrates, and organic acids, which are secreted into the rhizosphere by the roots along concentration gradients [2]. For soil microbial communities, these exudates are considered to be a rich reserve of labile C and N that can be readily absorbed without the need to employ exo-enzymes. Since the rhizosphere contains vast reserves of such nutrient rich substrates, the microbial activity tends to be several folds higher than in the bulk soil [3]. Also, up to 40% of the dry biomass made by crop plants is accounted for by the C released through root exudation [4]. It is pertinent to note that the microbial community structure and diversity in the rhizosphere is vastly determined by the expanse and chemical composition of the root exudates, which vary between plant species and between growth stages [5]. Apparently, the bacteria can exert favorable effects on the plant only if it is able to compete well with other microorganisms during their quest for nutrients from the substrate reserve.

10.3 Plant Growth–Promoting Rhizobacteria

1. Generally, PGPR function in three important ways:
2. They synthesize specific metabolites for the crops.
3. They facilitate the nutrients absorption from the soil.
4. They inhibit or suppress pathogens from infecting the plant.

Plant growth development by PGPR is expedited both directly and indirectly (Figure 10.1). The PGPR exhibit both synergistic and antagonistic

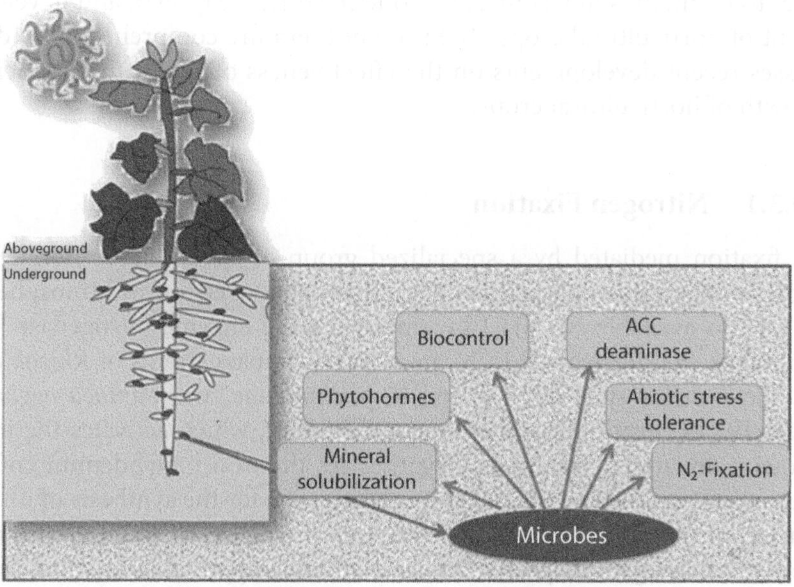

Figure 10.1 Conceptual model for mode of action of plant growth promoting rhizobacteria.

interactions while promoting plant growth [6] and markedly influence soil quality thereby transforming a sick barren soil into a fertile and productive one. Restoration of soil quality using PGPR has been an important and actively exploited facet for enhanced crop production in many regions of the world [7].

The symbiotic and non-symbiotic PGPR directly influence the growth of plants through secretion of an array of hormones like abscisic acid, auxins, gibberellins, cytokinins, IAA, and ethylene. Besides, several PGPR hydrolyze 1-aminocyclopropane-1-carboxylate (ACC) into α-ketobutyrate and ammonia, thereby reducing ethylene levels in the micro-rhizosphere environment and consequently enhancing plant growth. PGPR also positively influence plant growth by solubilizing insoluble minerals especially those containing P, Zn, etc. Furthermore, enhancing the resilience to nutrient stress tolerance helps in stabilizing soil structure and accumulation of organic matter. Enhanced nutrient cycling and availability to plants by PGPR can markedly reduce the amount of chemical fertilizers that go into the soil system.

PGPR indirectly influence plant growth by suppressing phyto-pathogens, by producing either siderophores (small iron binding molecules) or antibiotics. Besides, some PGPR have the capacity to produce HCN and or cell wall degrading enzymes like ß-1,3-glucanase and chitinase that severely inhibit phyto-pathogens. These interactions are vital to maintain soil fertility, and concurrently, the growth and development of horticultural crops. The present literature comprehensively discusses recent developments on the effectiveness of PGPR in enhancing growth of horticultural crops.

10.3.1 Nitrogen Fixation

N_2 fixation mediated by a specialized group of symbiotic or non-symbiotic prokaryotes is a process that involves conversion of atmospheric N (N_2) to available N (NH_3) by employing the enzyme nitrogenase [8]. Symbiotic PGPR involved in N_2 fixation encompass strains of *Rhizobium* sp., *Beijerinckia* sp., *Azoarcus* sp., *K. pneumoniae*, and *Pantoea agglomerans* [9]. N_2 fixation involves a gene called *nif*, which activates the iron protein, donates electrons, and synthesizes the iron molybdenum cofactor, including other regulatory genes imperative for the synthesis of nitrogenase and its activity [10]. Inoculation of biological N_2-fixing PGPR not only enhances growth-promotion and disease suppression but also helps in maintaining N level in arable soils [11].

10.3.2 Production of Phytohormones

Phytohormones or plant growth regulators (PGRs) are organic substances that positively influence various biological developments in plants at even very low applications. Besides the levels of hormonal signs, the dynamics in the levels of phytohormones can markedly influence plant growth and development.

The increased number of lateral roots and root hairs observed in plants inoculated with *A. brasilense* was attributed to the production of auxins, cytokinin-like, and gibberilin-like substances by the bacteria since similar results could be mimicked consequent to application of a mixture of gibberlic acid (GA), kinetin, and Indole-3-acetic acid (IAA) [12]. Besides, increased phytohormone production has also been implicated for the enhanced growth of plants after *Azospirillum* inoculation [13]. Some examples of PGPR that produce hormones, which play an important role in growth promotion, are *P. agglomerans*, *Rhodospirillum rubrum*, *Pseudomonas fluorescens*, *Paenibacillus polymyxa*, *Bacillus subtilis*, *Rhizobium leguminosarum*, *Pseudomonas* sp., and *Azotobacter* sp. [14, 15].

10.3.3 Production of Enzymes That can Transform Crop Growth

Ethylene is recognized as a "ripening hormone"; it stimulates adventitious root and root hair formation, breaks the dormancy, and stimulates germination of seeds. It is a strong plant hormone that influences the various phases of crop growth and development. However, if the ethylene levels remains elevated after germination, then root elongation is suppressed [16]. It is generally understood that many plant growth–promoting bacteria may stimulate plant growth by mitigating the concentration of ethylene in crops. This is recognized to the activity of the enzyme ACC deaminase, which hydrolyzes ACC, the direct precursor of ethylene in crops. The products of this hydrolysis, α-ketobutyrate and ammonia, can be utilized by the bacterium as a resource of carbon and nitrogen for growth and development [17]. In this way, the bacterium reduces the ethylene concentration and arrests some of the potentially detrimental consequences of high ethylene levels in crops [18].

The PGPR possessing ACC deaminase are available in various soils and offer assurance as a bacterial inoculum for development of crop growth, mostly under adverse ecological conditions such as phytopathogens, flooding, heavy metals, high salt, and drought. Application of plants with ACC deaminase-containing bacteria may assist plant growth by reducing harmful effects of stress. In environment, ACC deaminase has been frequently found in soil bacteria that colonize plant roots [19]. Many of these PGPR are identified by their capability to grow on minimal media containing ACC as its lone nitrogen source. In this way, *Achromobacter piechaudii*, *Alcaligenes* sp., *Bacillus pumilus*, *B. cereus*, *B. megaterium*, *Burkholderia* sp., *B. phytofirmans*, *Enterobacter aerogenes*, *E. sakazakii*, *E. cloacae*, *Kluyvera ascorbata*, *K. ascorbata*, *Pseudomonas putida*, *P. fluorescens*, *Sinorhizobium* sp., and *Variovorax paradoxus* are all observed to be capable to utilize ACC as the single nitrogen source for their growth [20].

10.3.4 Microbial Antagonism

PGPR can mitigate the development of a variety of plant pathogens in several of means like preventive accessible Fe through siderophore production, competing for nutrients, space, antibiosis, and producing lytic enzymes [21]. Among them, fluorescent pseudomonads are extensively stated for their wide-ranging antagonistic activity against numerous plant pathogens. Nitrogen fixing strains of *Rhizobium meliloti* have been noticed for siderophores production [22] in iron stress circumstances and thereby added a benefit to eliminate the pathogen, groundnut charcoal rot (*Macrophomina phaseolina*) [23]. *Azotobacter*, *Azospirillum*, *Enterobacter*, *Bacillus*, *Paenibacillus*, *Streptomyces*, and *Pseudomonas* are observed as the virulent genera of rhizobacteria acting against the plant pathogens like tomato mottle virus, tobacco necrosis virus, *Myzus persicae*, *Fusarium avenaceum*, *Acyrthosiphon kondoi*, and *Rhizoctonia bataticola*. PGPR strains, such as *Azospirillum* sp. and *P. fluorescens*, *Azotobacter* sp., and AM fungi, like *Gigaspora margarita*, *Glomus mossae*, and *G. fasciculatum*, are documented as the maximum efficient microbes to suppress wilt disease of brinjal *in vitro*. The microbial inoculants when used as consortia demonstrated very good potential in the conquest of diseases with the characteristic increase in shoot height, number of leaves, chlorophyll content, and thereby easing overall yield than when inoculated individual organism [24]. However, inoculation of these PGPR isolates did not impact the beneficial native rhizosphere microflora including the siderophore-producing bacterial strains and fluorescent pseudomonads.

10.3.5 Solubilization of Minerals

One of the varieties of process by which PGPR maintain crop development is by solubilization of insoluble minerals. Among the nutrients, phosphorus is the second most significant macronutrient following to nitrogen in preventing plant development. More than 40% of the global soils are limited in phosphorus and the acid weathered soils of sub-tropical and tropical regions are mainly susceptible to dearth of phosphorus [25]. A study reported that 98% of Indian soils require phosphatic fertilizers either in the form of biological or chemical fertilizer. Practice of phosphorus fertilizer application is followed though a majority of the soil phosphorus reaction, wherein produces are only barely soluble. In such circumstances, microbes offer a natural rescue structure capacity of solubilizing the insoluble inorganic phosphorus and sort it accessible to the plants. Phosphorus solubilization by PGPR has been well recognized. This group includes bacteria, some actinobacteria, and fungi. These microorganisms solubilize the inaccessible forms of inorganic phosphorus such as tricalcium, aluminum, iron, and rock phosphates into soluble forms by release of a variety of organic acids like succinic, malic, citric, fumaric, and gluconic and glyoxalic acids [26]. These phosphorus solubilizing microbes include different groups of organisms, which not only include phosphorus from insoluble forms but also cause a large fraction. The bacterial genuses with this competency are *Bacillus, Pseudomonas, Azosirillum, Rhizobium, Arthrobacter, Burkholderia, Serratia, Acinetobacter, Enterobacter, Erwinia*, and *Flavobacterium* [27, 28]. Atmospheric nitrogen fixation with phosphorus solubilization by diazotrophs is remarkable, and to our knowledge, such illustrations of diazotrophic phosphate solubilizer are *Pantoea agglomerans* [29] and *Swaminathania salitolerans* [30]. By production of organic acids, the phosphate solubilizing bacteria have the capability to solubilize insoluble inorganic phosphate [31] made it accessible to crops. On the other hand, production of phosphatase enzyme in the culture media indicated that there could also be a probable role of phosphatase in solubilizing inorganic phosphate.

Micronutrients such as Zn, Mn, and Fe are found to be scarce in most of the soils with Zn as a most vital nutrient all over the world [32]. Zn, the micronutrient necessary for crop growth, is an important constituent of over 300 enzymes and plays structural, catalytic, and co-catalytic roles in several crop systems. For mitigation of Zn deficiency for plants, their application is done mostly in soluble form as zinc sulfate, and the soluble form of Zn applied to the soil gets converted into various inaccessible forms due to the soil pH response. These alterations are based on other nutrients availability and the type of soil. Zn is mainly converted into zinc phosphate in maximum P fertilizing soils, while zinc carbonate in high calcareous soils reacts with Mn and Fe oxide minerals. Application of Zn solubilizing bacteria as a biofertilizer in crop production technology is certainly useful for a country like India having high zinc deficiency soils. A term called zinc solubilizing bacteria (ZSB) was framed for those bacteria that are capable of solubilizing the insoluble zinc minerals in agar medium as well as in soil [33–35]. Similarly, potassium solubilizing bacteria such as *Bacillus edaphicus, Bacillus mucilagenosus, Pseudomonas* sp., *Burkholderia* sp., and *Paenibacillus* sp. are examples of microbes that are used in biofetilizers. Through secretion and production of organic acids, potassium solubilizing bacteria are capable to solubilize potassium rock. These are aerobic bacteria which play a vital role in improving soil structure by development and stabilization of water stable soil aggregates. In addition, this Gram-positive bacterium can produce substances that improve crop growth or suppress root pathogens [36]. Sheng *et al.* [37] investigated silicon and potassium mobilization by silicate mineral solubilizing bacteria, *Bacillus globisporus.* In liquid cultures, the strain observed enhanced growth on the biotite than on muscovite and feldspar. The biotite is the most excellent potassium source for growth of the strain and gluconic acid seemed to be the most active component for the solubilization of the insoluble silicate minerals. Although these microbes mobilize the potassium resulting in improved soil fertility, studies on the subject of their use as a biofertilizer are meager.

10.3.6 Siderophore and Hydrogen Cyanide (HCN) Production

Iron is a vital constituent for all living beings. For iron uptake, it was recommended that crops can be beneficial from the siderophores produced by several PGPR. The shortage of bioavailable iron in soil habitats and on crop surfaces generates a livid competition among microbes. Even though iron is one of the most copious minerals on globe, in the soil, it is comparatively inaccessible for direct absorption by microbes [38]. Siderophores are small, high-affinity iron chelating compounds secreted by microbes such as bacteria, fungi, and various grasses. Under iron-limiting circumstances, plant growth–promoting bacteria produce low-molecular-weight compounds called siderophores (Greek: "iron carrier") to competitively acquire ferric ion. Microorganisms liberate siderophores to scavenge iron from these mineral phases by creation of soluble Fe^{3+} complexes that can be take up by active transportation mechanisms. Most of siderophores are non-ribosomal peptides, although some are biosynthesized separately [39]. Siderophores are also vital for some pathogenic microbes for their acquirement of iron. Among the siderophores, Enerobactin is the strongest binders to Fe^{3+}. Commonly reported siderophore producing strains are *Enterobacter, Pseudomonas, Rhodococcus, Bacillus, Azospirillum,* and *Azotobacter* [40].

For the reason that of the plenty of siderophore producing microbes in soils, along with their excellent chemical stability and Fe binding ability, these substances may provide considerably to an improved mobility of Fe in the soil and in the rhizosphere in particular, making it accessible for crops. However, microbial siderophores may act as vital component of Fe for higher plants in calcareous and alkaline soils [41], where iron accessibility is mainly limited.

10.3.7 Cyanide (HCN) Production

Cyanide (HCN) production is one of the possible ways by which rhizobacteria may suppress pathogens growth in soil. Among the most destructive plant pests, plant-parasitic nematodes are causing significant economic losses to agronomic crops worldwide. HCN is a significant substance produced from *Pseudomonas* sp. against root knot nematode and can act as a valuable model system for studying plant-parasitic nematode control [42]. Multitrophic level interactions arbitrate the capability of fungal pathogens to cause disease and the ability of bacterial antagonist to curb plant disease. For bacterial antibiotic HCN biosynthesis, pathogen metabolite provides the negative signal which can decide the relative significance of biological control mechanisms existing to antagonists and which may also influence fungus-bacterium ecological exchanges [43]. Direct correlations are found between HCN production *in vitro* and plant protection in the tomato/*Fusarium oxysporum* f. sp. *radicis-lycopersici* and cucumber/*Pythium ultimum* plant pathosystems, which were earlier reported. *B. subtilis* ubiquitous inhabitant of soil is widely documented as a powerful biocontrol agent. Besides, it produces different biologically active compounds with a broad spectrum of activities [44]. Multiple plant growth–promoting *B. megaterium* from rhizosphere of tea crop is able to help in the plant growth promotion, *viz.*, IAA production, siderophore production, phosphate solubilization, production of antifungal metabolite, and reduction of disease intensity [45].

10.3.8 Plant Growth–Promoting Rhizobacteria on Growth of Horticultural Crops

The application of synthetic chemicals has absolutely resulted in the getting of improved crop production. On the other hand, nowadays, there has been a great demand on farmers and consumers to decrease or eradicate the usage of man-made chemicals in horticulture, since vegetables and fruits are eaten as raw. This concern has encouraged searching for better replacements which are eco-friendly and cheaper. It is well established that PGPR play a vital part in improving soil health and crop through multifaceted growth promoting traits (Table 10.1).

Table 10.1 Plant growth–promoting rhizobacteria and their mechanisms of action on horticultural crops (last 3 years).

1. Vegetables crops				
Sl. no.	**Name of the crop**	**Organism involved**	**Mode of action**	**Reference**
1	Tomato	*Pseudomonas putida* and *Rothia* sp.	Induction of systemic resistance	[47]
2	Tomato	*Bacillus megaterium*	Growth promotion, salt tolerance	[48]
3	Tomato	*Azospirillum brasilense*	IAA production	[49]
4	Tomato	*B. subtilis*, and *B. amyloliquefaciens*	Non-ribosomal peptides	[50]
5	Cucumber	*Azospirillum brasilense*	IAA production	[49]
6	Cucumber	*Bacillus* sp.	Cytokinin synthesis	[51]
7	Cucumber	*Pseudomonas stutzeri*, *B. subtilis*, *Stenotrophomonas maltophilia*, and *Bacillus amyloliquefaciens*	IAA, N fixation, antibiosis	[52]
8	Chickpea	*Bacillus amyloliquefaciens*, *Paenibacillus polymyxa*, *Pseudomonas chlororaphis*, *Pseudomonas fluorescens*, *B. subtilis*, and *Pseudomonas* sp.	Growth promotion, antibiosis	[53]
9	Ladies finger	*Enterobacter* sp. UPMR18	Increase antioxidant enzyme activities and upregulation of ROS pathway genes	[54]

(Continued)

Table 10.1 Plant growth–promoting rhizobacteria and their mechanisms of action on horticultural crops (last 3 years). (*Continued*)

Sl. no.	Name of the crop	Organism involved	Mode of action	Reference
10	Potato	*Pseudomonas fluorescens*	IAA, GA, Siderophore, and phosphate solubilization	[55]
11	Cassava	*B. subtilis*	IAA production	[56]
12	Potato	*Azospirillum* sp. TN10, *Agrobacterium* sp. TN14, *Pseudomonas* sp. TN36, *Enterobacter* sp. TN38 and *Rhizobium* sp. TN42	N fixation, IAA production	[57]
13	Chilli	*Streptomyces* sp. and *Bacillus* sp.	Growth promotion, increase in capsaicin content	[58]
14	Brinjal	*B. subtilis* and *Cellulosimicrobium cellulans*	Antagonism against *Alternaria alternata*, *Penicillium digitatum*, and *Fusarium oxysporum*	[59]
15	Lettuce	Azospirillum brasilense	IAA production	[49]
16	Carrot, lettuce, Tomato	*Rhizobium* sp.	Sideophore production	[51]
17	Cauliflower, Tomato, Brinjal, Chilli	*Bacilus cereus*, *Pseudomonas rhodesiae*	Growth promotion	[60]

(*Continued*)

Table 10.1 Plant growth–promoting rhizobacteria and their mechanisms of action on horticultural crops (last 3 years). (*Continued*)

Sl. no.	Name of the crop	Organism involved	Mode of action	Reference
2. Spices, plantation, and medicinal crops				
1	Ginger	*Bacillus amyloliquefaciens* (GRB 35)	IAA, Ammonia, HCN Siderophore production, mineral solubilization	[61]
2	Turmeric	*Bacillus* spCL3, *Burkholderia thailandensis* CL4, *Bacillus cereus* CL7	IAA, NH_3 siderophore production and phosphate solubilization	[62]
3	Turmeric	*B. cereus*	Antagonism against *Pythium aphanidermatum*	[63]
4	Pepper	*Rhizobium* sp.	Siderophore production	[51]
5	Cumin	*Azospirillum*, *Bacillus megaterium*, *Pseudomonas fluorescence*	GA production, growth promotion	[64]
6	Tea	*Brevibacillus agri* TTD5, *B. megaterium* NT5, *Sporosarcina koreensis* BT22, *Aneurinibacillus aneurinilyticus* TTD21	PGP traits and antifungal activity, Antifungal activity against *G. cingulata*, *N. sphaerica*, *R. solani*, *P. theae*, and *F. oxysporum*	[65]
7	Tea, coffee and coconut	*Azotobacter* sp.	N_2 fixation	[51]

(*Continued*)

Table 10.1 Plant growth–promoting rhizobacteria and their mechanisms of action on horticultural crops (last 3 years). (*Continued*)

Sl. no.	Name of the crop	Organism involved	Mode of action	Reference
8	Coconut	*Pseudomonas putida* KnSF208, *Bacillus licheniformis* RSB14, *Bacillus megaterium* TEB2, *Bacillus megaterium* TSB16	PGP traits	[66]
9	Cocoa	*Bacillus cereus* ASB3, *Bacillus subtilis* VEB4, *Bacillus licheniformis* KGEB16, *Pseudomonas putida* KDSF23	PGP traits	[66]
10	Vetiver	*Bacillus* sp.	Growth promotion	[67]
11	Datura	*Pseudomonas plecoglossicida*, *Lysinibacillus fusiformis*, *Bacillus* sp.	Growth promotion	[68]
3. Fruit crops				
1	Apple	*Bacillus* T8, *Pseudomonas* BA-8	PGP traits	[69]
2	Grape	*Delfia tsuruhatensis*	PGP traits	[70]
3	Citrus	*Bacillus* sp., actinobacteria, and lactic acid bacteria	N_2 fixation, IAA production, P-solubilization,	[71]
4	Banana	*P. fluorescens* PS006	bacteriocins, 1-aminocyclopropane-1-carboxylic acid (ACC) deaminase	[72]

(*Continued*)

Table 10.1 Plant growth–promoting rhizobacteria and their mechanisms of action on horticultural crops (last 3 years). (*Continued*)

Sl. no.	Name of the crop	Organism involved	Mode of action	Reference
5	Banana	*Bacillus amyloliquefaciens* NJN-6	*Against Fusarium oxysporum* f. sp. *cubense*, biofilm formation	[73]
6	Banana	*Bacillus* sp. and *Pseudomonas* sp.	Nitrogenase and phosphate solubilisation	[74]
7	Banana	*Bacillus* sp.	IAA production, GA production, phosphate solubilization, siderophore production, and HCN production	[75]
8	Cherry	*Bacillus mycoides*	Growth promotion	[76]
9	Sapota	*Azotobacter* sp.	Growth promotion	[77]
10	Litchi	*P. jessenni, P. synxantha*	Growth promotion	[78]
11	Litchi	*Bacillus licheniformis* CKAI	Growth promotion	[79]
12	Strawberry	*Phyllobacterium* sp.	Siderophore production	[80]

10.4 Conclusion and Future Perspectives

It is a widely acknowledged information that certain rhizosphere associated bacteria, known to as PGPR, stimulate plant development and health. Versatile role of PGPR is now being extensively implemented for reducing abiotic stress and for resilient natural soil against variety of harmful heavy metals. Moreover, PGPR produce secondary metabolites that may stimulate plant resistance against attack of pathogen. Since, PGPR is likely associated to change in local and systemic physiology in plants to help defense under adverse environmental conditions and is also a vital part of the plant living ecosystem. Future investigation in biology of rhizosphere will depend on the progress of molecular methodologies to improve our understanding of rhizosphere and to accomplish a combined management of soil microbial dynamics. New alternatives should be examined for the use of bioinoculants horticultural such as fruits, vegetables, and flowers.

The need of today's world is improved production of the plant as well as health of soil to get in an eco-friendly manner. Hence, the study has to be aimed on the new concept of rhizosphere engineering based on positively partitioning of the exotic biological molecules, which create a distinctive setting for the communication between plant and microorganisms [46]. Rhizosphere engineering also can involve the choice by crops of favorable microbial populations. In case, some plant species choose for and maintain populations of antibiotic-producing strains that play a main role in soils naturally suppressive to soil-borne fungal pathogens. Rhizobacteria have been engineered to interfere with the production of stress-induced hormones such as ethylene, which retards root growth and lytic enzymes, and to produce antibiotics and active against soil borne pathogens. Novel molecular tools will keep on to provide a more absolute knowledge of the complex biochemical interactions that take place in the rhizosphere, ensuring that strategies to engineer the rhizosphere are safe, supportive to productivity, and significantly improve the sustainability of farming systems.

References

1. Schlaeppi, K. and Bulgarelli, D., The plant microbiome at work. *Mol. Plant Microbe Interact.*, 28, 212, 2015.
2. Kang, B.G., Kim, W.T., Yun, H.S., Chang, S.C., Use of plant growth-promoting rhizobacteria to control stress responses of plant roots. *Plant Biotechnol. Rep.*, 4, 179, 2010.

3. Kundan, R., Pant, G., Jado, N., Agrawal, P.K., Plant growth promoting rhizobacteria: mechanism and current prospective. *J. Fertilizers Pesticides.*, 6, 2, 2015.
4. Lynch, J.M. and Whipps, J.M., Substrate flow in the rhizosphere. *Plant Soil*, 129, 1, 1990.
5. Brimecombe, M.J., De Leij., F.A., Lynch, J.M., The effects of root exudates on rhizospheremicrobial populations, in: *The rhizosphere: Biochemistry and organic substances at the soil-plant interface*, R.Z. Pinton, P. Varanini, Nannipieri (Eds.), pp. 95–137, CRC Press, Boca Raton, 2001.
6. Bhardwaj, D., Ansari, M.W., Sahoo, R.K., Tuteja, N., Biofertilizers function as key player in sustainable agriculture by improving soil fertility, plant tolerance and crop productivity. *Microbial. Cell Fact.*, 13, 1, 2014.
7. Gabriela, F., Casanovas, E.M., Quillehauquy, V., Yommi, A.K., Goni, M.G., Roura, S.I., Barassi, C.A., *Azospirillum* inoculation effects on growth, product quality and storage life of lettuce plants grown under salt stress. *Scientia Horticul.*, 195, 154, 2015.
8. Shridhar, B.S., Review: nitrogen fixing microorganisms. *Int. J. Microbial. Res.*, 3, 46, 2012.
9. Ahemad, M. and Kibret, M., Mechanisms and applications of plant growth promoting rhizobacteria: current perspective. *J. King Saud Univ. Sci.*, 26, 1, 2014.
10. Reed, S.C., Cleveland, C.C., Townsend, A.R., Functional ecology of free-living nitrogen fixation: a contemporary perspective. *Annu. Rev. Ecol. Environ. Syst.*, 42, 489, 2011.
11. Sarathambal, C., Ilamurugu, K., Balachandar, D., Chinnadurai, C., Gharde., Y., Characterization and crop production efficiency of diazotrophic isolates from the rhizosphere of semi-arid tropical grasses of India. *Appl. Soil Ecol.*, 87, 1, 2015.
12. Damam, M., Kaloori, K., Gaddam, B., Kausar, R., Plant growth promoting substances (phytohormones) produced by rhizobacterial strains isolated from the rhizosphere of medicinal plants. *Int. J. Pharm. Sci. Rev.*, 37, 130, 2016.
13. Harari, A., Kigel, J., Okon, Y., Involvement of IAA in the interaction between *Azospirillum brasilense* and *Panicum miliaceum* roots. *Plant Soil*, 110, 275, 1988.
14. Sureshbabu, K., Amaresan, N., Kumar, K., Amazing multiple function properties of plant growth promoting rhizobacteria in the rhizosphere soil. *Int. J. Curr. Microbiol. Appl. Sci.*, 5, 661, 2016.
15. Prathap, M. and Ranjitha, K.B.D., A Critical review on plant growth promoting rhizobacteria. *J. Plant Pathol. Microbiol.*, 6, 1, 2015.
16. Jackson, M.B., *Ethylene in root growth and development, in: The Plant Hormone Ethylene*, A.K. Matoo and J.C. Suttle (Eds.), pp. 159–181, CRC Press, Boca Raton, 1991.
17. Honma, M. and Shimomura, T., Metabolism of 1-aminocyclopropane-1-carboxylic acid. *Agrl. Biol. Chem.*, 42, 1825, 1971.

18. Glick, B.R., Penrose, D.M., Li., J., A model for the lowering of plant ethylene concentrations by plant growth-promoting bacteria. *J. Theor. Biol.*, 190, 63, 1998.
19. Belimov, A.A., Kunakova, A.M., Safronova, V.I., Employment of rhizobacteria for the inoculation of barley plants cultivated in soil contaminated with lead and cadmium. *Microbiol.*, 73, 99, 2004.
20. Sarathambal, C., Khankhane, P.J., Gharde, Y., Kumar, B., Varun, M., Arun, S., The effect of Plant growth promoting rhizobacteria on the growth, physiology, and Cd uptake of *Arundo donax* L. *Int. J. Phytoremediation.*, 19, 360, 2017.
21. Meena, M.K., Gupta, S., Datta, S., Antifungal Potential of PGPR, their growth promoting activity on seed germination and seedling growth of winter wheat and genetic variability among bacterial isolates. *Int. J. Curr. Microbial. Appl. Sci.*, 5, 235, 2016.
22. Wang, X., Mavrodi, D.V., Ke, L., Mavrodi, O.V., Yang, M., Thomashow, L.S., Zheng, N., Weller, D.M., Zhang, J., Biocontrol and plant growth-promoting activity of rhizobacteria from Chinese fields with contaminated soils. *Microbial. Biotechnol.*, 8, 404, 2015.
23. Malviya, J. and Singh, K., Characterization of novel plant growth promoting and biocontrol strains of fluorescent Pseudomonads for crop. *Int. J. Med. Res.*, 1, 235, 2012.
24. Umadevi, P., Anandaraj, M., Sailas, B., Endophytic interactions of *Trichoderma harzianum* in a tropical perennial rhizo-ecosystem. *Res. J. Biotechnol.*, 12, 22, 2017.
25. Vance, C.P., Symbiotic nitrogen fixation and phosphorus acquisition. Plant nutrition in a world of declining renewable sources. *Plant Physiol.*, 127, 390, 2001.
26. Venkateswarlu, B., Balloli, S.S., Ramakrishna, Y.S. (Eds.), *Organic farming in rainfed Agriculture*, Central research institute for dry land agriculture, Hyderabad, 2007.
27. Richa, S., Subhash, C., Singh, A., Isolation of microorganism from soil contaminated with degraded paper in Jharna village. *J. Soil Sci. Environ. Manag.*, 4, 23, 2013.
28. Sarathambal, C. and Ilamurugu, K., Phosphate solubilising diazotrophic bacteria associated with rhizosphere of weedy grasses in India. *Ind. J. Weed Sci.*, 46, 364, 2014.
29. Verma, S.C., Jagdish, K., Ladha, B., Anil, K., Tripathi, A.K., Evaluation of plant growth promoting and colonization ability of endophytic diazotrophs from deep water rice. *J. Bacteriol.*, 91, 27, 2001.
30. Loganathan, P. and Nair, S., *Swaminathania salitolerans* gen. nov. sp. nov., a salt-tolerant nitrogen-fixing and phosphate solubilising bacterium from wild rice (*Porteresia coarctata*). *Int. J. Syst. Evol. Microbiol.*, 54, 1185, 2004.
31. Alexander, M., *Introduction to Soil Microbiology*, 2nd ed, pp. 338–339, Wiley Eastern Limited, New Delhi, 1977.

32. Alloway, B.J., *Zinc – The vital micronutrient for healthy, high-value crops*, pp. 1–135, International Zinc Association, Brussels, Belgium, 2001.
33. Dinesh, R., Srinivasan, V., Hamza, S., Sarathambal, C., Anke Gowda, S.J., Ganeshamurthy, A.N., Gupta, S.B., Aparna Nair, V., Subila, K.P., Lijina, A., Divya, V.C., Isolation and characterization of potential Zn solubilizing bacteria from soil and its effects on soil Zn release rates, soil available Zn and plant Zn content. *Geoderma*, 321, 173, 2018.
34. Sarathambal, C., Thangaraju, M., Paulraj, C., Gomathy, M., Assessing the zinc solubilization ability of *Gluconacetobacter diazotrophicus* in maize rhizosphere using labelled (65)Zn compounds. *Ind. J. Microbiol.*, 50, 103, 2010.
35. Saravanan, V.S., Madhaiyan, M., Osborne, J., Thangaraju, M., Sa, T.M., Ecological occurrence of *Gluconacetobacter diazotrophicus* and nitrogen-fixing Acetobacteraceae members: their possible role in plant growth promotion. *Microb. Ecol.*, 55, 130, 2008.
36. Parmar, P. and Sindhu, S.S., Potassium solubilisation by Rhizosphere Bacteria: influence of nutritional and environmental conditions. *J. Microbial. Res.*, 3, 25, 2013.
37. Sheng, X.F., Zhao, F., He, L.Y., Qiu, G., Chen, L., Isolation and characterization of silicate mineral-solubilizing *Bacillus globisporus* Q12 from the surfaces of weathered feldspar. *Can. J. Microbiol.*, 54, 1064, 2008.
38. Rathore, P., A review on approaches to develop plant growth promoting rhizobacteria. *Int. J. Recent Sci. Res.*, 5, 403, 2015.
39. Saha, M., Sarkar, S., Sarkar, B., Sharma, B.K., Bhattacharjee, S., Tribedi, P., Microbial siderophores and their potential applications: a review. *Environ. Sci. Pollut. Res.*, 23, 3984, 2016.
40. Tian, F., Ding, Y., Zhu, H., Yao, L., Du, B., Genetic diversity of siderophore-producing bacteria of tobacco rhizosphere. *Braz. J. Microbiol.*, 40, 276, 2009.
41. Bar-Ness, E., Hadar, Y., Chen, Y., Shanzer, A., Libman, J., Iron uptake by plants from microbial siderophores. *Plant Physiol.*, 99, 1329, 1992.
42. Aly, H., Kamalay, J., Walter, N., Okubara, P.A., Taylor, C.G., Characterization of the *Pseudomonas* genus of bacteria for plant parasitic nematode control. *ASM Conference on Pseudomonas*, seattle Washington, 2007.
43. Duffy, B., Keel, C., Défago, G., Potential role of pathogen signalling in multitrophic plant-microbe interactions involved in disease protection. *Appl. Environ. Microbiol.*, 70, 1836, 2004.
44. Ramette, A., Frapolli, M., Défago, G., Moënne-Loccoz, Y., Phylogeny of HCN synthase encoding hcnBC genes in biocontrol Fluorescent Pseudomonads and its relationship with host plant species and HCN synthesis ability. *Mol. Plant Microbe In.*, 16, 525, 2003.
45. Chakraborty, U., Chakraborty, B., Basnet, M., Plant growth promotion and induction of resistance in *Camellia sinensis* by *Bacillus megaterium*. *J. Bas. Microbiol.*, 46, 186, 2006.

46. Ryan, P.R., Yves, D., Linda, S., Thomashow, David, M., Weller Rhizosphere engineering and management for sustainable agriculture. *Plant Soil,* 321, 363, 2009.
47. Bano, A. and Muqarab, R., Plant defence induced by PGPR against *Spodoptera litura* intomato (*Solanum lycopersicum* L.). *Plant Biol.*, 19, 406, 2017.
48. Zameer, M., Zahid, H., Bushra, T., Qurban, A., Idrees, A.N., Saleem, M., Shahid, J., PGPR Potentially Improve Growth of Tomato Plants in Salt-Stressed Environment. *Tur. J. Agri. Food Sci. Technol.*, 4, 455, 2016.
49. Mangmang, J.S., Rosalind, D., Rogers, G., Optimal plant growth-promoting concentration of *Azospirillum brasilense* inoculated to cucumber, lettuce and tomato seeds varies between bacterial strains. *Isr. J. Plant Sci.*, 62, 145, 2015.
50. Ameen, A., François, K., Philippe, J., Behavior of plant growth promoting bacteria in tomato rhizosphere: Colonization of rhizosphere as influenced by PGPR and root exudates composition. *World Congress on Beneficial Microbes: Food, Pharma, Aqua & Beverages Industry, Ferment. Technol,* vol. 3, 2015.
51. Vejan, P., Abdullah, R., Tumirah, K., Ismail, S., Amru, N.B., Role of Plant Growth Promoting Rhizobacteria in Agricultural Sustainability—A Review. *Molecules,* 21, 573, 2016.
52. Islam, S., Akanda, A.M., Prova, A., Islam, M.T., Hossain, M.M., Isolation and identification of plant growth promoting rhizobacteria from cucumber rhizosphere and their effect on plant growth promotion and disease suppression. *Front. Microbiol.*, 6, 1360, 2016.
53. Imran, A., Mirza, M.S., Shah, T.M., Malik, K.A., Hafeez, F.Y., Differential response of kabuli and desi chickpea genotypes toward inoculation with PGPR in different soils. *Front. Microbiol.*, 6, 859, 2015.
54. Habib, S.H., Kausar, H., Saud, H.M., Plant growth-promoting rhizobacteria enhance salinity stress tolerance in okra through ROS-scavenging enzymes. *Biomed. Res. Int.*, 1, 1–10, 2016.
55. Kesaulya, H., Baharuddin, A., Zakaria, B., Syatrianty, A.S., Isolation and Physiological Characterization of PGPR from Potato Plant Rhizosphere in Medium Land of Buru Island. *Procedia Food Sci.*, 3, 190, 2015.
56. Natthiya, B., Kanjana, T., Mathukorn, S., Dusit, A., Sutruedee, P., The FTIR spectroscopy investigation of the cellular components of cassava after sensitization with plant growth promoting rhizobacteria, *Bacillus subtilis* CaSUT007. *Afr. J. Microbiol. Res.*, 6, 603, 2016.
57. Naqqash, T., Hameed, S., Imran, A., Hanif, M.K., Majeed, A., van Elsas, J.D., Differential Response of potato toward inoculation with taxonomically diverse Plant Growth Promoting Rhizobacteria. *Front. Plant Sci.*, 7, 144, 2016.
58. Datta, M., Dipak, P., Sankar, N.S., Sengupta, C., Plant Growth Promoting Rhizobacteria Improve the Production and Enhancement of Alkaloid Content in Chilli. *Front. Environ. Microbiol.*, 1, 24, 2015.
59. Sarbadhikary, S.B. and Mandal, N.C., Field application of two plant growth promoting rhizobacteria with potent antifungal properties. *Rhizosphere,* 3, 170, 2017.

60. Kalita, M., Moonmee, B., Dey, T., Gogoi, K., Pallavi, D., Unni, B.G., Ozah, D., Indira, S., Developing novel bacterial based bioformulation having PGPR properties for enhanced production of agricultural crops. *Indian J. Exp. Biol.*, 53, 56, 2015.
61. Dinesh, R., Anandaraj, M., Kumar, A., Bini, Y.K., Subila, K.P., Aravind, R., Isolation, characterization, and evaluation of multi-trait plant growth promoting rhizobacteria for their growth promoting and disease suppressing effects on ginger. *Microbiol. Res.*, 173, 34, 2015.
62. Kumar, A., Singh, M., Singh, P.P., Singh, S.K., Singh, P.K., Pandey, K.D., Isolation of plant growth promoting rhizobacteria and their impact on growth and curcumin content in *Curcuma longa* L. *Biocatal. Agric. Biotechnol.*, 8, 1, 2016.
63. Vinayarani, G. and Prakash, H.S., Growth Promoting Rhizospheric and Endophytic Bacteria from *Curcuma longa* L. as Biocontrol Agents against Rhizome Rot and Leaf Blight Diseases. *Plant Pathol. J.*, 34, 218, 2018.
64. Pavankumar, D.S., Maruthi Prasad, B.N., Umesha, K., Shivanna, M., Shankarappa, T.H., Halesh, G.K., Influence of plant growth promoting rhizobacteria and plant growth regulators on growth and yield of black cumin (*Nigella sativa* L.) VAR. NS-44. *J. Pharmacog. Phytochem.*, 7, 1–4, 2018.
65. Dutta, J. and Thakur, D., Evaluation of multifarious plant growth promoting traits, antagonistic potential and phylogenetic affiliation of rhizobacteria associated with commercial tea plants grown in Darjeeling, India. *PLoS ONE*, 12, e0182302, 2017.
66. Rajeela, K.T.H., Gupta, A., Gopal, M., Hegde, V., Thomas, G.V., Evaluation of Combinatorial capacity of coconut and cocoa plant growth promoting rhizobacteria (PGPR) with biocontrol agent *Trichoderma harzianum*. *Curr. Inves. Agri. Curr. Res.*, 3, 404–408, 2018.
67. Noorasyikin, M.N. and Zainab, M., PGPR by bacillus in Vetiver Grass's Root enhance strength of soil-root matrix system toward slope stabilization in Malaysia. *J Eng Appl. Sci.*, 12, 3051, 2017.
68. Rahmoune, B., Abdelkader, M., Madjda, K.S., Lakhdar, K., Andrew, S., Alexander, E., Joachim, K., Jürgen, P.J.T., van Dongen, Isolation and characterization of three new PGPR and their effects on the growth of *Arabidopsis* and *Datura* plants. *J. Plant Interactns.*, 12, 526–532, 2017.
69. Nazmi, C.S. and Pırlak, L., Effect of Plant Growth Promoting Rhizobacteria (PGPR) on Branching and Growing of Apple Sapling. *Erwerbs-Obstbau*, 59, 309, 2017.
70. Rolli, E., Ramona, M., Stefano, S., Erika, C., Francesca, M., Ameur, C., Sara, B., Leonardo, V., Claudia, S., Daniele, D., Root-associated bacteria promote grapevine growth: from the laboratory to the field. *Plant Soil*, 410, 369, 2017.
71. Giassi, V., Kiritani, C., Kupper, C., Bacteria as growth-promoting agents for citrus rootstocks. *Microbiol. Res.*, 190, 46, 2016.
72. Gamez, R.M., Rodríguez, F., Ramírez, S., Gómez, Y., Agarwala, R., Landsman, D., Mariño-Ramírez, L., Genome sequence of the banana plant

growth-promoting rhizobacterium *Pseudomonas fluorescens* PS006. *Genome Announc*, 4, e00329, 2016.

73. Yuan, J., Zhang, N., Qiwei, H., Raza, W., Li, R., Jorge, M.V., Shen, Q., Organic acids from root exudates of banana help root colonization of PGPR strain *Bacillus amyloliquefaciens* NJN-6. *Sci. Rep.*, 5, 13438, 2015.
74. Gechemba, O.R., Mugweru, J., Huxley, M.M., Nancy, L.M.B., Viviene, N.M., Potentially beneficial rhizobacteria associated with banana plants in Juja Kenya. *Res., Rev. J. Microbiol. Biotechnol.*, 19, 181–188, 2016.
75. Apastambh, A.R., Tanveer, K., Baig, M.M.V., Isolation and Characterization of Plant Growth Promoting Rhizobacteria from Banana Rhizosphere. *Int. J. Curr. Microbiol. App. Sci.*, 5, 59, 2016.
76. Arikan, S. and Pirlak., L., Effects of plant growth promoting rhizobacteria (PGPR) on growth, yield and fruit quality of Sour Cherry (*Prunus cerasus* L.). *Erwerbs-Obstbau*, 58, 1–6, 2016.
77. Prasad, H., Paramjeet, S., Meena, K., Solanki, S.P.S., Effect of Organic Manures and Biofertilizer on Plant Growth, Yield and Quality of Horticultural Crop: A Review. *Int. J. Chem. Stu.*, 5, 217, 2017.
78. Visen, A., Singh, P.N., Bohra, M., Narayan, A., Singh, S.K., Sharma, A.K., Enhancement of rooting, growth and survivability of litchi (*Litchi chinenesis*) air layers with the inoculation of Arbuscular mycorrhizal fungi (AMF) and plant growth promoting rhizobacteria. *Environ. Ecol.*, 34, 1162, 2016.
79. Chawla, W. and Mehta, K., Effect of different growing media on survival and growth of transplanted litchi layers. *Asian J. Hort.*, 10, 257, 2015.
80. Flores-Félix, J.D., Silva, L.R., Rivera, L.P., Marcos-García, M., García-Fraile, P., Martínez-Molina, E., Mateos, P.F., Velázquez, E., Andrade, P., Rivas, R., Plants Probiotics as a Tool to Produce Highly Functional Fruits: The Case of *Phyllobacterium* and Vitamin C in Strawberries. *PLoS One*, 10, e0122281, 2015.

11

Rhizosphere Microbiome: The Next-Generation Crop Improvement Strategy

M. Anandaraj[1], S. Manivannan[2]* and P. Umadevi[3]

[1]*ICAR-Indian Institute of Spices Research, Kozhikode, Kerala, India*
[2]*Department of Horticulture, Central University of Tamil Nadu, Neelakudi, Thiruvarur, India*
[3]*Rice Breeding and Genetics Research Center, ICAR–Indian Agricultural Research Institute, Aduthurai, India*

Abstract

The microbial communities associated with plants have received much importance in recent past because of their capacity to enhance the crop production. All tissues in the plants have microbes, benefiting the host to avoid adverse environmental stresses. However, very meager research has been conducted to correlate its potential as one of the crop improvement strategies. Research signifies the role of host species specifically the host genotype in driving the composition of microbial community, activity, and selection of partners. Studies have also found out the quantitative trait loci (QTL) governing the responses of plant to beneficial rhizosphere microbes. The understanding of new generation omics strategies, *viz.*, metagenomics, meta transcriptomics, and metaproteomics to screen for active microbial community of the rhizosphere, metabolomics, and proteomics to elucidate the beneficial microbe action on crop plant, is a prerequisite. The knowledge on the community pattern and its effect on crop yield and quality should support the improvement programs to exploit the beneficial indigenous microbial community and also the probiotically added biocontrol agents. We focus the technological platforms for studying rhizosphere microbiome and the major research findings of translational research toward crop improvement in this chapter. This chapter

**Corresponding author*: manivannan@cutn.ac.in

Bibhuti Bhusan Mishra, Suraja Kumar Nayak, Swati Mohapatra, and Deviprasad Samantaray (eds.) Environmental and Agricultural Microbiology: Applications for Sustainability, (243–256) © 2021 Scrivener Publishing LLC

encompasses several disciplines like horticulture, plant biotechnology, genetics, plant nutrition, and soil microbiology.

Keywords: Rhizosphere microbiome, omics platforms, crop improvement

11.1 Introduction

The term "rhizosphere" was coined by Lorenz Hiltner in the year 1904; his idea on rhizosphere was on plant nutrition is considerably influenced by the microbial composition of the rhizosphere. He envisioned that, in the rhizosphere, not only beneficial microbes are attracted by the root exudates but also the pathogens which are able to adjust to the specific root exudates. He hypothesized that the composition of rhizosphere microflora is the major factor for the resistance of plants toward pathogenesis. He visualized that the production of quality plant products is also dependent on the specific composition of the root microflora [1]. Based on this, a new school of thought on "Rhizosphere engineering" emerged and several conferences were held [2].

The plant microbiome also called as second genome of the plants denotes the entire microbial community the plant harbors [3]. The rhizosphere microbial diversity was suggested as bioindicators for plant productivity when the plants are grown under different environmental conditions [4]. The term rhizosphere microbiome denotes the collective microbial community present in the soil adhering to the rhizosphere of the plant. The rhizosphere microbial community is said to harbor a multitude of diverse microbes including beneficial, neutral, and pathogenic-microbes. The beneficial microbes are the main players in plant growth disease suppression and abiotic stress tolerance. Though the potential of the rhizosphere microbiome toward the plant fitness is greatly recognized, the information recorded for majority of rhizosphere microorganisms is meager. To enhance the plant fitness to various biotic/abiotic stresses, it is important to understand the nature of microorganism present in the rhizosphere microbiome and the mode of action toward the plant fitness. The mode of action of these microbes at the rhizosphere includes phytohormone production [5], nutrient mobilization to plants [6], increased tolerance to abiotic stresses [7], induction of the plant immune response system [8], manipulation of plant functional traits [9], modification of tissue ionome [10], and selective recruitment of beneficial microbes [11].

11.2 Rhizosphere Engineering

Rhizosphere engineering is nothing but reshaping the soil rhizosphere microbial community for sustainable productivity [12]. This could be done in two ways: one toward breeding of crops for better beneficial microbe association at rhizosphere and the other is toward probiotic application of beneficial microbes in soil. The most direct way to alter the microbiome is the probiotic application. Reshaping the plant rhizosphere microbiome by probiotic application of beneficial microorganisms to protect from pathogen infections is comparable with the application of probiotics in humans. Products with one or many species of bacteria or fungi are commercially available [13] for major crops. Rhizosphere engineering has also been demonstrated successfully in export oriented low volume high value spice crops. In India, seed coating [14] and biocapsules [15] for delivering efficient strains of bioagents were developed for spices and patent application was filed. These technologies have been commercialized and available to the farmers. Biocapsule technology is one of the 42 technologies of Indian Council of Agriculture (ICAR) exhibited by the entrepreneurs in the Rashtrapathi Bhavan and visited by Hon'ble President of India. In this technology, the carrier material is reduced by 100 times as 1-k talc formulations can be delivered in the form of 10 capsules with longer shelf life. As the microbes are isolated under traditional culturing conditions, the success depends on its adaptation to the changing soil chemical environment, *viz.*, pH, nutrient dynamics, and texture. There is a need for the inoculant to integrate themselves with the existing native microbiome so as to integrate in the food web. Some inoculants fail in the long-term persistence in soil as they are easily consumed by predators or outcompeted for resources by native microbes. The most effective organism forms association with the beneficial organism in a way of both population and functional abundance which emulates selective recruitment and strong network of native microbes in the soil [11]. Better inoculants can also be made with targeted functions through new gene editing tools [15], satisfying the consumer attitudes toward gene modifications. Rapid colonization of roots by a host-adapted microbiome may prevent pathogen establishment [17]. Hence, populations of beneficial microbes can be augmented by external application to boost flagging populations.

The second way to engineer the rhizosphere is achieved by breeding or genetically engineering the plant traits. Using several inbred lines of tomato, three quantitative trait loci (QTL) were identified which were associated with disease suppression by a strain of *B. cereus*. Phenotypic variation of resistance to *Pythium torulosum* among recombinant inbred lines of tomato was found to be associated with the intensity of growth of *B. cereus* on the seeds [18]. These results indicated that the presence of particular gene or loci in host plant and its possible utilization in enhancing the beneficial associations among the plants and rhizosphere soil microorganisms. The transgenic plants that were designed to secrete specific signal molecules showed that plants communicate with microorganisms in the rhizosphere in a more specific and defined manner [19]. The root exudates and root architecture are the important factors which determine the rhizosphere microbiome and dynamics; the engineering of plant using Clustered Regularly Interspaced Short Palindromic Repeats (CRISPR) and other gene editing tools will also be a promising option in the future [20].

11.3 Omics Tools to Study Rhizosphere Metagenome

The beneficial microbes in the rhizosphere play a prominent role in plant growth and disease suppression reactions. But the nature of microorganisms, community abundance and mechanisms toward increasing the plant fitness are less known. Therefore, development and application of meta-omics tools will bring out enormous understanding on this association. The knowledge obtained from these platforms will pave the way to translational research toward sustainable agriculture.

11.3.1 Metagenomics

Recent high-throughput sequencing methods, *viz.*, Roche 454, Illumina HiSeq, SOLid, Ion Torrent, and PacBio RSII, enables identification, relative quantification, and population as well as functional dynamics of microbial community in the soil sample and thereby provides information on community ecology. But the attempts on soil rhizosphere whole-genome metagenomics are very limited. Novel integrated bioinformatics platforms such as MEtagenome Analyzer (MEGAN) and *Metagenomic* Rapid Annotations using Subsystems Technology (MG-RAST) offer ways to find the community level taxomic affiliation, functional enrichment, and the interaction network. MG-RAST [21] is one of the prominent platforms which supports deposition and analysis of metagenomic datasets.

About 33 metagenome datasets derived from different soil sites, *viz.*, forest, desert, grass land, Arctic, and mangrove sediment through whole-genome metagenome shotgun sequencing using Roche 454 and Illumina platforms [22]. With the use of integrated bioinformatics tools, the phylogeny and functional characteristic of the microbial population were analyzed. Along with the profiling of microbial community from each soil type, an array of metagenomics biomarkers with 46 taxa and 33 metabolic modules were derived as indicators for differentiating the soil communities.

The comparative metagenomics was employed to compare the Loktak (the largest freshwater lakes of India) soil metagenomic data with available metagenomes of other four aquatic habitats (from pristine to highly polluted eutrophic habitats). The microbes, *viz.*, *Bradyrhizobium*, *Candidatus koribacter*, *Candidatus solibacter*, *Pedosphaera*, *Anaeromyxobacter*, *Sorangium*, *Opitutus*, and *Acidobacterium* genera, were selectively dominant in fresh water and this selective microbial enrichment was found to be the major phenomenon of bioremediation at Loktak Lake [23]. The diversity of bacteria and the dynamics in population under ambient CO_2 (a-CO_2) and elevated CO_2 + temperature (e-CO_2T) in low land rice rhizosphere was studied using whole-genome metagenomics approach [24]. The dominant bacterial communities were found to be *Firmicutes*, *Proteobacteria*, *Actinobacteria*, *Acidobacteria*, and *Planctomycetes*. Whereas the genera that are related to methane production, *viz.*, *Methanosphaera*, *Methanobacterium*, *Methanothermococcus*, and *Methanothermus*, were not present in a-CO_2. The enzymes involved in methanotrophy and acetoclastic methanogenesis pathways were with abundant reads in e-CO_2T compared to CO_2. The rich bacterial diversity and abundances of C and N decomposing bacteria in the rhizosphere were recorded under e-CO_2T that further suggested the possible exploration microbes for in nutrient cycling, environment management, and sustainable agriculture.

The gray mangroves rhizosphere microbiome metagenomics in the Red Sea using 454 GS FLX Titanium technology revealed the dominance by Proteobacteria, Bacteroidetes, and Firmicutes, with specific high abundance of methanogens and sulfate reducers, though many other specific groups were found to be enriched in the rhizosphere compared to the bulk soil. MG-RAST functional analysis observed the enrichment of aromatic compounds and potassium metabolisms along with the enrichment of pathway that utilize osmolytes [25].

Ascomycota was demonstrated to be the dominant phylum (76%–85%) in fungi in the rhizosphere metagenome of gray mangroves of the Red Sea in another study [26]. They also detected several commercially used fungi, *viz.*, secreted cellulases and cellulosome producers in the datasets using MG-RAST platform.

The population and functional dynamics of fungal and bacterial communities present in soil was characterized using metagenomic approach from vineyards in Central Chile and the native forest soil prior to the establishment of the vineyard as comparative datasets [27]. Analysis showed bacteria with high abundance than fungi in both habitats. Functional diversity on reads corresponding to genes coding for metabolism of amino acids, nucleotides, fatty acids, and secondary metabolism was enriched in forest soils, while potassium metabolism related genes was enriched vineyard soils.

In black pepper, rhizosphere microbiome was elucidated by using metagenomic tools and the results showed the "trichorhizosphere" with the presence of differential microbial communities recruited by the probiotic addition of *Trichoderma harzianum* [11]. The illumina hiseq sequenced soil metagenome, when analyzed with double approach, *viz.*, stand-alone and MG-RAST, yielded similar results for taxonomy and functional abundance. Statistical analysis of metagenomic profiles (STAMP) showed statistically higher proportion of *Acidobacteriaceae bacterium*, *Candidatus koribacter versatilis* in *Trichoderma* inoculated sample, uncultured bacteria in control and *Fusarium oxysporum*, *Talaromyces stipitatus*, *Pestalotiopsis fici* in *Trichoderma* inoculated sample, *Rhizophagus irregularis*, *Pseudogymnoasus pannarum* (Human pathogenic fungi), *Oidiodendran* in control sample, respectively. The relative abundance for the specific functional features showed the high abundance of heme and hemin uptake, iron acquisition, metabolism of aromatic compounds in *Trichoderma*-treated soil metagenome and with the less abundanceon phages, and pathogenicity islands and prophages than untreated soil (control).

11.3.2 Metaproteomics

Though metagenomics brings knowledge on the microbial diversity, the DNA abundance correlates sparsely with protein abundance to know exactly on soil microbial functions. To overcome this, the metaproteomics approach comes with entire protein profile from environmental samples which aid our understanding on soil microbial activity and to obtain a deeper understanding of root and root associated microbial interactions. It may be defined as the characterization of the expressed proteins by a microbial community in the given environmental sample. This platform works on the on liquid chromatography–mass spectrometry (LC-MS/MS)–based analysis of samples and conversion of the raw spectral data into peptide sequence followed by protein identification using databases. The progress of MS-based techniques and bioinformatics tools is making the metagenomic platform as a successful strategy to understand the rhizosphere soil microbial activity but the need for developing suitable protein extraction methods [28, 29] is the important research area to work on. This is particularly due to the sample complexity (rhizosphere samples with root exudates), high level of organism diversity (archea, Eukryotes), and wide range of protein abundance levels. Sometimes, the integration of data obtained from different extraction methods is needed to achieve significant coverage of total proteins in the soil sample.

Various studies have demonstrated the proteins expressed by the specific interaction between plant and soil microbes at the rhizosphere. The comparison between the profiles of metaproteomics of rhizosphere of plant sugarcane and ratoon sugarcane revealed that induced catabolic diversity, the expression of soil proteins derived from the microbes, plants, and fauna by the ratoon sugarcane. Among the soil proteins, majority (24.77%) were expressed from bacteria. The up-regulated l proteins of microbial origin were found to be signal transduction and membrane transport proteins [30]. The changes on soil protein abundance upon continuous monoculture of *Rehmanniaglutinosa* were determined using metaproteomics platform [31]. The lettuce (*Lactuca sativa*) rhizosphere upon inoculation of microbial consortium was found to have enhanced amount of proteins for virulence stress/defense response and energy metabolism, in presence of pathogenic strain of *Fusarium oxysporum* [32].

11.3.3 Metatranscriptomics

Metagenomics provides insight into the genes present in rhizosphere environment that acts as reference material for studying microbial gene expression, whereas the exploration of microbial gene expression can be achieved by Metatranscriptomics. Metatranscriptomics provides the information on global content of gene transcripts of the microbial community in a given environmental sample [33]. The primary goal of metatranscriptomics is to find out the over- and under-represented genes in the cDNA and is the right strategy in finding out the genes responsible for the disease suppression mechanism and the plant fitness at molecular level. This approach allows the researchers to identify/discover novel genes and its role, thereby bringing the active community members in the rhizosphere soil [34]. Here again, the extraction of soil RNA is major task due to the humic compounds interference, mix up of mRNA with other RNA types, and the short half-life of RNA [35].

A rhizosphere soil comparative metatranscriptomics was done with wheat grown under disease suppressive and non-suppressive soils for the wheat pathogen *R. solani* AG8 [36]. The mRNA annotation and differential expression found the dominant taxa in the rhizosphere as *Pseudomonas* spp.

and *Arthrobacter* spp. in non-suppressive samples and *Buttiauxella* spp. and *Stenotrophomonas* spp. in the disease suppressive samples. Suppressive soil metatranscriptome profile had higher expression of polyketide cyclase and many cold shock proteins, whereas the non-suppressive soil showed higher expression of antibiotic genes involved in phenazine biosynthesis and pyrrolnitrin. The reactive oxygen species (ROS) detoxifying genes and superoxide radicals related genes were found expressed in the non-suppressive rhizosphere samples than in suppressive soils which was attributed to be as response to the *R. solani* AG8 infection of wheat roots.

The entire active microbiomes were analyzed in bulk soil and rhizospheres of wheat, pea, and oats using metatranscriptomics [37]. In general, the rhizosphere microbiomes are different from bulk soil microbiomes and it differs between the plant species. The pea (legume) plant showed much stronger effect on the rhizosphere than the cereals (wheat and oat) with significantly different rhizosphere community. Nematodes and bacterivorous protozoa were found enriched in all rhizospheres except the pea rhizosphere which was highly enriched with fungi. The rhizosphere colonization was found toward selected metabolism, including cellulose degradation (cereals), H_2 oxidation (pea), and methylotrophy (all plants).

11.3.4 Ionomics

Ionome is defined as the mineral nutrient and trace element composition of an organism, representing the inorganic component of cellular and organismal systems [38]. Along with nucleic acids, proteins, and metabolites, the mineral elements are essential as building blocks of the living cell and in almost every process of an organism. The ionomics is one of the functional genomics strategies intended for rapid identification of plant genes and gene networks involved in regulating the intake and accumulation of mineral nutrients from the soil [39]. The multivariable ionomic signatures for different physiological state in plants have been established [40]. This was achieved by studying *Arabidopsis thaliana* ionome profiling, which were growing in different media condition. These multivariable ionomic signatures are potent enough to find specific physiological, soil environment, and geographical region [41] in which the crop has been grown. Recent studies report the ionome signatures of non-model, important horticultural crops [10, 42–44] from the active metabolic part of the plant through Inductively Coupled Plasma Mass Spectroscopy (ICPMS) analysis.

The collective function of rhizosphere microbiome in making the nutrients available to the plant depends upon the plant germplasm and type of microbial population present in the soil. Ionome signature derived from Okra, Palak, and Radish was different after application of microbes in their growth media [45]. Further, same dosage of the identical microbial population might not produce similar ionome signature [46], yield, and quality [47] within the two different germplasm of same species. Further, one of the quality parameters of ginger, i.e., volatile oil content, is found to be enhanced due to the application of PGPRs. However, the change in volatile oil content was also germplasm specific [48]. Hence, the ionome profiles would serve as the markers for identification of a specific microorganism to the specific germplasm toward its fitness. This can be validated by the ionome signatures as a function of the selective microbial application in soil.

11.4 As Next-Generation Crop Improvement Strategy

The agricultural scientists have got two important crop improvement platforms in general: (1) utilizing the genetic diversity and selecting/breeding for superior plant genotypes; and (2) manipulating the cultivation conditions through cultural practices to get the targeted/desired performance/plant fitness. The first approach has yielded many deliverables. The second approach is a manipulative approach. The component in this approach which is of utmost important and efficient as next-generation improvement strategy is the rhizosphere microbiome. Growing affordable sequencing technologies and taxonomic databases now make the profiling and identification of non-model crop-associated rhizosphere microbiomes as a widely accessible research method [11]. The ionome profiles of plants are also available nowadays [41, 49] which gives enormous information for the horticulturist and plant physiologist on the plant metabolism upon probiotic application of beneficial microbes. In future, the major ionome pattern exerted by the specific organism on its addition might serve as the indicator to test its action as well as to test the effectiveness of newer organism.

To achieve the selective effects of the rhizosphere microbiome on plants identifying the variation in the root exudates and manipulation of root exudation in agricultural cultivars are important. The main bottle neck to advance this concept is the *in situ* study of root exudation. However, recent developments, *viz.*, anion exchange membranes, are used to capture exudates to compare bulk soil and rhizosphere soil accounting for compounds derived from soil [50]. Exudates from live roots can be collected [51]. The suggestive evidence on the interaction of soil microbe and the root exudates in future will open up enormous translational research in rhizosphere engineering. The host genotype plays significant role in the formation and activity of the microbial community [52] and it is heritable [53]. This concept of identifying genes that are modulated by the beneficial microbes is easily achievable as the new generation omics methods like proteomics brings information proteins that are involved in induced systemic résistance in plants during the host–beneficial microbe interaction [54]. For instance, the *Phyllobacterium brassicacearum* STM196 strain mediates NRT2.5 and NRT2.6 gene and stimulates growth and also antagonizes high nitrate inhibition of lateral root development in *Arabidopsis thaliana* [55].

11.5 Conclusion

The novel omics technological platforms are the game changers toward using the rhizosphere microbiome into translational research for next-generation crop improvement paradigm. With the effective application of these tools and strategies, it may be possible to identify efficient genotype—PGPR combination which can shape the microbial population to promote the plant fitness in wide range of soil conditions. Genes imparting beneficial interactions can be reintroduced into elite germplasm from the microbes. The rhizosphere fitness against biotic and abiotic stresses may be increased by the genome editing.

References

1. Hartmann, A., Rotballer, M., Schimid, M., Loenz Hiltner, a pioneer in rhizhosphere microbial ecology and soil bacteriology research. *Plant Soil*, 312, 7, 2008.
2. Dessaux, Y., Hinsinger, P., Lemanceau, P., Rhizosphere: so many achievements and even more challenges. *Plant Soil*, 321, 1, 2009.
3. Mendes, R., Paolina Garbeva, P., Jos, M., Raaijmakers, J.M., The rhizosphere microbiome: significance of plant beneficial, plant pathogenic, and human pathogenic microorganisms. *FEMS Microbiol Rev.*, 37, 634, 2013.
4. Wagg, C., Jansa, J., Schmid, B., van der Heijden, M.G.A., Below ground biodiversity effects of plant symbionts support aboveground productivity. *Ecol. Lett.*, 14, 1001, 2011.
5. Ping, L. and Boland, W., Signals from underground: Bacterial volatiles promote growth in Arabidopsis. *Trends Plant Sci.*, 9, 263, 2004.

6. Janos, P., Plant responsiveness to mycorrhizas differs from dependence upon mycorrhizas. *Mycorrhiza*, 17, 75, 2007.
7. Redman, R.S., Sheehan, K.B., Stout, R.G., Rodriguez, R.J., Henson, J.M., Thermotolerance conferred to plant host and fungal endophyte during mutualistic symbiosis. *Science*, 298, 1581, 2002.
8. Jain, A., Singh, S., Sarma, B.K., Singh, H.B., Microbial consortium mediated reprogramming ofdefense network in pea to enhance tolerance against *Sclerotinia cclerotiorum*. *J. Appl. Microbiol.*, 112, 537, 2011.
9. Friesen, M.L., Porter, S.S., Stark, C., von Wettberg, E.J., Sachs, J.L., Martinez-Romero, E., Microbially Mediated Plant Functional Traits. *Annu. Rev. Ecol. Evol. Syst.*, 42, 23, 2011.
10. Manivannan, S., Ionome assisted safe food production for remunerative organic horticulture, in: *Doubling Faramers Income Through Horticulture*, K.L. Chadah, S.K. Singh, P. Kalia, W.S. Dhillon, T.K. Behera, J. Prakash (Eds.), pp. 517–520, Daya Publishing House, Astral International Pvt, Ltd, New Delhi, 2017.
11. Umadevi, P., Anandaraj, M., Vivek, S., Benjamin, S., *Trichoderma harzianum* MTCC 5179 impacts the population and functional dynamics in microbial community in rhizosphere of black pepper (*Piper nigrum* L.). *Braz. J. Microbiol.*, 49, 463, 2017.
12. Ahkamia, A.H., White III, R.A., Handakumburaa, P.P., Janssona, C., Rhizosphere engineering: Enhancing sustainable plant ecosystem productivity. *Rhizosphere*, 3, 233, 2017.
13. Calvo, P., Nelson, L., Kloepper, J.W., Agricultural uses of plant biostimulants. *Plant Soil*, 383, 3, 2014.
14. Anandaraj, M., Bini, Y.K., Johny, A.K., A seed coating composition and a process for its preparation, 2013. Indian Patent 4465/CHE/2013.
15. Anandaraj, M., Bini, Y.K., Dinesh, R., A novel method of storing and delivering PGPR/ microbes through biocapsules, Indian Patent 3594/CHE/ 2013, 2013.
16. Hutchison, C.A., Chuang, R.-Y., Noskov, V.N., Assad-Garcia, N., Deerinck, T.J., Design and synthesis of a minimal bacterial genome. *Science*, 351, 6280, 2016.
17. Umadevi, P., Anandaraj, M., Benjamin, S., Endophytic interactions of *Trichoderma harzianum* in a tropical perennial rhizo-ecosystem. *Res. J. Biotechnol.*, 2, 22, 2017.
18. Smith, K.P., Handelsman, J., Goodman, R.M., Genetic basis in plants for interactions with disease-suppressive bacteria. *PNAS*, 96, 4786, 1999.
19. Oger, P.M., Mansouri, H., Nesme, X., Dessaux, Y., Engineering root exudation of lotus toward the production of two novel carbon compounds leads to the selection of distinct microbial populations in the rhizosphere. *Microb. Ecol.*, 47, 96, 2004.
20. Nogales, A., Nobre, T., Valadas, V., Ragonezi, C., Döring, M., Polidoros, A., Arnholdt-Schmitt, B., Can functional hologenomics aid tackling current challenges in plant breeding? *Brief Funct Genomics*, 15, 288, 2015.

21. Meyer, F., Parman, D., D'Souza, M., Olson, R., Glass, E.M., Kubal, M., Paczian, T., Rodriguez, A., Stevens, R., Wilke, A., Wilkening, J., Edwards, R.A., The metagenomics RAST server- a public resource for the automatic phylogenetic and functional analysis of metagenomes. *BMC Bioinformatics*, 9, 386, 2008.
22. Xu, Z., Hansen, M.A., Hansen, L.H., Jacquiod, S., Sørensen, S.J., Bioinformatic Approaches Reveal Metagenomic Characterization of Soil Microbial Community. *PLoS ONE*, 9, e93445, 2014.
23. Puranik, S., Pal, R.R., More, R.P., Purohit, H.J., Metagenomic approach to characterize soil microbial diversity of Phumdi at Loktak Lake. *Wat. Sci. Tech.*, 74, 2075, 2016.
24. Bhattacharyya, P., Roy, K.S., Das, M., Ray, S., Balachandar, D., Karthikeyan, S., Nayak, A.K., Mohapatra, T., Elucidation of rice rhizosphere metagenome in relation to methane and nitrogen metabolism under elevated carbon dioxide and temperature using whole genome metagenomic approach. *Sci. Total Environ.*, 542, 886, 2016.
25. Alzubaidy, H., Essack, M., Malas, T.B., Bokhari, A., Motwalli, O., Kamanu, F.K., Jamhor, S.A., Mokhtar, N.A., Antunes, A., Simoes, M.F., Alam, I., Bougouffa, S., Lafi, F.F., Bajic, V.B., Archer, J.A.C., Rhizosphere microbiome metagenomics of gray mangroves (*Avicennia marina*) in the Red Sea. *Gene*, 576, 626, 2016.
26. Simoes, M.F., Antunes, A., Cristiane, A., Ottoni, C.A., Amini, M.S., Alam, I., Alzubaidy, H., Mokhtar, N., John, A.C., Archer, J.A.C., Bajic, V.B., Soil and Rhizosphere Associated Fungi in Gray Mangroves (*Avicennia marina*) from the Red Sea — A Metagenomic Approach. *Genomics Proteomics Bioinf.*, 13, 310, 2015.
27. Castañeda, L.E. and Barbosa, O., Metagenomic analysis exploring taxonomic and functional diversity of soil microbial communities in Chilean vineyards and surrounding native forests. *Peer J.*, 5, e3098, 2017.
28. Keiblinger, K.M., Wilhartitz, I.C., Schneider, T., Roschitzki, B., Schmid, E., Eberl, L., Soil metaproteomics - comparative evaluation of protein extraction protocols. *Soil Biol. Biochem.*, 54, 14, 2012.
29. Bastida, F., Hernández, T., García, C., Metaproteomics of soils from semiarid environment: functional and phylogenetic information obtained with different protein extraction methods. *J Proteomics*, 101, 31, 2014.
30. Lin, W., Wu, L., Lin, S., Zhang, A., Zhou, M., Lin, R., Metaproteomic analysis of ratoon sugarcane rhizospheric soil. *BMC Microbiol.*, 13, 135, 2013.
31. Wu, L., Wang, H., Zhang, Z., Lin, R., Zhang, Z., Lin, W., Comparative metaproteomic analysis on consecutively Rehmannia glutinosa monocultured rhizosphere soil. *PLoS ONE*, 6, e20611, 2016.
32. Moretti, M., Minerdi, D., Gehrig, P., Garibaldi, A., Gullino, M.L., Riedel, K., A Bacterial-Fungal Metaproteomic Analysis Enlightens an Intriguing Multicomponent Interaction in the Rhizosphere of *Lactuca sativa*. *J Proteome Res.*, 11, 2061, 2012.

33. de Menezes, A., Clipson, N., Doyle, E., Comparative metatranscriptomics reveals widespread community responses during phenanthrene degradation in soil. *Environ. Microbiol.*, 14, 2577, 2012.
34. Kim, Y. and Liesack, W., Differential assemblage of functional units in paddy soil microbiomes. *PLoS ONE*, 10, e0122221, 2015.
35. Carvalhais, L.C., Dennis, P.G., Fan, B., Fedoseyenko, D., Kierul, K., Becker, A., von Wiren, N., Borriss, R., Linking plant nutritional status to plant-microbe interactions. *Plos One*, 8, e68555, 2013.
36. Hayden, H.L., Savin, K.W., Wadeson, J., Gupta, V.V.S.R., Mele, P.M., Comparative Metatranscriptomics of Wheat Rhizosphere Microbiomes in Disease Suppressive and Non-suppressive Soils for *Rhizoctonia solani* AG8. *Front Microbiol.*, 9, 859, 2018.
37. Turner, T.R., Ramakrishnan, K., Walshaw, J., Heavens, D., Alston, M., Swarbreck, D., Osbourn, A., Grant, A., Poole, P.S., Comparative metatranscriptomics reveals kingdom level changes in the rhizosphere microbiome of plants. *ISME J.*, 7, 2248, 2013.
38. Salt, D., Baxter, I., Lahner, B., Ionomics and the study of the plant ionome. *Annu. Rev. Plant Biol.*, 59, 709, 2008.
39. Rus, A., Baxter, I., Kumar, M., Gustin, J., Lahner, B., Yakubova, E., Salt, D., Natural Variants of AtHKT1 Enhance Na Accumulation in Two Wild Populations of Arabidopsis. *PLoS Genet.*, 2, 1963, 2007.
40. Baxter, I., Ionomics: studying the social network of mineral nutrients. *Curr. Opin. Plant Biol.*, 12, 381, 2009.
41. Nadia, D., Manivannan, S., Muddarsu, V.R., Umadevi, P., Sujata, U., Ionome signatures discriminates the geographical origin of jackfruits (*Artocarpus heterophyllus* Lam.). *Food Chem.*, 339, 1278, 2021.
42. Chettri, S., Manivannan, S., Muddarsu, V.R., Nutrient and Elemental Composition of Wild Edible Ferns of the Himalaya. *Amer. Fern J.*, 108, 93, 2018.
43. Yamuna, P., Upadhyay, S., Bhatt, S.S., Sharma, L., Manivannan, S., Chanbisana, C., Nutritional Compositions of *Baccaurea sapida* and *Eleaocarpus sikkimnesis of* Sikkim Himalaya. *IJCMAS*, 7, 2101, 2018.
44. Lepcha, T.T., Upadhyay, S., Manivannan, S., Bhutia, K.D., Sharma, L., Muddarsu, V.R., Proximate and nutritional analysis of Culantro (*Eryngium foetidum*). *J. Pharmacogn. Phytochem.*, 3129, SP1, 2018.
45. Yadav, V., Effect of Next Generation Organic Inputs on Okra (*Abelmoschus esculantus* (L.) Moench), Palak (*Beta vulgaris var bengalensis*) and Radish (*Rapahanus sativus* L.), in: *Thesis*, Sikkim University, Gangtok India, 2017.
46. Lepcha, C., Organic Nutrient Management in Tomato, in: *Thesis*, Sikkim University, Gangtok India, 2015.
47. Tamang, D.L., Manivannan, S., Yadav, V., Effect of PGPR and organic inputs on growth and yield of commercial ginger varieties of Sikkim. *Fourth National Conference on Plant Growth-Promoting Rhizobacteria (PGPR) for Sustainability of Agriculture and Environment*, Mizoram University, Aizawl, Mizoram, India, 11th- 12th May, 2018.

48. Tamang, D.L. and Manivannan, S., Volatile oil contents as affected by PGPR and other organic inputs in commercial ginger varieties of Sikkim. *Symposium on Spices and Aromatic Crops(SYMSAC-IX)- Spices for doubling the farmer's income*, SASRD, Nagaland University, Medziphema, India, 15th-17th March, 2018.
49. Doyu, M., Designing an ionome data base for horticultural crops of Sikkim Himalaya, in: *Thesis*, Sikkim University, Gangtok India, 2018.
50. Shi Alan, E.S., Maureen O'Callaghan, R., DeAngelis, K.M., Jones, E.E., Stewart, A., Firestone, M.K., Condron, L.M., Effects of selected root exudate components on soil bacterial Communities. *FEMS Microbiol Ecol.*, 77, 600, 2011.
51. Phillips, R.P., Erlitz, Y., Bier, R., Bernhardt, E.S., Below ground responses to climate change New approach for capturing soluble root exudates inforest soils. *Funct. Ecol.*, 22, 990, 2008.
52. Zachow, H.C., Cardinale, M.M., Erlacher, M.A., Grube, R., Berg, T.G., The impact of the host genotype on the structure of the seed and rhizosphere microbiome: sugar beet and lettuce as model plants. *Workshop on Implementing Plant – Microbe Interaction in Plant Breeding*, 2015.
53. Pfeifer, J.A., Spor, A., Koren, O., Jin, Z., Tringe, S.G., Dangl, J.L., Buckler, E.S., Ley, R.E., Diversity and heritability of the maize rhizosphere microbiome under field conditions. *PNAS*, 110, 6548, 2013.
54. Umadevi, P. and Anandaraj, M., Proteomics analysis of the tripartite interaction between black pepper, *Trichoderma harzianum* and *Phytophthora capsici* provides insights into induced systemic resistance mediated by *Trichoderma* spp. *Eur J Plant Pathol.*, 154, 607, 2019.
55. Mantelin, S., Desbrosses, G., Larcher, M., Tranbarger, T.J., Cleyet-Marel, J.-C., Touraine, B., Nitrate-dependent control of root architecture and N nutrition are altered by a plant growth-promoting *Phyllobacterium* sp. *Planta*, 223, 591, 2006.

12

Methane Emission and Strategies for Mitigation in Livestock

Nibedita Sahoo[1], Swati Pattnaik[2]*, Matrujyoti Pattnaik[3] and Swati Mohapatra[4]

[1]*Department of Animal Nutrition, West-Bengal University of Animal and Fishery Sciences, Kolkata, India*
[2]*Department of Microbiology, Orissa University of Agriculture and Technology, Odisha, India*
[3]*Department of Public Health, ICMR - Regional Medical Research Centre, Bhubaneswar, Odisha, India*
[4]*Department of Biotechnology, Amity University, Noida, Uttar Pradesh, India*

Abstract

Livestock encompasses animals reared for labor and essential commodities like dairy, meat, and leather for commercial purposes. Ruminants like cows, goats, and sheep form a significant population of livestock. The rumen, also known as a fermentation vat, is the site of enteric fermentation and methane production and consists of methanogens (a group of archaea) that combine carbon dioxide and hydrogen to produce methane. Methane is a significant contributor of greenhouse effect and has been reported to reach 1,774 ppb from earlier attention. Data from the United States Environmental Protection Agency suggests that in 2014, cows reared for beef were the major contributors to global methane emissions at 71% with dairy cows and other sources of emissions contributing to 24% and 5%, respectively. Livestock nutritional strategies like incorporation of high cereal diet, biohydration of unsaturated fatty acids, enhancement of propionic acid production, protozoal inhibition, or supplementation with ionophores, fats, organic acid, probiotics, acetogens, and bacteriocins have been considered to keep the methane emission by livestock in check. In addition, research on improvement of vaccines against rumen methanogens and animal breeding and selection for inhibition of methane production is also underway.

**Corresponding author*: pattnaikswati4@gmail.com

Bibhuti Bhusan Mishra, Suraja Kumar Nayak, Swati Mohapatra, and Deviprasad Samantaray (eds.) Environmental and Agricultural Microbiology: Applications for Sustainability, (257–274) © 2021 Scrivener Publishing LLC

***Keywords*:** Livestock, methane emission, rumen methanogenesis, rumen methanogens, mitigation strategies, dietary manipulation

12.1 Introduction

Global warming is a wide-reaching environmental, economical, and social risk, and it is well documented that livestock production contributes to it by arising natural byproduct likes methane. Methane production is a matter of great concern as it contributes to 18% of overall warming. It has been accounted that methane is 25 times greater than carbon dioxide; it remains in the atmosphere for long years and increases 7% each year [1, 2]. Methane is classified under trace gases and the worldwide projected concentration is 1774 ± 1.8 parts per billion (ppb) [3]. Methane is oxidized to carbon monoxide (CO), carbon dioxide (CO_2), and water (H_2O) by photochemical reactions. By this means, the reaction consumes the hydroxyl radical (OH) and involves ozone (O_3) like gases. Methane generation affects earth's radiative balance in numerous ways. Methane oxidation generates CO_2 and water vapor as other important greenhouse gases. Moreover, it contributes to overall temperate through its infrared absorption spectrum.

Livestock contributes 100–120 kg of methane per year which is total of 2,300–2,760 kg of CO_2 as the reports showcase. Worldwide, there are about 1.5 billion cattle and all of them emit about 87 million metric tons of methane per year. Natural digestive process and manure management in livestock operations generally leads to high methane production. Methane emission affects energy gain and productivity through extensive dietary energy losses of ruminant. Enteric methane emission is processed by rumen and animal's hindgut by methanogens.

Livestock involves deforestation and desertification for creation of grasslands which leads to release of carbon in form of methane from cultivated soils. On the other hand, the rising of methane production is related with human populations which lead to high demand of livestock. Ruminants like cattle, sheep, buffalo, and goats have different digestive systems; they can change unusable plant materials into nourishing food and fiber and emit methane, the most potent greenhouse gas which affects to global climate. Global climate change by ruminants has become a great concern these days for its negative effects on human population and society. Research on isolation, characterization of rumen methanogens and improvement of methane release by various dietetic manipulations in the rumen has attracted great interest these days.

12.2 Contribution of Methane from Livestock

Continuous increase of methane concentration is strongly associated with raising populations and anthropogenic sources [4]. Fossil fuel, enteric fermentation, animal wastes, production of manure, paddy rice farming, and combustion of organic matters are the major anthropogenic sources of methane which contributes 70% of methane production. Out of which, two-thirds of the anthropogenic sources are agricultural sources.

According to Rolfe and Zeil (2001) [5], the total cattle population is accountable for 73% of methane emissions of all livestock worldwide and major population of livestock is sited in tropical region and fed chiefly low-quality diets and represents 10%–12% of loss of gross energy (GE) through methane. Livestock is major contributor of methane release and contributing about 80–115 mT per year which is comparable to 15%–20% of whole anthropogenic sourced methane emission [6]. Reports available that reported approximately 7591 million metric tonnes (Tg) of enteric methane were produced from cattle and buffaloes, small ruminants (sheep and goat), and pigs, respectively [7]. However, in the year of estimated that buffalo and camel contributed 6.2–8.1 Tg and 0.9–1.1 Tg and hindgut of pigs and horses released around 0.9–1.0 Tg and 1.7 Tg, correspondingly. Cattles produce 85%–90% of methane from fecal excretion and enteric fermentation and 95% of rumen methane is emitted via eructation. Enteric fermentation contributes up to 25% of anthropogenic methane emission [8]. Methane is produced through enteric fermentation at the rumen of ruminants and in the lower gastrointestinal tract of non-ruminants.

In ruminant animals, volatile fatty acid is formed where hydrogen is act as an intermediary and converts to methane. Methanogenic bacteria and protozoa utilize of carbohydrates in the gut of ruminants and produce microbial protein, volatile fatty acids, CO_2, and CH_4 with hydrogen (H_2). Methane from enteric fermentation of rumen contribute a loss of feed energy about 20–150 kJ/MJ intake in ruminants [9].

12.3 Methanogens

Methanogens constitute of bacteria that are strictly anaerobic and are members of the domain archaea. Methanogens contribute to the atmospheric methane production of around 70% which is found in both man-made and natural environments. Biologically produced methane puts negative impact on environment when released into the atmosphere. Conventionally, it was believed that methanogens could only produce methane by coupling the oxidation of fermented products with the reduction of carbon dioxide.

Among the methanogens, the largest producer of the greenhouse gas methane is the ruminant methanogens. Methane is usually produced by methanogenic archaea. As methanogens are accountable for methane production in ruminants, an effort is being made to identify and characterize these microbes.

12.3.1 Rumen Microbial Community

The micro-biota existing in the rumen is very complex. Majority of those microbial organisms in the rumen have not been identified yet. In addition to these, methanogenic archaea, protozoa, bacteriophages, and fungi contribute to the diversity and function of rumen in animals.

CO_2 and H_2 act as electron acceptor and donor in the rumen of cattle which are generally byproducts of fermentation. Organic matter degradation is achieved by the complete involvement of different groups of microbes. A plant molecule such as proteins and carbohydrates which is present in the rumen is degraded to monomers by primary anaerobic fermenters. The monomers produced from the degradation are converted by primary and secondary fermenters to fatty acids, CO_2 and H_2. Methanogens utilize the end products of fermentation as substrates to synthesize methane.

12.3.2 Methanogens Found in Rumen

Methanomicrobium mobile, *Methanobacterium bryantii*, *Methanobacterium formicicum*, *Methanobrevibacter millerae*, *Methanoculleus olentangyi*, *Methanobrevibacter ruminantium*, *Methanobrevibacter smithii*, *Methanosarcina barkeri*, *Methanobrevibacter olleyae*, *Methanobrevibacter boviskoreani*, *Methanobacterium beijingense*, *Methanoculleus marisnigri*, *Methanoculleus bourgensis*, and Methanosarcina mazei are the methanogens that have been isolated and cultured from the rumen of the animals by different isolation methods [10–12]. Genetic variations of methanogens in the rumen are considerably high even within the same ruminant species but most of them are not isolated as they are non-cultivable.

12.3.3 Enrichment of Methanogens from Rumen Liquor

Rumen liquor is collected from cattle before their feeding. An enrichment medium is used in the isolation of methanogens from samples. The medium contains NaCl and bile salt for proper growth and development. At first, a blend of CO_2 and H_2 (20:80), 0.5% w/v sodium formate were taken and the mixtures are utilized as substrates. The media is distributed in CO_2 flushed serum bottles and autoclaved thoroughly. Filter sterilized vitamins and antibiotics solutions are added after sterilization. Five percent of collected rumen sample is poured into each bottle by using syringe and then incubated at 39°C for 90 days in the dark.

12.3.4 Screening for Methane Production

Gas is taken out from the serum bottles after incubation by a gas tight syringe and analyzed using gas chromatography. The standard gas for estimation of methane is composed of 50% each of methane and carbon dioxide. The methane gas is identified on the basis of the retention time of standard and the response factor obtained was used to calculate the percentage of methane in the sample [13].

12.3.5 Isolation of Methanogens

Methane positive bottles are used with inoculate to make roll tubes [14]. The culture purity and cell morphology are confirmed microscopically. The microbial characteristics to utilize methanol and ethanol as substrates of growth are screened. Microbial growth is measured by optical density with the help of a spectrophotometer at 660 nm. In addition, 10% of cultures are inoculated in BY medium [15] containing different concentrations of sodium chloride and bile salt to check their sensitivity. The isolates are then grown at 39°C for 21 days.

12.3.6 Molecular Characterization

DNA toolkit is used to extract and purify DNA. PCR is done using 16S rRNA (Met86F and Met1340R)-based and mcrA gene-based primers. The reaction mixture comprises of 10X Taq buffer F, dNTPs primers, $MgCl_2$, and nuclease-free water. Thermal cycling steps were adopted [16, 17]. Respective amplified PCR products are sequenced and BLAST search is performed to obtain the homology with published methanogen sequences in NCBI GenBank database.

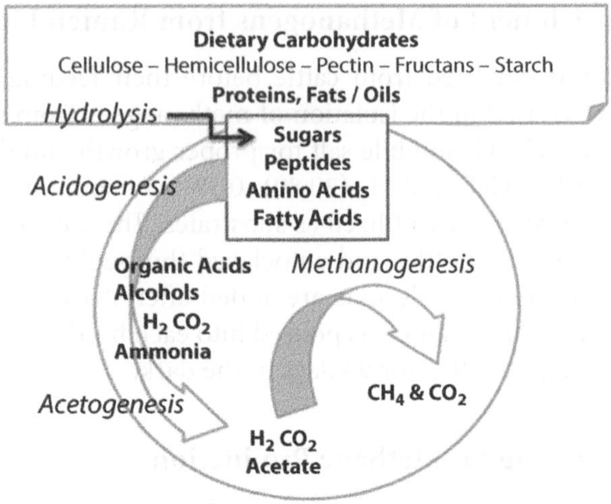

Figure 12.1 Processes of methanogenesis in general.

12.4 Methanogenesis: Methane Production

Methanogenesis is the process of production of methane by methanogens with simple substrate such as the acetate and carbon dioxide as terminal electron acceptors at low reduction potential (Figure 12.1). They may also use carbon from organic sources such as formic acid, methanol, and dimethyl sulfide for the process [12].

12.4.1 Pathways of Methanogenesis

Methanogens produce methane from substrates mainly $H_2 + CO_2$, formate, and acetate [18–20]. The $H_2 + CO_2$, acetate, and formate are acquired from carbohydrate fermentation. It is crucial to have a clear observation of their metabolism to orderly make the pathways of the methane production in the rumen of animals.

12.4.2 Pathway of CO_2 Reduction

The common substrate of methanogens for methane production is H_2+ CO_2. The detailed pathway of the methane formation is already described by Liu and Whitman (2008) [21]. The methane is produced by the reduction of CO_2 by H_2 as an electron donor by various intermediates with various cofactors and enzymes.

12.4.3 CO_2 Reduction to Formyl-Methanofuran

The reduction of CO_2 to formyl-methanofuran consists of two steps. The first step is the binding of CO_2 with methanofuran which is a CO_2 reduction factor and gives an intermediate product called N-carboxymethanofuran. The second step is H_2-dependent reduction to formyl-methanofuran. Ferredoxin is involved in the process which can accept an electron directly to form a reduced state. The methanofuran (MFR) is composed of a C4-substituted furfurylamine ring is present in all known methanogens at the level of 0.5–2.5 mg kg^{-1} of cell dry weight [10]. The formation of formyl-MFR is catalyzed by formyl-methanofuran dehydrogenase.

12.4.4 Conversion of the Formyl Group from Formyl-Methanofuran to Formyl-Tetrahydromethanopterin

An enzyme and a coenzyme take part in the process of formyl-methanofuran to formyl-tetrahydromethanopterin. The coenzyme tetrahydromethanopterin (H_4MPT) is a C1 carrier and consist of electron-donating methylene group. The formyl-transferase can transfer a formyl group. It depends on the salt concentration which coexist as monomer, dimer, or as a tetramer. The tetramers are generally thermostable but the monomers and dimers are the active forms.

12.4.5 Formation of Methenyl-Tetrahydromethanopterin

The process from formyl-tetrahydromethanopterin to methenyl-tetrahydromethanopterin is catalyzed by the enzyme methenyl-H4MPT cyclohydrolase (Mch). The enzyme stays in homotrimeric state but stable in aerobic conditions.

12.4.6 Reduction of Methenyl-Tetrahydromethanopterin to Methyl-Tetrahydromethanopterin

The methenyl-tetrahydromethanopterin is reduced to methylene-tetrahydromethanopterin and subsequently to methyl-tetrahydromethanopterin. In both the process, F_{420} is used as a coenzyme for hydride transfer as it is a low potential electron carrier. F_{420} is reduced to $F_{420}H_2$ by H_2 and F_{420}-reducing hydrogenase during the process of methanogenesis. The F_{420}-dependent methylene-H_4MPT dehydrogenase catalyzes reduction of methenyl-H_4MPT and $F_{420}H_2$ to methylene-H_4MPT. The methylene-H_4MPT reductase catalyzes the reduction from methylene-H_4MPT to methyl-H_4MPT.

12.4.7 Reduction of Methyl-Tetrahydromethanopterin to Methyl-S-Coenzyme M

The process of methyl-tetrahydromethanopterin to 2-mercaptoethanesulfonic acid and coenzyme M (HS-CoM) is catalyzed by HS-CoM methyltransferase (Mtr). The HS-Coenzyme M is methyl group carrier and forms methane by taking the methyl group from methylcobalamin to form Methyl-S-Coenzyme M.

12.4.8 Reduction of Methyl-S-Coenzyme M to CH_4

The reduction process from Methyl-S-Coenzyme M to methane is done by methyl-Coenzyme M reductase (MCR). The process involves coenzyme F430 as the prosthetic group and coenzyme B as an electron donor. A thiol group and an L-threonine phosphate group present in coenzyme B which is recognized by MCR. Thiol group replaces CH4 from methyl-Coenzyme M and L-threonine phosphate group binds to basic amino acids in MCR [12].

12.5 Strategies for Mitigation of Methane Emission

There are several options or approaches for alleviating methane production from animal. According to Seijan *et al.* (2011) [22], there are two methods to prevent methane production from animal, i.e., preventative and end of pipe method. In preventive method, through dietary manipulations, there is reduction of carbon and nitrogen input to animal which ultimately reduce methane production from animal. In end of pipe method, there is inhibition of methane production inside the animal body. According to Bunglavan (2014) [4], methane production in animal can be prevented by targeting methanogenic organism directly and substrate used by these organisms indirectly. Mostly, all strategies aim at improving productive efficiency by reducing methane production per animal or animal product.

12.5.1 Dietary Manipulation

12.5.1.1 Increasing Dry Matter Intake

Feed processing techniques increase feed intake of the animal. These techniques enhance feed ingredient feeding value by increasing digestibility of feed ingredients. These techniques are feed chopping, grinding, chemical treatment of feed residues such as acidic and alkali treatment, and supplementation of urea and molasses. Feed processing techniques reduce methane production by 10% [4]. Ruminal microbial efficiency was improved and ruminal pH was decreased with increasing dry matter (DM) intake. Additionally, the feed intake manipulation was altered the end-products of rumen fermentation. It is also reported that, 23% of GE losses in alfalfa hay–supplemented diet when DM intake was increased to 17 from 9 kg/d [23].

12.5.1.2 Increasing Ration Concentrate Fraction

Forage-to-concentrate ratio (F:C) in the ration has an impact on feed fermentation in the rumen and the ratio of acetate and propionate production. When forage-to-concentrate ratio increases, the methane production is also increased due to high production of acetate than propionate [24]. The increasing concentration fraction of a ration reduces methane emission in relation to energy intake [22]. According to Benchaar *et al.* (2001) [23], production of propionate and validate increased linearly and acetate production decreased when the proportion of concentrate is increased in the ration. Non-fiber carbohydrates (NFCs) in a ration increase amylolytic bacteria which change the ratio of short-chain fatty acid production in rumen and lead to high propionate and low acetate production [25]. Propionate production results in low hydrogen production and decreases methane production. Concentrates increase propionate production which decreases hydrogen formation in the rumen and inhibits methane producing bacteria, cellulose utilizing bacteria, and ciliate protozoa via pH reduction [25]. Concentrates also stimulate lactic acid bacteria to produce bacteriocin, which inhibits methane production. The curvilinear relationship between methane emission and concentrate supplementation and reported that methane emission was decreased to 2%–3% GE when concentrate was given at the rate of 80%–90% level [26]. However, high concentrate supplementation results in metabolic disorders such as ruminal acidosis, lower milk fat syndrome, and shorter reproductive life span of animal. Economically high concentrate feeding is expensive to maintain animal. Heavy grain production, harvesting, and transportation are results in emission of carbon dioxide and nitrous oxide, the global warming gases.

12.5.1.3 Supplementation of Lipid

Dietary supplementations of lipids such as fatty acid and oil have both *in vivo* and *in vitro* effect on rumen methanogens [24]. Lipids decrease organic matter which is fermented to methane. It also reduces the activity of methanogenic bacteria and has harmful effects on cellulose utilizing bacteria and protozoa [25]. Fatty acids bind with the cell membrane of methanogens and disturb transport mechanism of cell membrane [24]. On the other hand, Beauchemin *et al.* (2008) [27] conducted several trails on methane production and reported that methane production (g/kg DMI) was reduced to 5.6% when fat was increased to 1% on DM basis in beef cattle, dairy cows, and lambs. Supplementation of lipid at 1% level decreased methane production (g/kg DMI) to 3.8% [28]. Generally, methane acts as a hydrogen sink to get rid off of excess hydrogen because of unsaturation nature. They reduce the availability of hydrogen in rumen as their double and triple bonds are saturated by hydrogens [4]. Addition of 4.6% canola oil to a high-forage diet as the source of unsaturated fat decreased methane emission by 32% [29]. The saturated fatty acids break the cell membrane of methanogens which leads to death of organisms [25]. The most toxic saturated fatty acids are lauric acid (C12:0), followed by myristic acid (C14:0). According to Pereira *et al.* (2015) [25], medium-chain–rich fatty acids have pronounced effect on methane emission. Examples of some medium-chain fatty acids are coconut oil and canola oil that are rich in lauric acid or purified myristic acid.

12.5.1.4 Protozoa Removal

Protozoa removal is known as defaunation. There is an association between ciliated protozoa and methanogens where protozoa provides habitat to methanogens [4]. Protozoa also transfers hydrogen to methanogens which is required for methane formation from carbon dioxide reduction [24]. So, defaunation reduces methane production in an animal as methanogens have no habitat partner and less hydrogen available for methane production. Defaunating agents are copper sulfate, sodium lauryl sulfate, oil rich in PUFA, dioctyl sodium sulfosuccinate, etc. Complete elimination of protozoa from rumen results in 13% reduction of methane emission [30]. Qin *et al.* (2012) [31] found in their experiment that defaunation reduced methane production in all grain diet as compared faunation.

12.5.2 Feed Additives

12.5.2.1 Ionophore Compounds

Ionophore compounds are the antimicrobial agent used as a feed additive to improve the performance of the animal. Examples of some ionophore compounds are monensin and lasalocid. Monensin and lasalocid compounds are produced by various strains of *Streptomyces* sp. Chemically, they are carboxypolyether compounds. Ionophores inhibit methane precursors such as formate and hydrogen. They have no direct effect on methanogenic organisms as methanogenic organisms are resistant to ionophore compounds [25]. Monensins reduce gram-positive methanogens which are responsible for supplying substrate to methanogenesis [24]. Lasalosid is effective against hydrogen-producing bacteria and decreases methane production. Beauchemin *et al.* (2008) [27] conducted a trail in dairy cows and reported that monensin below 15 ppm did not affect methane production in gram per day or DM ingested. However, Odongo *et al.* (2007) [32] conducted a trail on beef and dairy cattle and reported that there was decline of methane formation about 4% and 10% in gram per day and 3% and 8% in gram per DM ingested at higher dose rate such as 24 to 35 ppm. The effect of monensin on methane emission in animal is dose dependant.

12.5.2.2 Halogenated Methane Compound

One of the methane inhibitor compounds is halogenated methane compound. Example of these compounds is chloroform, starch, chloral hydrate, carbon tetrachloride, methylene chloride, tricholroacetamide, hemiacetyl of chloral, methylene bromide, and bromochloromethane (BCM). They are methane inhibitors. BCM reacts with cobamine (coenzyme B) which is the reduced form of Vitamin B12 and inhibit methane production because coenzyme B has a role in the last step of methanogenesis [4]. The regulation of the "Montreal Protocol on Substances that Deplete the Ozone Layer" included BCM in their list. So, BCM should be used carefully. Ungerfeld *et al.* (2004) [33] stated that two compounds, i.e., 2-bromoethanesulfonate and 3-bromopropanesulfonate, might be effective in reducing methane production. In addition, 2-bromoethanesulfonate (BES) is a coenzyme M analog compound and 3-bromopropanesulfonate hinder methyl-CoM reductase enzyme. Mevastatin and lavastatin are the compounds which have the potentiality to inhibit 3-hydroxy-3-methylglutaryl coenzyme A (HMG-CoA) reductase [34]. So, HMG-CoA reductase inhibitors can be used to reduce methane production in animal. Mevastatin and lavastatin could restrain the rumen isolated *Methanobrevibacter* strains development and reduce methane production [34].

12.5.2.3 Organic Acid

Organic acids or dicarboxylic acids are propionate precursors or propionate enhancer. They have a major role in decreasing the number of reducing equivalent (hydrogen) in the rumen because they are used for propionate production, and in this process, reducing equivalents are used. Hence, organic acid can be used as an alternative hydrogen sink to reduce methane production. Among organic acid, malate is a potent antimethanogenic compound. Malate concentration in forages like alfalfa and Bermuda grass as DM is 2.9%–7.5% and 1.9%–4.5% of DM respectively. But the ratio varies with variety of forages and the stages of maturity [4]. Newbold *et al.* (2005) [35] compared 15 different propionate precursors and reported that in batch culture, fumarate and acrylate reduced methane excretion effectively. He also stated that as compared to acrylate, fumarate gave better result in artificial rumen. Wallace *et al.* (2006) [36] fed encapsulated fumaric acid in higher concentration (10% of the diet) to sheep and reported that there was 40 to 75% methane reduction. The use of organic acid in ruminant is limited because of the high cost. Green forages as a natural source of organic acid may be used as a tool in order to reduce methane emission from animal.

12.5.3 Microbial Feed Additives

As a microbial feed additive, acetogenic bacteria and yeast can be used to reduce methane production in an animal. Acetogenic bacteria or homoacetogens are H_2-utilizing bacteria and capable to use hydrogen as energy purpose for growth and formation of acetate from carbon dioxide [37]. Along with acetogenic bacteria, anti-methanogenic compound should be used because methanogenic bacteria are more potent than acetogenic bacteria in hydrogen utilization [4]. Yeast culture as probiotic is used in ruminant and it increases bacterial activity and stabilizes ruminal pH. Yeast cultures, *Saccharomyces cerevisiae*, modify fermentation process in the rumen and may reduce methane production [34]. However, more research work should be done to explore the use of microbial feed additive in an animal.

12.5.3.1 Vaccination

Many researchers investigated that vaccination is one of the methods to prevent methanogenesis in animals. Vaccination against rumen methanogens decreases the number or activity of the methanogenic organism in rumen and reduces methane emissions from the livestock.

The association between salivary immunoglobins and surface of methanogenic organism is important for the effectiveness of vaccination. Vaccination of animal results in salivary antibody production. These antibodies enter into rumen where they bind with the surface of methanogenic organism and reduce methane production. So, major target of vaccine against methanogenic organism is the membrane associated surface protein [25]. Wright *et al.* (2006) [38] reported that all rumen methanogens could not be isolated from the rumen. So, it is not possible to develop antibodies against that organism. Those non-isolated organisms may grow and replace the methanogenic organism in rumen. This presents a challenge in the production of an effective vaccine using prepared methanogen cells that can reduce enteric methane emissions. Recombinant vaccine is an option in order to prevent vaccine failure. Recombinant vaccines against cell surface protein of wide number of methanogenic organisms are an important tool to increase usefulness of vaccination. Cook *et al.* (2008) [39] reported that there was decline of *in vitro* methane production in cultured rumen liquid following administration of vaccine at higher doses. He first vaccinated hen with whole cell of three cultured methanogenic species and developed IgY antibodies in chicken egg. A vaccine generated from subcellular part of *Methanobrevibacter ruminantium* M1 strain and reported that there was agglutination of methanogenic organism and reduction of *in vitro* methane production [40]. In rumen, methanogenic organisms' population varies with the diet and geographic locations [41]. So, attention should be given toward development of a vaccine with a wide range of action against a number of methanogens. So, that there will be less chance of vaccination failure against methane production from animal.

12.5.3.2 Bacteriophages and Bacteriocins

As a biological control, bacteriophages and bacteriocins are one of the options to mitigate methane emission problem from animal. They reduce hydrogen availability for methane production by transferring it to the acetogenic or propiogenic bacteria and are capable to hinder the action of archaea methanogens [25]. Bacteriophages are an obligate microbial virus that infects and replicate within bacteria and archaea. Bacteriophage is composed of protein that encapsulates DNA or RNA genome material. Bacteriophage is also known as phages or bacterial virus. In ruminal fluid, bacteriophages are present in large number, i.e., > 109 particles/ml. They destroy or lyse their host such as bacteria and archaea during lytic phase of their development. Around, 750 complete genome sequenced bacteriophages are known present days. Out of 750, only six archaeal phages have been identified. Among them methanogenic bacteriophages are *Methanobacterium* phage psi M1, *Methanobacterium* phage psi M2 (a variant of M1), and *Methanothermobacter* phage psi M100. Other phages, i.e., *Siphoviridae* phages (siphophages), could inhibit methanogens like

Methanobrevibacter, *Methanobacterium*, and *Methanococcus* spp [41]. The major problems associated with the use of this method for methanogenesis prevention are rumen microorganisms quickly adopt to bacteriophages, and bacteriophages are host specific as there is high population of methanogen species in the rumen [25, 41].

The rumen has always a composite microbial community where bacteriocins play important role in regulation of rumen microbial ecosystem [41]. *Butyrivibrio* sp. strains and found that they exhibited more than 50% of bacteriocin-like activity. A *Lactococcus lactis* produces nisin which is a bacteriocin compound [25]. In his literature, he reported that when nisin concentration was increased from 30 µmol/L, the *in vitro* methane production was reduced upto 40% in a continuous culture system. Cookson *et al.* (2004) [42] identified a bacteriocin compound from *Streptococcus* sp. One of the active forms of bacteriocin compound from *Streptococcus* sp., i.e., bovicin HC5 reduced *in vitro* methane production about 50% [25, 43]. Sar *et al.* (2004) [44] used nisin along with nitrate as an alternative electron acceptor and reported that there was reduction of methane production in sheep.

12.5.4 Animal Breeding and Selection

Animal breeding is practiced to improve animal productivity and performance. It is evidenced that improvement of animal performance reduces methane production per unit of product. Breeding for enhancement of productivity reduces methane production by increasing utilization of feed energy and reducing maintenance requirements [45]. So, breeding of animal with high performing animal will counteract the problem of methane production from the animal. With similar potency, some animals per unit of intake are low methane emitters as compared to others. This may be due to the genetic difference among animals in methane production. Methane production at the same stages of lactation, i.e., 60 and 150 days on holstein cow from Northern Hemisphere and New Zealand region feeding same diet and found that there was reduction in methane production per kg of dry food stuff supplement, about 15% in cow from Northern Hemisphere than New Zealand region [46].

12.6 Conclusion

Increasing demand of livestock and 70–120 kg of methane emission per year are great concerns as well as challenging for the global climate change. The rumen methanogens, methane producers in livestock, utilize H_2, as substrate for methane emission. By decreasing the H_2-producing protozoa and fibrolytic rumen microbes or by increasing non-methanogenic cultures, decline of rumen methanogenesis is possible. It is essential and important as the livestock contributes 18% of the overall global warming. Incorporation of high cereal diet, biohydration of unsaturated fatty acids, enhancement of propionic acid production, protozoal inhibition, or supplementation with ionophores, fats, probiotics, acetogens, bacteriocins, and organic acids have been considered earlier to check the methane emission. In addition, advance research in vaccine development, bacteriophages, and bacteriocins against rumen methanogens and animal breeding selection is also underway. Manipulation of fermentation kinetics, rumen microbiome, diet, and adaptation to anti-methanogens are achievable for commercial applications on farms, growing crops and raising livestock with minimal environmental pollution.

References

1. Hongmin, D., Mangino, J., McAllister, T.A., Hatfield, J.L., Johnson, D.E., Keith, R.L., Lima, M.A., Romanovskaya, A., Intergovernamental Panel on Climate Change, in: *Emissions from livestock and manure management*, vol. 10, p. 1, 2006.
2. IPCC *Guidelines for National Greenhouse Gas Inventories*, Institute for Global Environmental Strategies (IGES), Hayama, Japan, 2006.
3. Forster, P., Ramaswamy, V., Artaxo, P., Berntsen, T., Betts, R., Fahey, D., Haywood, J., Lean, J., Lowe, D., Myhre, G., Nganga, J., Prinn, R., Raga, G., Schulz, M., Van Dorland, R. *et al.*, Changes in Atmospheric Constituents and in Radiative Forcing, in: *Climate Change 2007: The Physical Science Basis. Contribution of Working Group I to the Fourth Assessment Report of the IPCC*, S. Solomon *et al.* (Ed.), pp. 129–234, Cambridge University Press, Cambridge, UK, 2007, http://www.cambridge.org/catalogue/catalogue.asp?isbn=9780521705967.
4. Bunglavan, S.J., Methanogenesis and recent techniques for mitigation of methanogenesis in ruminants. *J. Livestock Sci.*, 5, 35, 2014.
5. Rolfe, J. and Zeil, V., Methane emissions from cattle production – Issues in meeting the Kyoto targets. Paper presented at *45th Annual Conference of the Australian Agricultural and Resource Economics Society*, 2001.
6. IPCC, *Contribution of working group I to the third assessment report of the intergovernmental panel on climate change*, Cambridge University Press, Cambridge, UK, 2001.
7. McMichael, A.J., Powles, J.W., Butler, C.D., Uauy, R., Food, livestock production, energy, climate change, and health. *Lancet*, 370, 1253, 2007.

8. Wuebbles, D.J. and Hayhoe, K., Atmospheric methane and global change. *Earth Sci. Rev.*, 57, 177, 2002.
9. Singh, G.P., Nagpal, A.K., Sainj, N., Methane production in relation to productivity of livestock and environment: a review. *Indian J. Anim. Sci.*, 75, 143, 2005.
10. Sirohi, S.K., Neha, P., Singh, B., Puniya, A.K., Rumen methanogens: a review. *Indian J. Microbio.*, 50, 253, 2010.
11. Zhou, X., Meile, L., Kreuzer, M., Zeitz, J.O., The Effect of Saturated Fatty Acids on Methanogenesis and Cell Viability of *Methanobrevibacter ruminantium*. *Archaea*, 2013, 106916, 2013. http://dx.doi.org/10.1155/2013/106916.
12. Qiao, J., Tan, Z., Wang, M., Potential and existing mechanisms of enteric methane production in ruminants. *Sci. Agric.*, 70, 430, 2014.
13. Kumar, S., Dagar, S.S., Puniya, A.K., Isolation and characterization of methanogens from rumenof Murrah buffalo. *Ann. Microbiol.*, 62, 345, 2012.
14. Hungate, R.E., A roll-tube method for cultivation of strict anaerobes. *Methods Microbiol.*, 3, 117, 1969.
15. Joblin, K.N., Methods in gut microbial ecology for ruminants. Methanogenic archaea, in: *Methods in Gut Microbial Ecology for Ruminants*, H.P.S. Makkar and C.S. McSweeney (Eds.), pp. 47–53, Springer, Netherlands, 2005.
16. Wright, A.D.G., Auckland, C.H., Lynn, D.H., Molecular diversity of methanogens in feedlot cattle from Ontario and Prince Edward Island, Canada. *Appl. Env. Microb.*, 73, 4206, 2007.
17. Luton, P.E., Wayne, J.M., Sharp, R.J., Riley, P.W., The mcrA gene as an alternative to 16S rRNA in the phylogenetic analysis of methanogen populations in landfill. *Microbio.*, 148, 3521, 2002.
18. Oppermann, R.A., Nelson, W.O., Brown, R.E., *In vivo* Studies of methanogenesis in the bovine rumen: dissimilation of dcetate. *J. Gen. Microbiol.*, 25, 103, 1961.
19. Hungate, R.E., Smith, W., Bauchop, T., Yu, I., Rabinowitz, J.C., Formate as an intermediate in the bovine rumen fermentation. *J. Bactiol.*, 102, 389, 1970.
20. Rea, S., Bowman, J.P., Sam, P., Carolyn, P., Andre Denis, G.W., *Methanobrevibacter millerae* sp. nov. and *Methanobrevibacter olleyae* sp. nov., methanogensfrom the ovine and bovine rumen that can utilize formate for growth. *Inter. J. Sys. Evol. Microb.*, 57, 450, 2007.
21. Liu, Y. and Whitman, W.B., Metabolic, phylogenetic, and ecological diversity of the methanogenic archaea. *Ann. New York Acad. Sci.*, 1125, 171, 2008.
22. Sejian, V., Lal, R., Lakritz, J., Ezeji, T., Measurement and prediction of enteric methane emission. *Int. J. Biometeorol.*, 55, 1, 2011.
23. Benchaar, C., Pomar, C., Chiquette, J., Evaluation of dietary strategies to reduce methane production in ruminants: A modelling approach. *Can. J. Anim. Sci.*, 81, 563, 2001.
24. Hook, S.E., Wright, A.G., McBride, B.W., Methanogens: Methane Producers of the Rumen and Mitigation Strategies. *Archaea*, 2010, 945785, 2010.

25. Pereira, L.G.R., Machado, F.S., Campos, M.M., Júnior, R.G., Tomich, T.R., Reis, L.G., Coombs, C., Enteric methane mitigation strategies in ruminants: a review. *Rev Colomb Cienc Pecu.*, 28, 124, 2015.
27. Beauchemin, K.A., Kreuzer, M., O'Mara, F., McAllister, T.A., Nutritional management for enteric methane abatement: a review. *Aust J. Exp. Agric.*, 48, 21, 2008.
26. Sauvant, D. and Giger-Reverdin, S., Empirical modelling meta-analysis of digestive interactions and CH4 production in ruminants, in: *Energy and Protein Metabolism and Nutrition*, vol. 561, I. Ortigues-Marty, N. Miraux, W. Brand-Williams (Eds.), Wageningen Academic, Wageningen, The Netherlands, 2007.
28. Martin, C., Morgavi, D.P., Doreau, M., Methane mitigation in ruminants: from microbes to the farm scale. *Animal*, 4, 351, 2009.
29. Beauchemin, K.A. and McGinn, S.M., Methane emissions from beef cattle: effects offumaric acid, essential oil, and canola oil. *J. Anim. Sci.*, 84, 1489, 2006.
30. Morgavi, D.P., Forano, E., Martin, C., Newbold, C.J., Microbial ecosystem and methanogenesis in ruminants. *Animal*, 4, 1024, 2010.
31. Qin, W.Z., Li, C.Y., Kim, J.K., Jul, J.G., Song, M.K., Effect of defaunation on ruminal fermentation characteristics and methane production by rumen microbes *in vitro* when incubated with starchy feed sources. *Asian-Aust. J. Anim. Sci.*, 25, 1381, 2012.
32. Odongo, N.E., Bagg, R., Vessie, G., Dick, P., Rashid, M.M., Hook, S.E., Gray, J.T., Kebreak, E., France, J., McBride, B.W., Long-term effects of feeding monensin on methane production in lactating dairy cows. *J. Dairy Sci.*, 90, 1781, 2007.
33. Ungerfeld, E.M., Rust, S.R., Boone, D.R., Liu, Y., Effects of several inhibitors on pure cultures of ruminal methanogens. *J. Appl. Microb.*, 97, 520, 2004.
34. Mitsumori, M. and Sun, W., Control of rumen microbial fermentation for mitigating methane emissions from the rumen. *Asian-Aust. J. Anim. Sci.*, 21, 144, 2008.
35. Newbold, C.J., López, S., Nelson, N., Ouda, J.O., Wallace, R.J., Moss, A.R., Propionate precursors and other metabolic intermediates as possible alternative electron acceptors to methanogenesis in ruminal fermentation *in vitro*. *Brit. J. Nut.*, 94, 27, 2005.
36. Wallace, R.J., Wood, T.A., Rowe, A., Price, J., Yanez, D.R., Williams, S.P., Newbold, C.J., Encapsulated fumaric acid as a means of decreasing ruminal methane emissions. *Int. Cong. Ser.*, 1293, 148, 2006.
37. Kahraman, O., Ozbilgin, A., Alatas, M.S., Citil, O.B., Strategies to reduce methane production in ruminants. *Animal Sci.*, LVIII, 144, 2015.
38. Wright, A.D.G., Toovey, A.F., Pimm, C.L., Molecular identification of methanogenic archaea from sheep in Queensland, Australia reveal more uncultured novel archaea. *Anaerobe*, 12, 134, 2006.

39. Cook, S.R., Maiti, P.K., Chaves, A.V., Benchaar, C., Beauchemin, K.A., McAllister, T.A., Avian (IgY) anti-methanogen antibodies for reducing ruminal methane production: *in vitro* assessment of their effects. *Aust. J. Exp. Agric.*, 48, 260, 2008.
40. Wedlock, D.N., Pedersen, G., Denis, M., Dey, D., Janssen, P.H., Buddle, B.M., Development of a vaccine to mitigate greenhouse gas emissions in agriculture: vaccination of sheep with methanogen fractions induces antibodies that block methane production *in vitro*. *New Ze. Veter. J.*, 58, 29, 2010.
41. McAllister, T.A. and Newbold, C.J., Redirecting rumen fermentation to reduce methanogenesis. *Aust. J. Exp. Agric.*, 48, 7, 2008.
42. Cookson, S.L., Noel, S.J., Kelly, W.J., Attwood, G.T., The use of PCR for the identification and characterisation of bacteriocin genes from bacterial strains isolated from the rumen or caecal contents of cattle and sheep. *FEMS Microb. Eco.*, 48, 199, 2004.
43. Lee, S.S., Hsu, J.T., Mantovani, H.C., Russell, J.B., The effect of bovicin HC5, a bacteriocin from *Streptococcus bovis* HC5, on ruminal methane production *in vitro*. *FEMS Microb. Lett.*, 217, 51, 2002.
44. Sar, C., Santoso, B., Mwenya, B., Gamo, Y., Kobayashi, T., Morikawa, R., Kimura, K., Mizukoshi, H., Takahashi, J., Manipulation of rumen methanogenesis by the combination of nitrate and β-1-4 galactooligosaccharides of nisin in sheep. *Ani. Feed Sci. Tech.*, 115, 129, 2004.
45. Chagunda, M.G.G., Romer, D.A.M., Roberts, D.J., Effect of genotype and feding regime on enteric methane, non-milk nitrogen and performance of dairy cows during the winter feeding period. *Livest. Sci.*, 122, 323, 2009.
46. Waghorn, G.C., Woodward, S.L., Tavendale, M., Inconsistencies in rumen methane production-effects of forage composition and animal genotype. *Int. Cong. Series.*, 1293, 115, 2006.

13

Liquid Biofertilizers and Their Applications: An Overview

Avro Dey

Microbiologist, CORDET, IFFCO, Paradeep, India

Abstract

Application of microorganisms in agricultural realm is of great importance owing to exo- and endo-metabolites synthesis and ultimately escalates plant growth and yield. The indigenous rhizospheric soil bacteria acclaimed as plant growth–promoting rhizobacteria (PGPR) act upon plant growth enhancing agent. The elementary idea of biofertilizers with the addition of microbes for the growth of plant is useful for the general practice of the farmers. Biofertilizers are essential and have often harsh impact to the soil and plants than the chemical counterparts. Solid carrier-based and liquid-based biofertilizers are the categories of biofertilizers. Liquid-based biofertilizer is more beneficial, cost effective, and ease in application. The useful effects of liquid biofertilizers without compromising with the environment are the daily needs in this 21st century, where urbanization and industrialization are growing rapidly to fulfil the needs of the society. The manufacture and production of liquid biofertilizers commercially for the betterment of the society and mankind is of utter importance, thus making the agriculture and production of different types of crops with more yield and less use of chemical fertilization.

Keywords: Plant growth–promoting rhizobacteria (PGPR), biofertilizers, liquid biofertilizers, chemical fertilizers, soil nutrients

13.1 Introduction

Microorganisms such as bacteria, fungi, algae, viruses, and protozoa dwell in the soil as it avails the best habitat for them. Population explosion coupled

Email: dey.avro@gmail.com

Bibhuti Bhusan Mishra, Suraja Kumar Nayak, Swati Mohapatra, and Deviprasad Samantaray (eds.) Environmental and Agricultural Microbiology: Applications for Sustainability, (275–292) © 2021 Scrivener Publishing LLC

with industrialization and urbanization has affected the water bodies and physicochemical parameters of the soil. Mineralization of organic and inorganic elements present in the soil is carried out by different microorganisms which are in process of availability to the aquatic and terrestrial environment. Several species of flora and fauna with diversity are found in the surface waters, rivers, lakes, and maritime waters. Industrialization is leading to the effluent from the industries such as allochthonous substances which are sometimes biodegraded by the bacterial population in the rivers and water bodies. These microorganisms also play a valuable role in the environment by controlling pollution and other important activities.

Microorganisms play an important role in the agriculture, pharmaceuticals, industrial, clinical, and food sectors, and in this chapter, the importance of microorganisms in the agricultural realm has been discussed in detail. The abundant presence of microorganisms in the soil and water always helps in the betterment of crops and plants. The trick of microorganisms to adapt to different mechanisms helps in development of crop plants [1, 2]. In recent years, the elementary idea of biofertilizers has given us a view of more production and less use of chemical fertilizers and pesticides. Plant growth–promoting rhizobacteria are the beneficial groups of bacteria present in the root region having an abundance of organic matter due to the root exudes, which are responsible for promoting plant growth. Plant health and soil fertility are influenced by the group of root-associated bacteria by interacting with the plant root in single or through biofilms [3]. PGPR helps as plant growth promoters which are direct mechanisms and acts as biological control agents by an indirect mechanism. Direct mechanisms of PGPR gives an idea about the use of nitrogen by the plants which is done by the process of nitrogen fixation; these microorganisms also help in breaking down the complex form of phosphorous to the bioavailable form which can easily be taken by the plants; microorganism helps in chelating and sequestration of iron by siderophores which is essential in the plant growth [4]. Additionally, soil microbes also produce many plant hormones like auxins, gibberellins, and cytokines for the growth of the plant. The plant ethylene level is lowered by using ACC deaminase that accumulates during biotic and abiotic stresses [5–11]. The well-equipped root colonizers like *Bacillus* sp. and *Pseudomonas* sp. produce a good broad spectrum antifungal molecule, which are beneficial against various phytopathogens, thus acting as effective biocontrol agents [12]. Furthermore, many mechanisms including the production of antibiotics, siderophores [4, 13], and HCN production [14] are added to the combat against soil borne pathogens through competing for habitat and nutrient. To increase the productivity coupled with an eco-friendly environment and cost effective nature, the inclusion of highly specific bacteria as bio-inoculants is necessary for the agriculture system.

PGPR is the burning interest of the researchers leading to the commercialization of the microorganisms for the growth of crops in an organic way to increase the existing population of beneficial microorganisms in the absence of pesticides and chemical fertilizers. Inoculating the seeds with the medium increases the availability of nutrients, solubilizing potassium, phosphorous, chelating iron and copper, oxidizing sulfur, and fixing nitrogen. Nitrogen can also be cycled from the organic material by a specific group of bacteria; these specific groups can also fix the atmospheric nitrogen in the soil [15]. Basing on the foresaid favoring concepts and knowledge on PGPR, the concept of biofertlizer came into existence to overcome the problematic issues of chemical fertilizers and pesticides.

13.1.1 Chemical Fertilizer and its Harmful Effect

In this 21st century, the unrestricted use of synthetic fertilizers for the vast production of crops is leading to the damage of the soil texture and health by hardening the soil, decreased fertility, polluting air and water, and release of toxic gases (greenhouse), which is hazardous to human health and environment. Chemical fertilizers are also harmful to the plants and soil in the long run as they have higher salt content. The fertile soil having essential soil nutrients and minerals are depleted by the uninterrupted use of chemical fertilizers. The soil fertility and other useful nutrients cannot be replenished by the use of chemical fertilizers; however nitrogen, potassium, and phosphorous are replenished [16]. The intense use of chemical fertilizers and certain nutrients results in soil degradation by making an imbalance in the supply of nutrients and also causes loss of equilibrium in stable soil. The hardening of soil is caused mainly due to the overuse of phosphorous which cannot be dissolved in water; similarly, alkaline fertilizers like sodium-nitrate develop alkalinity in soil reducing its fertility and making it barren [17].

Plant grows faster with the help of chemical fertilizers but they lack the time to develop good root growth, strong stems, nutritious fruits, and vegetables. The survival rate of these plants is much low due to their susceptibility towards pests and diseases as they lack a good immune system and enough resistance. Fertilizer burn (root burn) is also a major concern along with the use of excessive fertilizers as chemical fertilizers do not allow enough water intake for plants. Uses of excessive nitrogenous fertilizers (Urea) cause water pollution as it breaks down to nitrates and is easily soluble in water and penetrates through the soil to nearby rivers, wells and other water sources including groundwater. Nevertheless, it is to remember that the application of synthetic fertilizers destroys the natural soil microflora and fauna (which includes beneficial insects, fungus, bacteria, actinobacteria, verrucomicrobia, and some cyanobacteria) [18].

13.2 Biofertilizers "Boon for Mankind"

A substance containing living microorganism when applied to seeds, plant surfaces, or soil, colonizes the rhizosphere or the interior of the plant to increase the availability of the primary nutrients for the growth promotion is termed as biofertilizer. The plants can be grown healthy through the application of microorganisms present in a biofertilizer as it restores the soil's natural nutrient cycle and builds soil organic matter, thus enhancing the sustainability and the health of the soil. Microorganisms through the natural process fix nitrogen, solubilize phosphorous and potash, and stimulate plant growth through the synthesis of growth-promoting substances which is essential and adds nutrients to the plants. It is also expected that the reduced use of chemical fertilizer and pesticides is necessary with the use of biofertilizers, whereas biofertilizers are not replaceable material for the chemical fertilizer and their use.

Biofertilizers are sustainable and environmentally friendly organic agro-input. It has been recorded for a long time the use of *Rhizobium*, *Azotobacter*, *Azospirilium*, and blue-green algae (BGA) as biofertilizers. It is well known about the *Rhizobium* forming a symbiotic relationship with the root nodules of the leguminous plants [19]. Whereas *Azotobacter* is mainly used with crops like maize, wheat, cotton, potato, mustard, and other vegetable crops. Sorghum, millets, and sugarcane need *Azospirilium* inoculants as the best biofertilizer. For a paddy crop in both lowland and upland conditions, it is necessary to fix nitrogen for their growth which is mainly carried out by BGA belonging to cyanobacteria genus, *Nostoc* or *Anabaena* or *Tolypothrix* or *Aulosira*. Nitrogen up to 60 kg/ha/season and enrichment of soil with organic matter is carried out by *Anabaena* in association with water fern *Azolla*. In the coastal districts, a good practice is carried out of using seaweeds as manures as the seaweed is rich in potassium, phosphorous, trace elements, etc. Seaweeds are also used for breaking down clays. The bottom mud (contains abundant BGA) of the dried ponds in the tropical countries are used as manures, the mixture of BGA and seaweeds acts as a perfect biofertilizer [20].

13.3 Carrier-Based Biofertilizers

The preparation of biofertilizers is well established as carrier-based inoculation containing useful microorganisms. Microorganisms present in the carrier material helps in easy handling, high effectiveness, and long-term storage of biofertilizers. Bacterial inoculation process is the basic type of biofertilizer containing *Rhizobium*, nitrogen fixing rhizobacteria, PGPR, phosphate solubilizing bacteria, etc. In this chapter, type of carrier materials available for biofertilizers, and preparation in general of carrier-based inoculants will be described. Biofertilizer comprises bacteria having a close relationship with roots of the plants, mainly *Rhizobium* having a symbiotic relationship with legume roots. Successful inoculation of *Rhizobium* or rhizobacteria is achieved by placing the large population of the bacterial strains close to the emerging root that causes the maximum formation of nodules by the rhizobial strain. The rhizobial strain also occupies the major portion of the rhizosphere as the vital member of the rhizobacteria. There are two types of carrier-based systems in general practices as solid carrier-based biofertilizers and liquid biofertilizers.

13.3.1 Solid Carrier-Based Biofertilizers

Solid-based carrier system was first introduced to the general agricultural practices since ancient era. This implementation was gradually modified accordingly overtime. Nowadays, several different types of solid carriers are used for the growth and expansion of the beneficial microbes (mainly *Rhizobium*) present in the mixture, which are depicted in the Table 13.1 as given below.

13.3.2 Liquid Biofertilizer

Liquid formulation of the desired microorganism in their dormant form supplemented with the nutrients for their growth. In addition to the substances useful for the formation of resting spores and cyst, which increases the shelf life and tolerance to adverse conditions, is termed as liquid biofertilizer. The root exudates and the carbon present in the soil help in the germination of the active batch of cells from the dormant cells after reaching the soil.

Table 13.1 Solid carrier used for biofertilizers.

Carrier material	Inoculant bacterium	Characteristics
Sterile oxalic acid industrial waste	*Rhizobium* sp.	- Seed inoculation. - *Rhizobium* multiplication in a carrier in ambient temperature up to 90 days. - Grain yield, nodule number, and nitrogen content are increased after the carrier sterilization.
Alginate-perlite dry granule	*Rhizobium* sp.	- Soil inoculation. - Survival rate of *Rhizobium* strains for more than 180 days in dry granules form. - Without losing much viability the inoculant can be stored in a dry state.
Composted sawdust	*Bradyrhizobium*, *Rhizobium*, and *Azospirillum*	- Seed inoculation. - Good growth and survival of the inoculant strains.
Agriperlite, Expanded clay, Kaolin,Celite, Diatom, PorosilMP, Microcel, Vermiculite	*Agrobacterium radiobacter* K84	- Crown gall control - Improved formulation of K84 cells. - Effect of carrier storage temperature and carrier water content on survival of K84 was examined.
Cheese whey grown cells in peat	Rhizobium meliloti	- Seed inoculation - Better survival at various temperatures during storage, even under desiccation.
Mineral soils	*Rhizobium* sp.	- Seed inoculation - *Rhizobium* survived better at 4°C than at higher temperatures.

(*Continued*)

Table 13.1 Solid carrier used for biofertilizers. (*Continued*)

Carrier material	Inoculant bacterium	Characteristics
Coal, Charcoal, Lignite, Talcum Powder, Bentonite	*Rhizobium* sp., *Azotobacter* sp., *Azospirillum* sp., *Bacillus* sp., *Pseudomonads*	- Seed inoculation - Growth and survival of strains are well supported by these strains. - The count of the bacterial colonies is good at more than 10^7–10^9 bacterium per g till 12 months.
Granular inoculants amended with nutrients	*Bradyrhizobium japonicum*	- Seed inoculation - Betonite granules, illite, and smectite granules, or silica granules amended with glycerol, Na glutamate and inoculated with either peat or liquid *Bradyrhizobium japonicum* inoculants. - N content of the grains was increased and early nodulation of soyabean was found.
Soybean oil or peanut oil added with lyophilized cells	*Rhizobium* sp.	- Seed inoculation - Provide more protection than peat-based inoculants when rhizobia are inoculated on seeds and exposed to the condition of drought and high temperature.
Perlite	*Rhizobium* sp., *Bradyrhizobium* sp., *Bacillus*, *Pseudomonad*	- Seed inoculation - Bacterial survival is at its best when there is a combination of a sucrose adhesive with the perlite carrier - Produced a similar number of nodules, nodule dry weight, crop yield, and nitrogen content as peat-based inoculants.

(*Continued*)

Table 13.1 Solid carrier used for biofertilizers. (*Continued*)

Carrier material	Inoculant bacterium	Characteristics
Wastewater sludge	*Sinorhizobium meliloti*	- Seed inoculants - Result showed the suitability of using sludge as a carrier because it had the same or a higher potential than peat to support the survival of *S. meliloti*.
Wheat bran, sugarcane bagasse	*Rhizobium* sp./ *Bradyrhizobium* and rockphosphate solubilizing fungus *Aspergillus niger*	- Soil inoculants. - The number of cultured microorganisms was the highest with peat, followed by bran and sugarcane baggas.
Nutrient-supplemented pumice	*Rhizobium* sp.	- Seed inoculants - During sowing the inoculants is easily mixed with the seeds, easy to handle and storage.

13.4 Sterilization of the Carrier

To maintain the desired number of inoculant bacteria on the carrier for long storage period the process of sterilization of carrier material is required. The carrier should be sterilized using Gamma-irradiation as the process does not change the physical and chemical properties of the material. Gamma-irradiated at 50 kGy (5 Mrads) on the carrier material packed in thin-walled polythene bags. Another way of carrier sterilization is autoclaving. Carrier material is packed in partially opened, thin-walled polypropylene bags, and autoclaved for 60 min at 121°C.

In case of liquid biofertilizers the liquid medium carrier is autoclaved or sterilized via fermenter for 30 min at 121°C and 15 lbs pressure.

13.5 Merits of Using Liquid Biofertilizer Over Solid Carrier-Based Biofertilizer

Biofertilizers having a good number of cells and the efficiency of the microorganisms to fix nitrogen or solubilize phosphates determine the solidity and durability of the biofertilizer. The liquid biofertilizers are believed to be the best alternative to synthetic fertilizers and conventional carrier-based biofertilizers in the modern agriculture due to their high moisture-retaining ability, longer shelf life than carrier-based biofertilizers, better survival on the seed and nodulation, and ease of handling, storage, and transportation, all favoring a sustainable agricultural system of high productivity.

Basing on these criteria and focusing on other similar criteria, it can be assumed that there is an advantage of using liquid biofertilizers rather than using solid carrier-based biofertilizers.

i. Shelf life of the microbes is estimated to be 15–24 months in the case of liquid biofertilizers, whereas it is low (8–12 months) in the case of solid base.
ii. There is no effect of high temperature (45°C) on its properties in liquid biofertilizers, but temperature plays a vital role in the solid carrier-based due to the carrier and its tolerance to temperature.
iii. Contamination level is negligible in liquid biofertilizers, whereas contamination is the running problem in solid-based system.
iv. The bacterial count in the solid carrier-based cannot be maintained for 12 months but the load of bacteria is higher (10^8 CFU/ml) and can be maintained for 15–24 months in the liquid culture medium.
v. Liquid biofertilizers are easy for the farmers to use rather than using solid carrier-based biofertilizers.
vi. The farming community uses the liquid biofertilizers 10 times less than the solid carrier-based product.
vii. Liquid biofertilizers are easier for quality control protocols in comparison to the solid carrier-based biofertilizers.
viii. Liquid biofertilizers are cost effective with respect to solid carrier-based biofertilizers.
ix. Liquid biofertilizers have a high export potential in comparison to solid carrier-based system.
x. Mass production time of the liquid biofertilizer is less with respect to solid carrier-based system.

13.6 Types of Liquid Biofertilizer

Biofertilizer is a substance containing living microorganism that, when applied to seeds, plant surfaces, or soil, colonizes the rhizosphere or the interior of the plant to increase the availability of the primary nutrients for the growth promotion. With the implementation of biofertilizers, the nutrients are added through natural processes like solubilizing phosphates, nitrogen fixation, and stimulating the plant growth through the synthesis of growth promoting substances, whereas the fertilizer directly increases soil fertility. Biofertilizers are grouped in different ways based on their nature and function (Table 13.2).

Table 13.2 Categorization of different biofertilizers.

Sl. no.	Groups	Examples
Nitrogen (N_2) Fixing Biofertilizers		
1.	Free living	*Azotobacter, Clostridium, Anabaena, Nostoc*
2.	Symbiotic	*Rhizobium, Frankia, Anabaena azollae*
3.	Associative Symbiotic	*Azospirillum*
Phosphate Solubilizing Biofertilizers		
1.	Bacteria	*Bacillus megaterium* var. *phosphoricum, Bacillus circulans, Pseudomonas striata, Pseudomonas putida*
2.	Fungi	*Penicillium* sp., *Aspergillus awamori*
Potassium Mobilizing Biofertilizers		
1.	Arbuscular mycorrhiza	*Glomus* sp., *Gigaspora* sp., *Acaulospora* sp., *Scutellospora* sp., *Scelorocytis* sp.
2.	Ectomycorrhiza	*Laccaria* sp., *Pisolithus* sp., *Boletus* sp., *Amanita* sp.
3.	Orchid mycorrhiza	*Rhizoctonia solani*
4.	Bacteria	*Frateuria aurentia* and *Bacillus* sp.
Biofertilizers for micronutrients		
1.	Silicate and Zinc Solubilizers	*Bacillus* sp.
Plant Growth-Promoting Rhizobacteria		
1.	Pseudomonas	*Pseudomonas fluorescens*

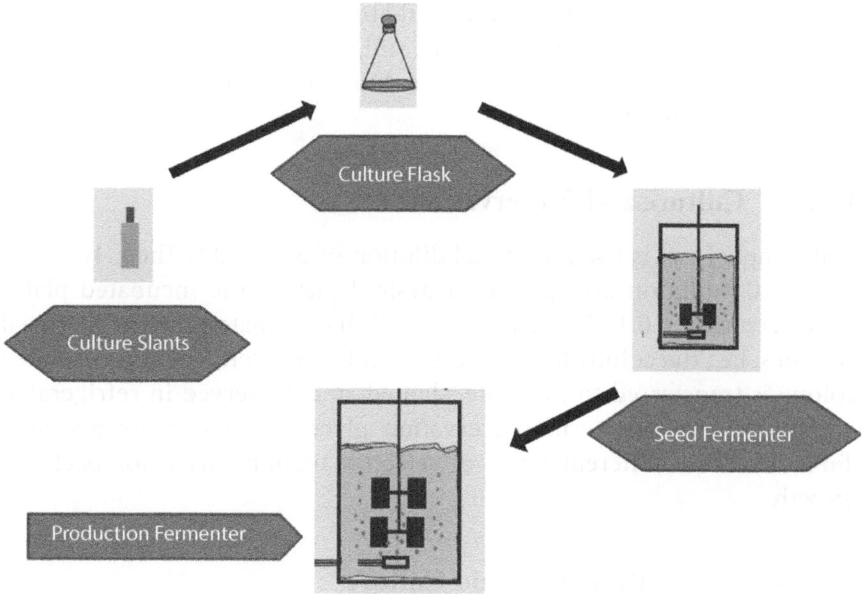

Figure 13.1 Schematic diagram showing mass production process of biofertlizer.

13.7 Production of Liquid Biofertilizers

The mass production or the commercial production of the liquid biofertilizers is carried out by the fermentation process [21] as shown in Figure 13.1. The process is carried out in closed vessels known as a fermenter.

13.7.1 Isolation of the Microorganism

The microorganism is isolated from the local soil. Top layer of the soil is removed, and 10 g of soil sample is collected from the location. The soils are stored in the refrigerated condition in sealed poly bags after they are powdered and sieved through a fine mesh with air dried up to 20% moisture level.

13.7.2 Preparation of Medium and Growth Condition

The desired concentrations of the ingredients were mixed in distilled water and sterilized at 120 ± 1°C (115 ± 0.1 kPa) for 15 min in an autoclave. A membrane filtration is required for sterilization where the ingredients are passed through a 0.22-μm membrane filter. Moreover,

1.5%–2.0% agar is added with the liquid broth after the pH is adjusted to form the solid medium. Plates, slants, still, and cultures were maintained in a BOD at 35 ± 0.1°C and shake cultures were maintained at 100 ± 0.5rpm at 35 ± 0.1°C.

13.7.3 Culture and Preservation

Soil sample (1 g) is used for serial dilution of up to 10^{-6}. Then, 100 µl of 10^{-4} to 10^{-6} dilution are spread on desired plates. The incubated plates were kept at 35 ± 0.1°C for 24–96 h in a BOD incubator and the bacterial colonies, i.e., the colony forming units (CFU) are seen. Then, the isolated colony is transferred to the plate, slanted, and preserved in refrigerated condition. For long-term preservation glycerol stocks were adopted. Table 13.3 lists different types of selective medium used for bacterial growth.

13.7.4 Preparation of Liquid Culture

Liquid culture is prepared in a small 5-L flask. The medium is sterilized for 20 min at 121°C and 15-kPa pressure and then cooled up to 35°C. The bacterial inoculum is then transferred to the liquid medium through a loop full of bacteria from the stored plates. The liquid medium is then placed in the BOD incubator cum rotary shaker at 100 ± 0.5 rpm and 30 ± 0.1°C for the desired growth time.

Table 13.3 Bacteria used as liquid biofertilizers and their selective growth medium.

Sl. no.	Microorganism	Medium	Growth period
1.	*Azotobacter* sp.	Jensen's nitrogen free medium	24–48 h
2.	Phosphate solubilizing bacteria	Pikovskaya medium	24 h
3.	Potash mobilizing bacteria	Alexandrov's medium	4–5 days
4.	*Rhizobium* sp.	YEMA medium	3–4 days
5.	Zinc solubilizing bacteria	Zinc silicate medium	48–72 h

13.7.5 Fermentation and Mass Production

The closed vessel with appropriate arrangement for aeration, agitation, temperature, and pH control, and drain or overflow vent to separate the waste biomass of cultured microorganisms along with their products is termed as fermenter (bioreactor). The fermentation technology stands on six different parameters which if not fulfilled hampers the growth and development of the microorganism. The parameters are temperature, pH, agitation, pressure, aeration, and turbidity. These parameters are generally maintained in a liquid biofertilizer producing industry for the growth of microorganisms [22].

The desired medium inside the fermenter is first sterilized at 121°C and 15-kPa pressure for 20 min and cooled to 35°C. Then, the inoculums already prepared in 5-L flask are transferred to the cooled medium inside the vessel very aseptically, and then, the parameters are maintained inside the vessel. Temperature is maintained at 35°C–38°C with pH ranging accordingly. The culture is left for the desired time for the growth of the microorganism.

13.7.6 Formulation of the Liquid Biofertilizers

The Microorganisms before bottling and packing for commercial use need to be formulated to increase their self-life and performance. The liquid formulation is also essential for the product to increase its efficiency and growth. The materials required for formulation are as follows.

i. Preservative: A preservative is required for the preservation of the microbial cells present inside the product. Mainly glycerol (2%–5%) is added as a preservative.
ii. Adhesive: An adhesive is required for the adhesion of the microbial cells to the seeds and roots. Acacia powder (1%–2%) is added as an adhesive.
iii. Lubrication and Turbidity: A lubricant is necessary for the lubrication of the cells. Edible Vanaspati Oil (1%–2%) is added.
iv. Surfactant: To reduce the surface tension of the liquid medium and preventing atmospheric contaminants from entering inside the packing. Labolene (0.5%–0.8 %) is added.

13.8 Applications of Biofertilizers

Biofertilizers are used to enrich the soil nutrients by natural processes, the application of liquid biofertilizers are mainly carried out by three different methods (Table 13.4).

i. Seed Treatment: Crops mainly grown from the seeds are treated with liquid biofertilizers before field application. In this process 8 to 10kg of seeds are taken in a bowl and 250 ml of liquid biofertilizers with 1 L of water/rice starch and mixed gently. The seeds are kept in the solution for 15–20 min and then removed to a polythene sheet under the shed away from sunlight and heat. The treated seeds are left to dry for 20 min. After they are dried the seeds are used for sowing. Examples: beans, lentils, pulses, vegetables, etc.
ii. Root Treatment: Crops mainly grown from the seedling are selected for this method. Seedling needed for 1 acre of land is treated with 250 ml of liquid biofertilizer with 1 L of starch/water. The seedling roots are dipped in the solution and kept for 15–20 min. Then, the seedlings are kept under the shed for 20 min to dry up. Further these are planted in the soil. Examples: Rice, Wheat, barley, Brinjal, some vegetables, etc.
iii. Soil Treatment: Plants which are already grown and ready for flowering and fruits are treated with this method. About 500 ml of liquid biofertilizer is added to the 3–4 kg of cow dung/humus and kept for 15–20 days with 20% moisture. The microorganism grows rapidly in this source and becomes a good source of nutrients and other resources. This mixture is applied to the plant by removing the top layer of the soil near the root region of the plant. Then, 200–250 g of the mixture is added surrounding the plants root region; e.g., coconut, banana, mango, and other fruiting trees. But this type of treatment is not as good as the other two as it gives 60%–70% results comparing the other two with 80%–90% results.

Table 13.4 Different types of biofertilizers have different treatment types which are as follows.

Sl. no.	Product	Treatment	Advantages	Applications
1.	Azotobacter	Seed, Root, Soil.	It helps in increasing the greenery of the crops. Fix 30–40 kg/ha N and 15%–20% increase in yield.	Useful for all rabi crops, oilseeds, vegetables, fruits, etc.
2.	Phosphate Solubilizing Bacteria	Seed, Root, Soil.	It helps in strengthening of the roots which is useful in the accumulation of nutrients from the soil. 20–25 kg/ha P and 10%–20% increase in yield.	Useful for oilseeds, legumes, vegetables, flowers, fruits, etc.
3.	NPK Consortia	Seed, Root, Soil.	It helps in the strengthening of the stems and growth of the leaves, flowers and fruits/grains. Fixes 30–40 kg/ha N, 20–25 kg/ha P, 10–20 kg/ha K and 10%–20% increase in yield.	Useful for oil seeds, legumes, vegetables, flowers, fruits, etc.

(*Continued*)

Table 13.4 Different types of biofertilizers have different treatment types which are as follows. (*Continued*)

Sl. no.	Product	Treatment	Advantages	Applications
4.	Rhizobium	Seed, Root	Enhances the root nodules for better Nitrogen fixation	Leguminous Plants.
5.	Azospirillum	Seed, Root	It helps in increasing the greenery of the crops. Fix 20–25 kg/ha N and 5%–10% increase in yield.	Cereals like Rice, Maize, Wheat, Barley, etc.
6.	Zinc Solubilizing Bacteria	Seed, Root	It helps in the drought, increasing the drought resistance of the crop. It provides 02–05 Kg/ ha Z and 5% increase in yield.	Fruits, vegetables, cereals, flowers, oilseeds, etc.

13.9 Conclusion

Application of biofertilizer increases the soil fertility which is further enhanced by continuous use. By availing the minor and major nutrients from its complex forms to easily accessible form for plant consumption, microbial biofertilizers increases the yield in a simpler way. Agribiotechnology plays an important role in production of liquid biofertilizer which are efficient with soil microorganisms that upsurge the effective plant captivation, increasing soil quality, reducing chemical intakes, and increasing yields. The treated soils can be rehabilitated in natural way sustainability without hampering the soil itself and the environment.

References

1. Baliyarsingh, B., Nayak, S.K., Mishra, B.B., Soil Microbial Diversity: An Ecophysiological Study and Role in Plant Productivity, in: *Advances in Soil Microbiology: Recent Trends and Future Prospects, Microorganisms for Sustainability*, vol. 4, T.K. Adhya, B.B. Mishra, K. Annapurna, D.K. Verma, U. Kumar (Eds.), pp. 1–17, Springer Nature, Singapore, 2017.
2. Nayak, S.K. and Mishra, B.B. (Eds.), *Frontiers in soil and Environmental Microbiology*, CRC Press, Boca Raton, Unites States, 2020.
3. Nayak, S.K., Nayak, S., Patra, J.K., Rhizobacteria and its biofilm for sustainable agriculture: A concise review, in: *New and Future Developments in Microbial Biotechnology and Bioengineering: Microbial Biofilms*, M.K. Yadav and B.P. Singh (Eds.), pp. 165–175, Elsevier, Amsterdam, Netherlands, 2020.
4. Pahari, A., Pradhan, A., Nayak, S.K., Mishra, B.B., Bacterial Siderophore as a Plant Growth Promoter, in: *Microbial Biotechnology*, vol. 1, J.K. Patra, C. Vishnuprasad, G. Das (Eds.), pp. 163–180, Springer Nature, Singapore, 2017.
5. Glick, B.R., Patten, C.L., Holguin, G., Penrose, D.M., *Biochemical and Genetic Mechanisms used by Plant Growth Promoting Bacteria*, Imperial College Press, London, 1999.
6. Govindasamy, V., Senthil Kumar, M., Kumar, U., Annapurna, K., PGPR: Biotechnology for management of abiotic and biotic stresses in crop plants, in: *Potential microorganisms for sustainable agriculture*, D.K. Maheshwari and R.C. Dubey (Eds.), pp. 26–48, IK International Publication, India, 2008.
7. Govindasamy, V., Senthilkumar, M., Gaikwad, K., Annapurna, K., Isolation and characterization of ACC deaminase gene from two plant growth promoting rhizobacteria. *Curr. Microbiol.*, 57, 312, 2008.
8. Kumar, U., Vikas, S., Vithal, L., Pranita, B., Annapurna, K., Fluorescent pseudomonads in rapeseed mustard. *ICAR News*, 16, 12, 2010.
9. Annapurna, K., Ramadoss, D., Senthil Kumar, M., Govindasamy, V., Mageshawar, V., Kumar, U., Vithal, L., Pranita, B., Effect of plant growth promoting rhizobacteria in wheat. *ICAR News*, 16, 15, 2010.
10. Govindasamy, V., Senthil Kumar, M., Mageshwar, V., Kumar, U., Pranita, B., Vikas Sharma, V., Annapurna, K., *Bacillus and Paenibacillus spp.*: Potential PGPR for sustainable agriculture, in: *Plant growth and health promoting bacteria, Microbiology Monograph*, D.K. Maheshwari and R.C. Dubey (Eds.), pp. 333–364, Springer, Berlin, Heidelberg, 2011.
11. Kumar, U., Kumar, V., Ramadoss, D., Bose, P., Annapurna, K., Microbial consortium for increased nutrient uptake from basmati rice rhizosphere. *ICAR NEWS*, 18, 14, 2012.
12. Nayak, S.K., Nayak, S., Mishra, B.B., Antimycotic Role of Soil Bacillus sp. Against Rice Pathogens: A Biocontrol Prospective, in: *Microbial Biotechnology*, vol. 1, J.K. Patra, C. Vishnuprasad, G. Das (Eds.), pp. 29–60, Springer Nature, Singapore, 2017.

13. Kloepper, J.W., Leong, J., Teintze, M., Schroth, M.N., *Pseudomonas* siderophores: A mechanism explaining disease suppressive soils. *Curr. Microbiol.*, 4, 317, 1980, 1980.
14. Defago, G. and Haas, D., Pseudomonads as antagonists of soil borne plant pathogens: mode of action and genetic analysis, in: *Soil Biochemistry*, J.M. Bolley and G. Stortzky (Eds.), pp. 249–291, New York, Baul, 1990.
15. Goring, C.A.I. and Laskowski, D.A., The effects of pesticides on nitrogen transformations in soils, in: *Nitrogen in Agricultural Soils*, F.J. Stevenson (Ed.), pp. 689–720, American Society of Agronomy, Madison, Wisconsin, USA, 1982.
16. Nayak, S.K., Dash, B., Baliyarsingh, B., Microbial Remediation of Persistent Agro-chemicals by Soil Bacteria: An Overview, in: *Microbial Biotechnology*, vol. II, J.K. Patra, G. Das, H.S. Shin (Eds.), pp. 275–301, Springer, Singapore, 2018.
17. Obrzud, R.F. and Truty, A., *The hardening soil model -A Practical guidebook, Z Soil.PC 100701 report revised 21.10.2018*, Zace Services Ltd, Preverenges, Switzerland, 2018.
18. Dash, B., Nayak, S., Pahari, A., Nayak, S.K., Verrucomicrobia in soil: An agricultural prospective, in: *Frontiers in soil and Environmental Microbiology*, S.K. Nayak and B.B. Mishra (Eds.), pp. 37–46, CRC Press, Boca Raton, Unites States, 2020.
19. Pahari, A., Pradhan, A., Nayak, S.K., Mishra, B.B., Plant Growth Promoting Rhizobacteria (Pgpr): Prospects and Application, in: *Frontiers in soil and Environmental Microbiology*, S.K. Nayak and B.B. Mishra (Eds.), pp. 47–55, CRC Press, Boca Raton, Unites States, 2020.
20. Mishra, B.B., Nayak, S.K., Pahari, A., *Agriculturally Important Microorganisms: Mechanisms and Applications for Sustainable Agriculture*, New India Publishing Agency, New Delhi, India, 2021.
21. Namfon, P., Ratchanok, S., Chalida, D., Optimization of the liquid biofertilizer production in batch fermentation with byproduct from MSG. *AIP Conference Proceedings*, vol. 1823, p. 020074, 2017.
22. Mohod, S., Lakhawat, G.P., Deshmukh, S.K., Ugwekar, R.P., Production of Liquid Biofertilizers and its Quality Control. *Int. J. Emerg. Trend Eng. Basic Sci.*, 2, 158, 2015.

14

Extremozymes: Biocatalysts From Extremophilic Microorganisms and Their Relevance in Current Biotechnology

Khushbu Kumari Singh[1†] and Lopamudra Ray[1,2*†]

[1]School of Biotechnology, Kalinga Institute of Industrial Technology, Bhubaneswar, Odisha, India

[2]School of Law, Kalinga Institute of Industrial Technology, Bhubaneswar, Odisha, India

Abstract

Prokaryotic life does not seem to be restricted to a specific environment and has ruled our planet evolutionary history, evolving and inhabiting almost all environmental niches. Studies in the past few decades showed microbial communities can adapt to diverse environmental conditions which include extremes of pressure, temperature, pH, and salinity. These microorganisms, known as extremophiles have adopted different molecular mechanisms to survive such extreme conditions. Biocatalysts produced by such microorganisms are termed as extremozymes possessing unique properties and stability owing to new opportunities for biotransformation and development of economy. Extremophile-derived biocatalysts or extremozymes can perform catalytic activity under extreme conditions, suited to industrial processes, which were earlier thought impossible for enzyme activity. Lately, different extremophilic proteins such as thermophilic, halophilic, piezophilic, and acidophilic have been studied. The optimal stability and activity of these extremophilic enzymes make them a biocatalytic alternative for the present-day biotechnological applications. Also, these enzymes represent a keystone for environment-friendly and sustainable development of industrial technologies.

***Keywords*:** Extremozymes, thermophiles, halophiles, acidophiles, piezophiles, biotechnological applications

**Corresponding author*: lray@kiitbiotech.ac.in; lopamudra.ray@kls.ac.in

†*Equal contribution as first author*

Bibhuti Bhusan Mishra, Suraja Kumar Nayak, Swati Mohapatra, and Deviprasad Samantaray (eds.) Environmental and Agricultural Microbiology: Applications for Sustainability, (293–312) © 2021 Scrivener Publishing LLC

14.1 Introduction

The term extremophile was first coined by Robert MacElroy in the 1970s [1], although the remarkable discovery of extremophiles was made way back in 1966 when Thomas Brock noticed microorganisms growing in the extreme conditions of Yellowstone National Park (USA). Since then, thermophiles have been found worldwide in geothermal sites including hot springs, volcanic areas, Antarctic biotopes, and other locations that were earlier not believed to support life. Extremophiles are a group of microorganisms that can survive and adapt to extreme environmental variables such as temperature (as low as −2°C to 10°C and as high as 55°C to 121°C), salinity (2 to 5 M), hydrostatic pressure (>500 atmospheres), alkalinity (pH > 9), acidity (pH < 4), extremes of chemicals and heavy metals, and polyextremity or different osmotic barriers [2–5]. Depending on their habitat, extremophiles are categorized as acidophiles, alkalophiles, thermophiles, osmophiles, psychrophiles, radioresistance, and halophiles [6–8]. In Table 14.1, different forms of extreme conditions with their representative microorganisms growing under related environmental conditions have been represented [9].

In the last decade, a major push driving extensive research on extremophiles received more attention for the potential biotechnological and industrial applications [10–12]. More recently, an intensive study on these extremophilic microorganisms has been fueled by industries for the search of novel species and the molecular mechanism that helps them grow in such extreme environments expecting novel pathways and enzymes [13]. Enzymes being nature's biocatalysts help to accelerate the rate of a chemical reaction by lowering its activation energy. Enzymes have been used unknowingly in biotechnological processes for food and beverage production since ancient times. Numerous enzymes have been studied and identified in the past few decades, where most of these enzymes have been used for biotechnological and industrial applications, still, enzyme markets are insufficient to meet the industrial demands [14, 15]. This insufficiency of the enzymes is related to its instability at different industrial conditions [16]. Also, the enzymes used for industrial purposes employ ecological processes [17]. Therefore, the industry needs biocatalyst which can withstand the extreme conditions while processing, such as a wide range of temperature, pH, aeration with high reproducibility [18–20]. With the advance understanding of biotechnology, interest for stable enzymes has grown as a policy to attain a bio-based economy. The exploration of enzymes from extremophiles with improved stability and novel activities at extreme conditions and their industrial importance have been studied lately and it is believed that the stability of these enzymes can contribute to filling the gap between chemical and biological processes [21]. This chapter focuses on the pertinent role of extremophilic microorganisms and their potential in biotechnology and industrial applications.

14.2 Extremophiles: The Source of Novel Enzymes

Extreme environments as classified in Table 14.1 refer to different extreme environmental conditions at which microorganisms thrive. The intrinsic characteristics of the biocatalysts from these microorganisms have

Table 14.1 Different group of extremophiles and their microbial habitats.

Environmental feature	Classification	Description	Microbial habitations	Examples
Temperature	Hyperthermophiles	>80°	Hydrothermal vents	*Pyrolobus furiosus*
	Thermophiles	50°–80°	Hot-springs	*Cyanidium caldarium*
	Mesophiles	15°–4°	Acid mine drainage	*Acidithiobacillus ferrooxidans*
	Psycrophiles	<15°	Arctic soils, Deep oceans	*Arthrobacter russicus*
Salinity	Halophiles	2-5 M NaCL	Salt lakes, brine	*Tetragenococcus halophilus*
pH	Acidophiles	pH < 2	Acidic hot springs, sulfide mines	*Alicyclobacillus, Ferroplasma* sp.
	Alkaliphile	pH > 8	Soda lakes/ desserts	*Natronomonas pharaonis, Spirulina* sp. (pH 10.5)
Radiaoactivity	Radioresistant		Soil polluted area	*Thermococcus gammatolerans*
Toxic metals	Metalophiles	Endure higher metal concentrations	Polluted areas, volcanic areas	*Ralstonia metallidurans* CH34 (Zn, Co, Hg, Pb)

influenced the industrial market with a constant rise in their demand over the year. These biocatalysts being active and resistant at high temperatures are often favored for different polymers (chitin, starch, and cellulose) degradation. Under these extreme conditions, the solubility and subsequently substrates usability is increased. These biocatalysts besides being easily biodegradable are stable and efficient and are considered as a greener solution to many industrial challenges.

14.2.1 Thermophilic Extremozymes

Thermophiles are known to be the most studied extremophiles in the past four decades [40, 41]. A work on thermophiles has gained broad attention because of its potential to survive at a higher temperature attributed to the need for a stable and active enzyme at higher temperatures (41°C–122°C) [42]. A wide number of thermophilic enzymes particularly proteases and different polymer-degrading enzymes such as amylases, chitinases, cellulases, xylanases, pectinases, and phytases have been characterized and used for industrial applications (Table 14.2) [43]. Enzymes from thermophilic microorganisms possess certain benefits to compare to their mesophilic counterparts. These thermophilic enzymes can survive proteolysis and are resistant to extreme conditions like high pressure, salinity, organic solvents, and denaturing agents. Moreover, the use of these enzymes reduce the risk of contamination and undesired product formation, by reducing adhesiveness, and increasing the substrate solubility [44].

To maintain its stability and activity, thermozymes possess certain physical properties and electrostatic interactions. Several studies have been made to elucidate and compare the structural feature of thermozymes with that of its mesophilic counterparts, to understand the mechanism lying behind its thermostability [45]. After a comprehensive study, structural alignment, and homology modeling, it was concluded that the thermostability of the enzyme was acquired by enhancement of electrostatic charge, macromolecule core property and replacing the exposed reactive amino acids [46–48]. Thermostable enzymes such as lipases are being utilized by different biotechnological industries for esterification, transesterification, and organic biogenesis. Also, thermostable lipases are being utilized in paper, leather, milk, and different pharmaceutical industries [18, 19]. Thermostable proteases are used for the biogenesis of dipeptides, DNA, and starch purification [49, 50]. Thermozymes including cellulases and xylanases have been used in bleaching industries and have had an important application in bioremediation [51, 52]. Moreover, thermostable

Table 14.2 Biocatalysts isolated from different extremophiles and their foreseen application in biotechnological industries.

Category	Environmental source	Biocatalysts	Biotechnological applications	References
Thermophiles (50°–110°)	Hot spring, Yellow Stone National Park, USA	Proteases, lipases	Detergents formulation, hydrolysis in food and feed industries, removal of lipid and protein stains, transesterification of waste oils and alcohols for biodiesel production, wastewater management, flavor modification in food industries.	[15, 22–26]
		Amylases, glucosidases, cellulases, starch	Starch, cellulose and pectin processing, washing of cotton fabrics, the ripening of cheese, dough fermentation, saccharifying enzymes, oligosaccharide synthesis, clarifying wine and juices	[15, 24, 27–30]
		Xylanases	Hydrolysis of starch, paper bleaching, cheese ripening	[31]
		DNA polymerases, dehydrogenases	Genetic engineering and stereospecific reactions	[15]
Psychrophiles (0°–20°)	Antarctic ice and Arctic ocean	Amylases, lipases, proteases, dehydrogenases	Detergent formulation, bakery, polymer degrading agents, biosensors	[15, 23, 32]

(Continued)

Table 14.2 Biocatalysts isolated from different extremophiles and their foreseen application in biotechnological industries. (*Continued*)

Category	Environmental source	Biocatalysts	Biotechnological applications	References
Piezophiles (Hydrostatic pressure of 40MPa or more)	The deep ocean, Antarctic ice, Mariana Trench	Lipases	Food processing and antibiotic production	[15, 33]
Halophiles (2%–20% salt)	Salt lake, USA	Proteases, dehydrogenases	Peptide synthesis, cosmetic adhesives, pharmaceuticals	[22, 34, 35]
	Volcanic spring, Acid mine drainage, USA	Amylase, glucoamylase Proteases Oxidases Cellulases	Starch processing Feed component for animals Coal desulfurization Elimination of hemicellulose from animal feed material	[36, 37] [38, 39] [15] [22]
Alkaliphiles (pH > 10)	Soda lake, USA	Cellulases, proteases, amylases, lipases, cyclodextrin	Polymer degrading agents in detergents, fermentation of beer, wine, juice, and bread	[15, 38]

amylases from *Pyrococcus furiosus* are found to be applicable in mutational studies [53]. Also, the biodetergents being used today possess thermostable enzymes such as proteases, lipases, and cellulases that are unsusceptible to extreme conditions. Thermozymes avail not only their activity at a higher temperature but they also lack the catalytic activity confirmation at high temperatures [32]. Thermophilic enzymes are considered to be of great potential for biotechnological applications.

14.2.2 Psychrophilic Extremozymes

Cold environment like Antarctica and deep oceans covering 70% of the earth's surface, representing the major biome harbors plenty of psychrophilic microorganisms. These microbes are functional in a restricted temperature range with less energy demand. Psychrophiles have evolved to thrive at cold temperature by developing several biochemical mechanisms, including RNA chaperones, cold shock proteins, increased membrane fluidity preserving the semi-liquid state of membranes, and producing secondary metabolites, enzymes, antifreeze proteins active at a cold temperature [54, 55]. The most common adjunctive feature of psychrophilic extremozyme is their reaction rate which is higher at a lower temperature [56]. The thermal stability of psychrophilic enzymes at low temperatures can be explained by the rise in the flexibility of the molecule when compared with that of thermophilic extremozymes and mesophilic enzymes. The flexibility of psychrophilic enzymes at low temperatures is maintained due to the presence of more of α-helix compare to β-sheets in psychrophilic proteins [57].

Recently, attempts to decrease the energy consumption have made extremozymes from these psychrophilic microorganisms an interesting study for industrial applications. Psychrophilic enzymes are being used in diverse industries including food processing and molecular diagnostics to chemical synthesis [58, 59]. For example, psychrophilic extremozymes such as beta glycanases, proteases, and lipases are considered to have commercial potential to be used in detergent industries making laundry feasible at a lower temperature. The polymer degrading psychrophilic extremozymes such as cellulase, glucosidases, and amylases have gained interest by pulp, paper, biofuel, and textile industries and also for different bioremediation application [60]. Also, these psychrophilic enzymes are known to provide a potential benefit to food and feed industries, by avoiding spoilage of thermosensitive products saving their nutritional value and flavor [58, 61].

14.2.3 Halophilic Extremozymes

Extremely saline environments, like the Dead Sea and Great Salt Lake, are yet another extremophilic habitat inhabiting halophilic microorganisms. These halophiles can survive a hypersaline environment (at least 1 M NaCl) by maintaining osmotic balance and have evolved modifying its structural, chemical, and physiological parts allowing protein selectivity and stability [62, 63].

The halophilic enzymes have employed different adaptation mechanisms. These enzymes acquire more negative charge amino acids on their surface to be active at higher ionic strength and inhibit precipitation. Also, the negative charge on the surface of halophilic proteins helps them by reducing their tendency to precipitate at higher salt concentration, by decreasing their surface hydrophobicity. Accordingly, halophilic extremozymes when compared to its no halophilic counterpart exhibit low solubility in the presence of low surrounding salt concentration, limiting their applicability [64]. However, these enzymes are active and stable in media with little water activity, as they have enough water to possess proper charge distribution at the active site, to maintain the enzyme conformation and structure [65]. In the presence of high sodium and potassium chloride concentrations, these enzymes are also involved in various stabilization and solubility processes, increasing the enzyme's solubility in the presence of organic solvents [66, 67]. Halophiles such as *Marinococcus, Micrococcus, Bacillus,* and *Halobacillus* have reported the production of halophilic extremozymes like lipases, xylanases, and proteases [68, 69]. Also, halophilic lipases and esterase are known for their potential in polyunsaturated fatty acids (PUFAs) and biodiesel production [70, 71]. Moreover, to date, halophiles from the archaeal domain are considered to be the main source for halophilic extremozymes. For example, halophilic extremozymes are being used in stabilizing agents like betains, ecotines, and different polymers which are used for the manufacture of biodegradable plastics [72]. Also, retinal proteins like bacteriorhodopsin are used in holographic films and other light-sensitive applications [73]. Halophilic enzymes compressed in reverse micelles can be used for the development of novel applications like bioremediation for the accumulation of toxic and hazardous waste [65].

14.2.4 Alkaliphilic/Acidiophilic Extremozymes

Microorganisms surviving under extremes of pH are useful for industries requiring highly acidic or alkaline reaction conditions, for instance in industries dealing with detergent production. In addition, one of the most remarkable characteristics of these extremophilic microorganisms is their ability to maintain their neutral internal pH, such that the extreme growth environment does not influence the intracellular enzymes of these extremophilic microorganisms. Extracellular enzymes, however, are suited to the extremes of the pH environment. The need for stable and active enzymes at extreme pH by laundry industries have encouraged alkaliphilic bacterial and archeal screening to produce enzymes such as proteases, lipases, amylases, and other enzymes that are resistant under high pH and chelator concentration of detergents. Also, studies have been made by combining homology PCR and activity screening to screen alkaline proteases from archaeal and bacterial isolates, obtained from extreme environments [74]. Also, acidophilic biocatalysts such as amylases and glucoamylases resistant at low pH for polymer hydrolyzing applications are being screened [40, 75].

14.2.5 Piezophilic Extremozymes

Microorganisms that thrive at hydrostatic pressures such as in the deep ocean and volcano environment like *Pyrococcus abyssi* are known as piezophiles [76, 77]. Several studies showed *Sulfolobus solfataricus* adaptation to the piezophilic environment because of the presence of Sso7D protein [78, 79]. Peptidases from *Pyrococcus horikoshii* have been reported for its stability and activity at hydrostatic pressure [80]. Piezophiles even though after having great potential to serve biotechnological industries, very few research exists on biocatalysts from piezophiles. Alpha-amylase from piezophiles has been reported to produce trisaccharides in the presence of maltooligosaccharide as a substrate, offering great potential for food and feed industries [81, 82]. Piezophilic biocatalysts have shown high potential to be used in different industries including food, detergent, and chemical industries [2].

14.3 The Potential Application of Extremozymes in Biotechnology

Extremophiles are considered to be of great importance for the future development of biotechnological industries [83]. The expected potential of extremozymes has increased ascendingly with the positive screening of novel microbial strains and recognition of new compounds, pathways, and molecular mechanisms. The ever increasing rise in demands for biocatalysts and metabolites by industries has made progressive research in a direction to screen enzymes capable to withstand the extremes of industrial procedure and conditions [84]. For instance, extremozymes that are being commercially used include protease, amylase, lipase, cellulase, pectinase, pullulanase, chitinase, oxidase, peroxidase, glucoamylase, and many more as shown in Table 14.2. With the increase in energy consumption, the need for psychrophilic enzymes such as amylase, lipases, and proteases have increased and are now available with industries including Novozymes and Genecor [85]. Also, these psychrophilic enzymes because of its high activity and low structural stability are considered of great importance in food and feed industries [36]. In biotransformation industries together with volatile substrates, pharmacological industries, and cosmetic industries, psychrophilic extremozymes are counted to be of great potential owing to their high structural flexibility. Their low energy requisite and high flexibility offers a substantial advantage above mesophilic enzymes [86]. Moreover, these enzymes are also being used for agricultural applications to increase water management for plants that are under water-deficient pressure [87, 88]. Extremophiles are considered to be a boon for biotechnological applications. Their capability to produce enzymes that can sustain extremes of the industrial procedure is useful in making commercially valuable products. And one of the most important applications of these extremozymes is their usage in the bioremediation of lethal pollutants from water and sediments and for producing biological molecules for pharmacological and industrial purposes [89–92].

Extremozymes, being active and stable at a wide range of temperature, pH, ionic strength, salt concentration, and their ability to perform even in organic solvents which would otherwise degrade most of the other biocatalysts makes enzymes from extremophile offer a wide range of biotechnological prospects for biotransformation and biocatalysis [93, 94]. These extremozymes are being used for different commercial purposes [94]. Approximately, 65% of the worldwide production of extremozymes include detergent, starch, paper, and textile industries and rest 25% of the extremozymes are being used by food processing industries [95]. Thermostable enzymes such as α-amylase and amylopullulanase, due to its stability at high temperature and narrow pH range and is highly cost-effective, are used in the starch saccharification process [96]. Also, thermostable and alkali xylanases obtained from extremophiles have shown benefits in the pulp bleaching process by pre bleaching of pulps to decrease their chlorine need [97–99]. Hyperthermophilic archaeon *Thermococcus kodakaraensis* KOD1 has been reported for thermostable chaperonins CpkA and CpkB which are used by biotechnological industries for enzyme stabilization

[100]. Besides, these extremozymes usually have a high reaction rate, making them capable of removing xenobiotic contaminants and modulating the accumulation of heavy metals and pollutants from the environment. For example, a halophilic *Marinobacter hydrocarbonoclasticus* bacteria are known for its ability to breakdown aliphatic and aromatic hydrocarbon contaminants [101]. Also, an extremely enriched halophilic bacterial culture was cultivated by [102] for degrading toxic compounds such as benzene, toluene, and xylene. Further, extremophiles producing cellulolytic enzymes are also considered to be of great importance in biotechnological applications relating to brewing, cosmetics, textiles, paper, pulp food, and detergent industries. Extremophilic cellulases and proteases are also involved in cellulose modification to upsurge its color intensity and dirt removal from cotton garment industries [83, 103]. Xylanases from extremophiles also offers great biotechnological potential for industries by bioleaching of pulp and paper and thus decreasing the environmental halogen contamination [104, 105].

Additionally, one of the foremost biotechnological push on are likely in the area of protein engineering. Structural property identification of proteins responsible for thermal activity and stability plays a significant role in the development of proteins with required catalytic and thermal properties. For instance, an extremophile *Bacillus stearothermophilus* was mutated based on rational design to increase its thermostability [106]. The protein after the mutation was found to be thermostable and functional at 100°C and even with denaturing agents compared to the wild type protein. Further, a breakthrough in biotechnological industries are expected with the development of genetically engineered extremophiles. A genetically modified strain of *Deinococcus* has been studied for its potential in organopollutant degradation in radioactive and metal waste environment. This recombinant strain was genetically modified to express toluene dioxygenase which facilitated toluene, chlorobenezene, and indole oxidation in radioactive environment [107]. The unique and diversified properties of enzymes from extremophiles involving great reproducibility, high stability, low energy consumption, and economic viability represents importance of these enzymes compared to mesophilic enzymes in biotechnological applications by different industries [108, 109].

14.4 Conclusion and Future Perspectives

The paradigm moves of industries toward renewable source consumption have increased the need for biocatalysts which is speculated to increase the demand for thermostable enzymes in the future. Extremozymes, considering its stability and activity in extreme environments are being used as a potent source of novel enzymes. Particularly, thermophilic extremozymes because of its high resistivity under extreme temperature, organic solvents, and pH are considered as of great potential in biotechnological industries. Also, the economic potential of extremozymes has been seen in different industrial applications including food, agriculture, laundry, textile, paper, and pharmaceutical. The increase in demand for extremophilic biocatalysts by biotechnological industries with the development of new industrial procedures based on extremozymes has made extremophiles an interesting subject of research. The extremophiles are known as a viable source can be used in different bio-based applications toward the development of the biotech economy.

References

1. MacElroy, R.D., Some comments on the evolution of extremophiles. *Biosystems*, 6, 74, 1974.
2. Cavicchioli, R., Amils, R., Wagner, D., McGenity, T., Life and applications of extremophiles. *Environ. Microbiol.*, 13, 1903, 2011.
3. Deppe, U., Richnow, H.-H., Michaelis, W., Antranikian, G., Degradation of crude oil by an arctic microbial consortium. *Extremophiles*, 9, 461, 2005.
4. Navarro-González, R., Iniguez, E., de la Rosa, J., McKay, C.P., Characterization of organics, microorganisms, desert soils, and Mars-like soils by thermal volatilization coupled to mass spectrometry and their implications for the search for organics on Mars by Phoenix and future space missions. *Astrobiology*, 9, 703, 2009.
5. Seitz, K.H., Studdert, C., Sanchez, J., de Castro, R., Intracellular proteolytic activity of the haloalkaliphilic archaeon *Natronococcus occultus*. Effect of starvation. *J. Basic Microbiol.*, 37, 313, 1997.
6. Edwards, C. (Ed.), *Microbiology of extreme environments*, McGraw-Hill, University of California, 1990.
7. Jennings, D., Osmophiles, in: *Microbiology of extreme environments*, C. Edwards (Ed.), pp. 117–146, McGraw-Hill, University of California, 1990.
8. Ulukanli, Z. and DIĞRAK, M., Alkaliphilic micro-organisms and habitats. *Turk. J. Biol.*, 26, 181, 2002.
9. Madigan, M.T., Martinko, J.M., Parker, J. (Eds.), *Brock biology of microorganisms*, vol. 11, Prentice Hall, Upper Saddle River, New Jersey, 1997.
10. Cárdenas, J.P., Valdés, J., Quatrini, R., Duarte, F., Holmes, D.S., Lessons from the genomes of extremely acidophilic bacteria and archaea with special emphasis on bioleaching microorganisms. *Appl. Microbiol. Biotechnol.*, 88, 605, 2010.

11. López-López, O., Cerdan, M.E., Gonzalez Siso, M.I., New extremophilic lipases and esterases from metagenomics. *Curr. Protein. Pept. Sci.*, 15, 445, 2014.
12. Yildiz, S.Y., Radchenkova, N., Arga, K.Y., Kambourova, M., Oner, E.T., Genomic analysis of Brevibacillus thermoruber 423 reveals its biotechnological and industrial potential. *Appl. Microbiol. Biotechnol.*, 99, 2277, 2015.
13. Herbert, R.A., A perspective on the biotechnological potential of extremophiles. *Trends Biotechnol.*, 10, 395, 1992.
14. Demirjian, D.C., Morís-Varas, F., Cassidy, C.S., Enzymes from extremophiles. *Curr. Opin. Chem. Biol.*, 5, 144, 2001.
15. Van Den Burg, B., Extremophiles as a source for novel enzymes. *Curr. Opin. Microbiol.*, 6, 213, 2003.
16. Irwin, J.A. and Baird, A.W., Extremophiles and their application to veterinary medicine. *Ir. Vet. J.*, 57, 348, 2004.
17. Díaz-Tena, E., Rodríguez-Ezquerro, A., de Lacalle Marcaide, L.L., Bustinduy, L.G., Sáenz, A.E., Use of extremophiles microorganisms for metal removal. *Procedia Eng.*, 63, 67, 2013.
18. Eichler, J., Biotechnological uses of archaeal extremozymes. *Biotechnol. Adv.*, 19, 261, 2001.
19. Fujiwara, S., Extremophiles: Developments of their special functions and potential resources. *J. Biosci. Bioeng.*, 94, 518, 2002.
20. Haki, G. and Rakshit, S., Developments in industrially important thermostable enzymes: a review. *Bioresour. Technol.*, 89, 17, 2003.
21. Schill, N., Van Gulik, W., Voisard, D., Von Stockar, U., Continuous cultures limited by a gaseous substrate: development of a simple, unstructured mathematical model and experimental verification with Methanobacterium thermoautotrophicum. *Biotechnol. Bioeng.*, 51, 645, 1996.
22. Joseph, B., Ramteke, P.W., Thomas, G., Cold active microbial lipases: some hot issues and recent developments. *Biotechnol. Adv.*, 26, 457, 2008.
23. Sarmiento, F., Peralta, R., Blamey, J.M., Cold and hot extremozymes: industrial relevance and current trends. *Front Bioeng. Biotechnol.*, 3, 148, 2015.
24. Schmid, A.K., Reiss, D.J., Pan, M., Koide, T., Baliga, N.S., A single transcription factor regulates evolutionarily diverse but functionally linked metabolic pathways in response to nutrient availability. *Mol. Syst. Biol.*, 5, 282, 2009.
25. Pennisi, E., Biotechnology: in industry, extremophiles begin to make their mark. *Science*, 276, 705, 1997.
26. Schiraldi, C., Martino, A., Acone, M., Di Lernia, I., Di Lazzaro, A., Marulli, F., Generoso, M., Cartenì, M., De Rosa, M., Effective production of a thermostable α-glucosidase from Sulfolobus solfataricus in Escherichia coli exploiting a microfiltration bioreactor. *Biotechnol. Bioeng.*, 70, 670, 676, 2000.
27. Martino, A., Schiraldi, C., Fusco, S., Di Lernia, I., Costabile, T., Pellicano, T., Marotta, M., Generoso, M., Van der Oost, J., Sensen, C.W., Properties of the recombinant α-glucosidase from Sulfolobus solfataricus in relation to starch processing. *J. Mol. Catal., B Enzym.*, 11, 787, 2001.

28. de Pascale, D., Sasso, M.P., Di Lernia, I., Di Lazzaro, A., Furia, A., Farina, M.C., Rossi, M., De Rosa, M., Recombinant thermophilic enzymes for trehalose and trehalosyl dextrins production. *J. Mol. Catal., B Enzym.*, 11, 777, 2001.
29. de Pascale, D., Di Lernia, I., Sasso, M.P., Furia, A., De Rosa, M., Rossi, M., A novel thermophilic chimeric enzyme for trehalose production. *Extremophiles*, 6, 463, 2002.
30. Collins, T., D'Amico, S., Marx, J.-C., Feller, G., Gerday, C., Cold-adapted enzymes, in: *Physiology and biochemistry of extremophiles*, C. Gerday, and N. Glansdorff, (Eds.), pp. 165–179, ASM Press, Washington, D.C, 2007.
31. Egorova, K. and Antranikian, G., Industrial relevance of thermophilic Archaea. *Curr. Opin. Microbiol.*, 8, 649, 2005.
32. Champdore, Md., Staiano, M., Rossi, M., D'Auria, S., Proteins from extremophiles as stable tools for advanced biotechnological applications of high social interest. *J. R. Soc. Interface*, 4, 183, 2007.
33. Aurilia, V., Parracino, A., D'Auria, S., Microbial carbohydrate esterases in cold adapted environments. *Gene*, 410, 234, 2008.
34. Sauer, T. and Galinski, E.A., Bacterial milking: a novel bioprocess for production of compatible solutes. *Biotechnol. Bioeng.*, 57, 306, 1998.
35. Ruiz, D.M. and De Castro, R.E., Effect of organic solvents on the activity and stability of an extracellular protease secreted by the haloalkaliphilic archaeon Natrialba magadii. *J. Ind. Microbiol. Biotechnol.*, 34, 111, 2007.
36. Gerday, C., Aittaleb, M., Bentahir, M., Chessa, J.-P., Claverie, P., Collins, T., D'Amico, S., Dumont, J., Garsoux, G., Georlette, D., Cold-adapted enzymes: from fundamentals to biotechnology. *Trends Biotechnol.*, 18, 103, 2000.
37. Huang, H., Luo, H., Wang, Y., Fu, D., Shao, N., Yang, P., Meng, K., Yao, B., Novel low-temperature-active phytase from Erwinia carotovora var. carotovota ACCC 10276. *J. Microbiol. Biotechnol.*, 19, 1085, 2009.
38. Ueda, M., Goto, T., Nakazawa, M., Miyatake, K., Sakaguchi, M., Inouye, K., A novel cold-adapted cellulase complex from Eisenia foetida: characterization of a multienzyme complex with carboxymethylcellulase, β-glucosidase, β-1, 3 glucanase, and β-xylosidase. *Comp. Biochem. Physiol. B Biochem. Mol. Biol.*, 157, 26, 2010.
39. Parkes, R.J., Cragg, B.A., Banning., N., Brock, F., Webster, G., Fry, J.C., Hornibrook, E., Pancost, R.D., Kelly, S., Knab, N., Biogeochemistry and biodiversity of methane cycling in subsurface marine sediments (Skagerrak, Denmark). *Environ. Microbiol.*, 9, 1146, 2007.
40. Bertoldo, C. and Antranikian, G., Starch-hydrolyzing enzymes from thermophilic archaea and bacteria. *Curr. Opin. Chem. Biol.*, 6, 151, 2002.
41. Van Der Maarel, M.J., Van der Veen, B., Uitdehaag, J.C., Leemhuis, H., Dijkhuizen, L., Properties and applications of starch-converting enzymes of the α-amylase family. *J. Biotechnol.*, 94, 137, 2002.
42. Ruberti, E., L'evoluzione biologica: storia e sviluppi recenti. *Energia Ambiente e Innovazione*, 6, 1997.

43. Sunna, A. and Bergquist, P.L., A gene encoding a novel extremely thermostable 1, 4-β-xylanase isolated directly from an environmental DNA sample. *Extremophiles*, 7, 63, 2003.
44. Raddadi, N., Cherif, A., Daffonchio, D., Neifar, M., Fava, F., Biotechnological applications of extremophiles, extremozymes and extremolytes. *Appl. Microbiol. Biotechnol.*, 99, 7907, 2015.
45. Sterner, R. and Liebl, W., Thermophilic adaptation of proteins. *Crit. Rev. Biochem. Mol. Biol.*, 36, 39, 2001.
46. Paiardini, A., Gianese, G., Bossa, F., Pascarella, S., Structural plasticity of thermophilic serine hydroxymethyltransferases. *Proteins.*, 50, 122, 2003.
47. Bräsen, C., Urbanke, C., Schönheit, P., A novel octameric AMP-forming acetyl-CoA synthetase from the hyperthermophilic crenarchaeon *Pyrobaculum aerophilum*. *FEBS Lett.*, 579, 477, 2005.
48. Mayer, F., Küper, U., Meyer, C., Daxer, S., Müller, V., Rachel, R., Huber, H., AMP-forming acetyl coenzyme A synthetase in the outermost membrane of the hyperthermophilic crenarchaeon *Ignicoccus hospitalis*. *J. bacteriol.*, 194, 1572, 2012.
49. Jayakumar, R., Jayashree, S., Annapurna, B., Seshadri, S., Characterization of thermostable serine alkaline protease from an alkaliphilic strain *Bacillus pumilus* MCAS8 and its applications. *Appl. Biochem. Biotechnol.*, 168, 1849, 2012.
50. Bruins, M.E., Janssen, A.E., Boom, R.M., Thermozymes and their applications. *Appl. Biochem. Biotechnol.*, 90, 155, 2001.
51. de Pascale, D., Cusano, A.M., Autore, F., Parrilli, E., Di Prisco, G., Marino, G., Tutino, M.L., The cold-active Lip1 lipase from the Antarctic bacterium *Pseudoalteromonas haloplanktis* TAC125 is a member of a new bacterial lipolytic enzyme family. *Extremophiles*, 12, 311, 2008.
52. Unsworth, L.D., van der Oost, J., Koutsopoulos, S., Hyperthermophilic enzymes– stability, activity and implementation strategies for high temperature applications. *FEBS J.*, 274, 4044, 2007.
53. Rosenbaum, E., Gabel, F., Durá, M.A., Finet, S., Cléry-Barraud, C., Masson, P., Franzetti, B., Effects of hydrostatic pressure on the quaternary structure and enzymatic activity of a large peptidase complex from *Pyrococcus horikoshii*. *Arch. Biochem. Biophys.*, 517, 104, 2012.
54. Casanueva, A., Tuffin, M., Cary, C., Cowan, D.A., Molecular adaptations to psychrophily: the impact of 'omic'technologies. *Trends Microbiol.*, 18, 374, 2010.
55. De Maayer, P., Anderson, D., Cary, C., Cowan, D.A., Some like it cold: understanding the survival strategies of psychrophiles. *EMBO Rep.*, 15, 508, 2014.
56. Siddiqui, K.S. and Cavicchioli, R., Cold-adapted enzymes. *Annu. Rev. Biochem.*, 75, 403, 2006.
57. Dalmaso, G.Z.L., Ferreira, D., Vermelho, A.B., Marine extremophiles: a source of hydrolases for biotechnological applications. *Mar. Drugs*, 13, 1925, 2015.

58. Cavicchioli, R., Charlton, T., Ertan, H., Omar, S.M., Siddiqui, K., Williams, T., Biotechnological uses of enzymes from psychrophiles. *Microb. Biotechnol.*, 4, 449, 2011.
59. Gomes, J. and Steiner, W., The biocatalytic potential of extremophiles and extremozymes. *Food Technol. Biotechnol.*, 42, 223, 2004.
60. Horikoshi, K., Alkaliphiles: some applications of their products for biotechnology. *Microbiol. Mol. Biol. Rev.*, 63, 735, 1999.
61. Luisa Tutino, M., di Prisco, G., Marino, G., de Pascale, D., Cold-adapted esterases and lipases: from fundamentals to application. *Protein Pept. Lett.*, 16, 1172, 2009.
62. Delgado-García, M., Valdivia-Urdiales, B., Aguilar-González, C.N., Contreras-Esquivel, J.C., Rodríguez-Herrera, R., Halophilic hydrolases as a new tool for the biotechnological industries. *J. Sci. Food Agric.*, 92, 2575, 2012.
63. Jackson, C.R., Langner, H.W., Donahoe-Christiansen, J., Inskeep, W.P., McDermott, T.R., Molecular analysis of microbial community structure in an arsenite-oxidizing acidic thermal spring. *Environ. Microbiol.*, 3, 532, 2001.
64. Madern, D., Ebel, C., Zaccai, G., Halophilic adaptation of enzymes. *Extremophiles*, 4, 91, 2000.
65. Sellek, G.A. and Chaudhuri, J.B., Biocatalysis in organic media using enzymes from extremophiles. *Enzyme Microb. Technol.*, 25, 471, 1999.
66. Datta, S., Holmes, B., Park, J.I., Chen, Z., Dibble, D.C., Hadi, M., Blanch, H.W., Simmons, B.A., Sapra, R., Ionic liquid tolerant hyperthermophilic cellulases for biomass pretreatment and hydrolysis. *Green Chem.*, 12, 338, 2010.
67. Raddadi, N., Cherif, A., Daffonchio, D., Fava, F., Halo-alkalitolerant and thermostable cellulases with improved tolerance to ionic liquids and organic solvents from *Paenibacillus tarimensis* isolated from the Chott El Fejej, Sahara desert, Tunisia. *Bioresour. Technol.*, 150, 121, 2013.
68. Sutrisno, A., Ueda, M., Abe, Y., Nakazawa, M., Miyatake, K., A chitinase with high activity toward partially N-acetylated chitosan from a new, moderately thermophilic, chitin-degrading bacterium, *Ralstonia* sp. A-471. *Appl. Microbiol. Biotechnol.*, 63, 398, 2004.
69. Wołosowska, S. and Synowiecki, J., Thermostable β-glucosidase with a broad substrate specifity suitable for processing of lactose-containing products. *Food Chem.*, 85, 181, 2004.
70. Litchfield, C.D., Potential for industrial products from the halophilic Archaea. *J. Ind. Microbiol. Biotechnol.*, 38, 1635, 2011.
71. Schreck, S.D. and Grunden, A.M., Biotechnological applications of halophilic lipases and thioesterases. *Appl. Microbiol. Biotechnol.*, 98, 1011, 2014.
72. Madigan, M.T. and Orent, A., Thermophilic and halophilic extremophiles. *Curr. Opin. Microbiol.*, 2, 265, 1999.
73. Rodriguez-Valera, F., Biotechnological potential of halobacteria. *Biochem. Soc. Symp.*, 58, 135, 1992.

74. Koyuncu, B., Expression of recombinant acid protease (thermopsin) gene from Thermoplasma volcanium, in: *Thesis*, vol. 113, The Graduate School of Natural and Applied Science Middle East Technical University, Ankara, Turkey, 2006.
75. Serour, E. and Antranikian, G., Novel thermoactive glucoamylases from the thermoacidophilic Archaea Thermoplasma acidophilum, Picrophilus torridus and Picrophilus oshimae. *Antonie Van Leeuwenhoek*, 81, 73, 2002.
76. Boonyaratanakornkit, B.B., Park, C.B., Clark, D.S., Pressure effects on intra- and intermolecular interactions within proteins. *Biochim. Biophys. Acta.*, 1595, 235, 2002.
77. Takai, K., Miyazaki, M., Hirayama, H., Nakagawa, S., Querellou, J., Godfroy, A., Isolation and physiological characterization of two novel, piezophilic, thermophilic chemolithoautotrophs from a deep-sea hydrothermal vent chimney. *Environ. Microbiol.*, 11, 1983, 2009.
78. Fusi, P., Grisa, M., Mombelli, E., Consonni, R., Tortora, P., Vanoni, M., Expression of a synthetic gene encoding P2 ribonuclease from the extreme thermoacidophilic archaebacterium Sulfolobus solfataricus in mesophylic hosts. *Gene*, 154, 99, 1995.
79. Mombelli, E., Shehi, E., Fusi, P., Tortora, P., Exploring hyperthermophilic proteins under pressure: theoretical aspects and experimental findings. *Biochim. Biophys. Acta.*, 1595, 392, 2002.
80. Cavicchioli, R., Extremophiles and the search for extraterrestrial life. *Astrobiology*, 2, 281, 2002.
81. Abe, F. and Horikoshi, K., The biotechnological potential of piezophiles. *Trends Biotechnol.*, 19, 102, 2001.
82. Giuliano, M., Schiraldi, C., Marotta, M., Hugenholtz, J., De Rosa, M., Expression of Sulfolobus solfataricus α-glucosidase in Lactococcus lactis. *Appl. Microbiol. Biotechnol.*, 64, 829, 2004.
83. Kuhad, R.C., Gupta, R., Singh, A., Microbial cellulases and their industrial applications. *Enz. Res.*, 2011, 280696, 2011.
84. Sukumaran, R.K., Singhania, R.R., Pandey, A., Microbial cellulases-production, applications and challenges. *J. Sci. Ind. Res.*, 64, 832, 2005.
85. Kumar, L., Awasthi, G., Singh, B., Extremophiles: a novel source of industrially important enzymes. *Biotechnology*, 10, 121, 2011.
86. Margesin, R., Schinner, F., Marx, J.-C., Gerday, C. (Eds.), *Psychrophiles: from biodiversity to biotechnology*, Springer-Verlag, Berlin, 2008.
87. Marasco, R., Rolli, E., Ettoumi, B., Vigani, G., Mapelli, F., Borin, S., Abou-Hadid, A.F., El-Behairy, U.A., Sorlini, C., Cherif, A., A drought resistance-promoting microbiome is selected by root system under desert farming. *PloS one*, 7, 2012.
88. Rolli, E., Marasco, R., Vigani, G., Ettoumi, B., Mapelli, F., Deangelis, M.L., Gandolfi, C., Casati, E., Previtali, F., Gerbino, R., Improved plant resistance to drought is promoted by the root-associated microbiome as a water stress-dependent trait. *Environ. Microbiol.*, 17, 316, 2015.

89. Hotta, Y., Ezaki, S., Atomi, H., Imanaka, T., Extremely stable and versatile carboxylesterase from a hyperthermophilic archaeon. *Appl. Environ. Microbiol.*, 68, 3925, 2002.
90. Johnson, D.B., Biomining—biotechnologies for extracting and recovering metals from ores and waste materials. *Curr. Opin. Biotechnol.*, 30, 24, 2014.
91. Karasová-Lipovová, P., Strnad, H., Spiwok, V., Malá, Š., Králová, B., Russell, N.J., The cloning, purification and characterisation of a cold-active β-galactosidase from the psychrotolerant Antarctic bacterium *Arthrobacter* sp. C2-2. *Enzyme Microb. Technol.*, 33, 836, 2003.
92. Navarro, C.A., von Bernath, D., Jerez, C.A., Heavy metal resistance strategies of acidophilic bacteria and their acquisition: importance for biomining and bioremediation. *Biol. Res.*, 46, 363, 2013.
93. Adrio, J.L. and Demain, A.L., Microbial enzymes: tools for biotechnological processes. *Biomolecules*, 4, 117, 2014.
94. Karmakar, M. and Ray, R., Current trends in research and application of microbial cellulases. *Res. J. Microbiol.*, 6, 41, 2011.
95. Birgisson, H., Delgado, O., Arroyo, L.G., Hatti-Kaul, R., Mattiasson, B., Cold-adapted yeasts as producers of cold-active polygalacturonases. *Extremophiles*, 7, 185, 2003.
96. Satyanarayana, T., Rao, J.U.M., Ezhilvannan, M., α-Amylases, in: *Enzyme Technology*, A. Pandey, C. Webb, M. Fernandes, C. Larroche (Eds.), pp. 189–220, Springer-Verlag, New York, 2006.
97. Satyanarayana, T., Raghukumar, C., Shivaji, S., Extremophilic microbes: diversity and perspectives. *Curr. Sci.*, 89, 1, 78–90, 2005.
98. Archana, A. and Satyanarayana, T., Xylanase production by thermophilic Bacillus licheniformis A99 in solid-state fermentation. *Enzyme Microb. Technol.*, 21, 12, 1997.
99. Archana, A. and Satyanarayana, T., Purification and characterization of a cellulase-free xylanase of a moderate thermophile *Bacillus licheniformis* A99. *World J. Microbiol. Biotechnol.*, 19, 53, 2003.
100. Izumi, M., Fujiwara, S., Shiraki, K., Takagi, M., Fukui, K., Imanaka, T., Utilization of immobilized archaeal chaperonin for enzyme stabilization. *J. Biosci. Bioeng.*, 91, 316, 2001.
101. Gauthier, M.J., Lafay, B., Christen, R., Fernandez, L., Acquaviva, M., Bonin, P., Bertrand, J.-C., *Marinobacter hydrocarbonoclasticus* gen. nov., sp. nov., a new, extremely halotolerant, hydrocarbon-degrading marine bacterium. *Int. J. Syst. Evol. Microbiol.*, 42, 568, 1992.
102. Nicholson, C.A. and Fathepure, B.Z., Biodegradation of benzene by halophilic and halotolerant bacteria under aerobic conditions. *Appl. Environ. Microbiol.*, 70, 1222, 2004.
103. Young, P., Major microbial diversity initiative recommended. *Asm News*, 63, 417, 1997.

104. Kuddus, M. and Ramteke, P.W., Cold-active extracellular alkaline protease from an alkaliphilic Stenotrophomonas maltophilia: production of enzyme and its industrial applications. *Can. J. Microbiol.*, 55, 1294, 2009.
105. Yumoto, I., Bioenergetics of alkaliphilic *Bacillus* spp. *J. Biosci. Bioeng.*, 93, 342, 2002.
106. Cavicchioli, R. and Thomas, T., Extremophiles, in: *Encyclopedia of microbiology*, 2nd edn, vol. 2, pp. 317–337, Academic press, London, 2003.
107. Gurung, N., Ray, S., Bose, S., Rai, V., A broader view: microbial enzymes and their relevance in industries, medicine, and beyond. *Biomed. Res. Int.*, 2013, 329121, 2013.
108. Ferrer, M., Golyshina, O., Beloqui, A., Golyshin, P.N., Mining enzymes from extreme environments. *Curr. Opin. Microbiol.*, 10, 207, 2007.
109. Wang, F., Hao, J., Yang, C., Sun, M., Cloning, expression, and identification of a novel extracellular cold-adapted alkaline protease gene of the marine bacterium strain YS-80-122. *Appl. Biochem. Biotechnol.*, 162, 1497, 2010.

15

Microbial Chitinases and Their Applications: An Overview

Suraja Kumar Nayak[1*], Swapnarani Nayak[2], Swaraj Mohanty[3], Jitendra Kumar Sundaray[2] and Bibhuti Bhusan Mishra[4]

[1]*Department of Biotechnology, College of Engineering and Technology, Bhubaneswar, Odisha, India*

[2]*Fish Genetics and Biotechnology Division, ICAR-Central Institute of Freshwater Aquaculture, Bhubaneswar, Odisha, India*

[3]*Peptide and Proteomics Division, Defence Institute of Physiology and Allied Sciences (DIPAS), Defence Research and Development Organisation (DRDO), Delhi, India*

[4]*Department of Microbiology, College of Basic Science & Humanities, Orissa University of Agriculture and Technology, Bhubaneswar, India*

Abstract

Microorganisms synthesize an array of glycosyl hydrolytic enzymes quantitatively and chitinase is one among them. In general, the persistence of chitinase form microbial source is availing elemental carbon and nitrogen as precursors or nutrient sources and the insatiability of this enzyme plays a significant role mainly in parasitism against chitinaceous host. Out of different microbial chitinases availed in public domain, soil bacterial chitinases share more than half of genomes and mostly are of actinobacterial origin. The others are from proteobacteria, yeast, moulds, and few viral sources. With the advent of biotechnology, the production of chitinases is increased to many folds. Extracellular chitinases disintegrate chitin polymers to produce oligomers having several applications. The microbial chitinases are being broadly applied in various realms including agricultural, biomedical, pharmaceutical, industrial, and environmental. The wide application is well marked as biopesticides and biocontrol agents. Furthermore, the waste minimization with chitinase application is surplus to the environmental cleanup initiatives. The present chapter is inclusive of production of microbial chitinases and their detail applications in descriptive manner.

**Corresponding author*: surajnayak3@gmail.com

Bibhuti Bhusan Mishra, Suraja Kumar Nayak, Swati Mohapatra, and Deviprasad Samantaray (eds.) Environmental and Agricultural Microbiology: Applications for Sustainability, (313–340) © 2021 Scrivener Publishing LLC

***Keywords*:** Microbial chitinases, chito-oligosaccharides (COS), submerged fermentation, biocontrol agents, biopesticides, waste management

15.1 Introduction

Enzymes are organic compounds (biocatalysts) that catalyze all the chemical reactions of living beings with greater specificity and rate enhancements. As (bio)catalyst enzymes facilitate tremendous opportunity for industries to carry out efficient and economical biocatalytic conversions. Being proteins, enzymes possess properties such as specificity toward the reactions they catalyze and the substrates on which they act upon along with rapid action and minimal waste generation [1]. Advancements in biotechnology build on milestones in enzyme redesigning for substrate specificity and thermostability. Diverse applications and tailor made modifications paved the way for enzyme technology to meet the upcoming challenges in future and under-lens of researchers. Biotechnology has also paved new area and era of research in the field of genetics, protein engineering, and bioinformatics.

Enzymes have been isolated, purified, studied, and improved from prokaryotes and eukaryotes while the prokaryotic sources are the frequent plate to target due to wide biochemical diversity, feasibility in time and space for mass growth, and affluence in genetic manipulation [2]. Currently, there are more than 5,500 known enzymes divided into oxidoreductases, transferases, hydrolases, lyases, isomerases, and ligases based on the type of reaction it catalyzes. Based on target substances, specific names have been opted for enzyme (IUB 1961). Chitinases (EC 3.2.1.14) belong to hydrolases (EC 3) class and glycosyl hydrolase (GH) (EC 3.2.1) superfamily of enzymes are known to degrade chitin, which is a complex biopolymer resistant to degradation. It is the highest polymer occurs in fungal cell wall, shells of crustaceans, and insects' exoskeletons and other members of arthropodata [3]. Chitin and its biosimilars have a broad usage in medical and environmental realms among others. Amid a broad array of uses, chitin has its application in chitinase augmentation. Chitinases have great importance as they have wide spread applications in agriculture, biochemical processing, medicine, protein engineering, flavor enhancer in food and feed, cell wall degradation, and waste management in environment [4]. In recent years, the use of chitinases as a biocontrol agent is one of the attractive agriculturally and environmentally safe strategies [5]. In microbes, chitinases implicate cell wall morphogenesis, whereas in plants, it involved in body physiology and defense mechanism [6].

Although large numbers of microorganisms are identified to be involved in chitinase production, industrial chitinase production is restricted to few efficient strains including *Trichoderma harzanium* and *T. flavofuscum*. Industrial chitinase production broadly prepared through submerged fermentation, solid state fermentation, and immobilization techniques using different substrates and potent strains as described in detail later in this chapter. In recent year, the genetic engineering technology has being used to clone chitinase producing genes in suitable hosts for enhanced chitinase production. Studies have made possible with protein engineering to produce chitinases with desired functions for meeting the requirements in disorders treatment. Bioreactors with optimized nutritional and other parameters have been developed for industrial scale production with less percapita. Also, the recovery processes are difficult without full information about the required parameters [7]. Because of vast application of microbial chitinases, researches undertaken in last decades to give a long leap in chitinase production enhancement. The present chapter focuses on microbial chitinase, its types, different microbial sources, molecular biology of microbial chitinase enzyme, bioprocess technologies in biosynthesis, and their detail application in various realms.

15.2 Chitinases and Its Types

Chitinases are an important enzyme for low cost waste minimization. In 1911, Bernard for the first time described chitinase in plants, i.e., orchid [*Dactylorhiza incarnata* (L.) Soó (1962) erstwhile *Orchis incarnata* L.] bulbs, and it performs as diffusible antimycotic and thermosensitive factor. Similarly, in 1929, Karrer and Hofmann noticed this enzyme in terrestrial gastropod mollusks (*Helix pomatia* Linn.). In recent years Jeuniaux's scientific research led to renewed interests in chitinases. As a member of GH superfamily, it hydrolyzes the β-1,4-glycosidic bonds between the C1 and C4 of two consecutive *N*-acetyl-D-glucosamine in chitins. Complete biodegradation of chitin chain to free *N*-acetyl glucosamine (GlcNAc), i.e., the hydrolytic polymerization of chitin is due to composition of diverse enzymes/biocatalysts in chitinolytic systems [8, 9]. Presently, the nomenclature of chitinolytic enzymes is indecisive. On the basis of their mode of action, in 1992, International Union of Biochemistry and Molecular Biology (IUBMB) classified chitinase into two broad categories as endochitinases (EC 3.2.1.14) and exochitinases (EC 3.2.1.30). As the name suggests, endochitinases are targeting the internal sites and generates diacetylchitobiose, chitotriose, and chitotetraose which are low molecular mass multimers of glucosamine. On the other hand, exochitinase have been separated into chitobiosidases (EC 3.2.1.29) and β-*N*-acetylglucosaminidases (EC 3.2.1.52) which works for diacetylchitobiose release at the non-reducing end of chitin and works on the nonreducing end of multimers/oligomers obtained from endochitinases into GlcNAc monomers, respectively [10, 11]. In recent times, IUBMB put together chitobiase and β-N-acetylglucosaminidases into β-N-acetylhexosaminidases (EC.3.2.1.52) [12]. Often few chitinases exhibiting pronounced lysozyme (EC 3.2.1.17) activity corresponding to the cleavage of a glycosidic bond between the C1 of NAM and C4 of GlcNAc in peptidoglycans of bacteria. In few cases, transglycosidase activities also associated with exochitinases [13].

As stated earlier based on amino acid sequences and structural dissimilarity chitinases has been grouped into GH superfamilies 18(GH 18), 19(GH 19), and 20(GH 20) [14]. Scientific works on GH 18 and GH 19 express often parallel sequence and also have totally dissimilar 3D structure so concluded to have ancestral difference. GH 18 includes chitinases from bacteria, fungi, viruses, animals, insects, and plants. It consists of a number of conserved repeats of amino acids and enzyme core, which has 8 α-helices and 8 β-strands, creating a barrel positioned down α helices, in turn, forming a ring at the external [15]. Through retaining mechanism, the catalytic reaction (substrate-assisted catalysis) of the GH 18 takes place in which β-1,4-glycosidic associations are targeted (hydrolysis) for β-anomer production. Catalytic domains as a multi-domain arrangement and both a cysteine rich chitin-binding domain (CBD) and a serine/threonine rich glycosylated domain have been recognized as structural characteristics of chitinase diversification in prokaryotes as well as eukaryotes [16]. The resemblance between both bacterial and fungal chitinases leads to compare the catalytic domains in all the sources. *Streptomyces* chitinases and plant chitinases are included in GH 19 [17]. The catalytic mechanism is similar to lysozyme and chitosanase which is a general acid and base mechanism. GH 19 chitinases hydrolyze both GlcNAcs and glucosamines. GH 20 involves *N*-acetylglucosaminidases from bacteria (*Vibrio harveyi*, *Streptomycetes* sp.), certain fungi, and *N*-acetylhexosaminidase from amoeba (*Dictyostelium discoideum*) and human [9, 18].

Localization of the enzymes, length and types of aminoacids, isoelectric point (pI), pH, presence of inducers, and signal and *N*-terminal peptides further classified chitinases into five different classes. Class I chitinases experimentally show their presence in higher eukaryotes such as plants. Whereas class II is found in higher and lower eukaryotes like plants and fungi. It is also found in bacteria. At present, there is no information available on sequence similarity of either class I, class II, or class III. Not only sequence similarity but also immunological properties are observed in class I and class IV but they differ in peptide length. The biotic interaction of plants and microbes leads to detection of class V chitinases [19].

Despite of abundances, chitin is not accumulated in environment due to occurrence of bacterial chitinases. Their presence is well marked in cell walls of fungi as well as in cuticles of insect. Additionally, chitinases have shown significantly inhibitory capabilities to phytopathogens thus can be employed as biocontrol agents, i.e., alternative to chemical pesticides. There is a growing demand of chitin derivatives, for various industrial, clinical, and pharmaceutical purposes which are explained in detail in application section.

15.3 Sources of Microbial Chitinase

Bacteria, fungi, actinobacteria, viruses, insects, animals, and higher plants encoded with different kind of chitinases. Also, their presence varies from soil to sediment and water and it is for nutrition, morphogenesis, and defense against chitin-containing pathogens. Few researchers agreed that chitinolytic microorganisms from soil were usually more active correlated to water and sediment members and could be more appropriate for practical applications. According to reports, chitinolytic microorganisms are abundantly present in environments with high amounts of chitin (such as shrimp shells). Bacteria synthesize chitinases for the purpose of availing N and C as a source of energy by degrading chitin. Moreover, bacterial pathogenesis is chitinase dependent [20].

Chitinolytic bacteria are only 4% of all-over heterotrophic bacterial population and fungi are significantly lower than that of bacteria. It comprises of 25%–60% of the total fungi [21]. Microbial chitinases ranging from 20 to 120 kDa, and most lie in 20- to 60-kDa molecular mass [22]. Insect chitinases (~40–85 kDa) are larger in size than both plant (~25–40 kDa) and bacteria (~20–60 kDa). The maximum and minimum temperature of chitinase activity ranges in between 25°C and 65°C including some psychrophiles and thermophiles while 37°C ± 3°C is known to be the optimum temperature. Similarly, the optimum pH is 5.0–8.0 and few isolates were also exhibiting activity in acidic and alkaline pHs. Maximum chitinases from different organisms restricted pI in between 4.5 and 8.5.

15.3.1 Bacterial Chitinases

Bacteria are the major sources of microbial chitinases. Plants contain eight conserved cysteine residues in CBDs which are not found in bacteria [23].

The basic function of chitinase is binding of a non-catalytic chitin binding protein (CBP) to polymer of chitin. It is found that only four numbers of amino acids are conserved in the catalytic domain of plant class III chitinases and bacteria [24]. *Aeromonas* sp., *Arthrobacter* sp., *Bacillus* sp., *Beneckea* sp., *Clostridium* sp., *Chromobacterium* sp., *Cytophaga* sp., *Enterobacter* sp., *Erwinia* sp., *Flavobacterium* sp., *Klebsiella* sp., *Pseudomonas aeruginosa* K-187, *Serratia marcescens, S. griseus* HUT 6037, and *Vibrio* sp. can synthesize several different kinds of chitinases and have been isolated from both soil as well as water resources [25–28]. Moreover, *Clostridium* sp., a chitinolytic bacterial species, were present in the faeces of domestic and wild herbivores [Bison (*Bibos bonasus* Linn.), llama (*Llama vicugna paca* Linn.), Elk (*Elaphurus davidianus* Milne-Edwards), Sheep (*Ovis aries* Linn.), and Yak (*Bos grunniens* Linn.)]. Also found in the stomach of cows, providing a living environment in exchange of digesting this complex compound. Sturz and Robinson observed that in the sediments, aerobic heterotrophic bacteria dominate and play a decisive role in degradation of chitin [29].

Chitinases produced by *S. marcescens* maximally belong to GH 18. These are ChiA, ChiB, ChiC, and CBP (CBP21) having (β/α) eight TIM-barrel catalytic domain with maximum six sugar subsites [30]. Both ChiA and ChiB have multimodular organization, which simply means that these have an *N*-terminal chitin binding module (CBM) with a fibronectin-like fold in ChiA, whereas these have CBM5 in C-terminal. CBMs found in chitinases are characterized by presence of conserved exposed tryptophan residues able to interact with substrate [31]. *S. marcescens* Nima secretes an endochitinase, a novel *N*-acetylglucosaminidase and an exochitinase, have been reported [32]. Genes responsible for chitinases also have been reported in *S. marcescens* QMB 1466, *S. marcescens* KCTC2172, and *S. marcescens* BJL200 [10, 33, 34]. Sequence homology revealed that bacterial GH 18 chitinases are classified into A, B, and C subfamilies and are not widely recognized. Subfamily A chitinases have a third domain equivalent to the insertion of an $\alpha + \beta$ fold region between the seventh and eighth $(\alpha/\beta)_8$ barrel which is not found in subfamilies B and C [35]. *Serratia* sp. and *Bacillus* sp. along with others have been producing chitinases of four types [36]. A marine isolate *Alteromonas* sp. O-7 in chitin containing environment synthesizes chitinase A, chitinase B, chitinase C, and chitinase D at ~50°C temperature and 7.0 pH [37]. *Vibrio* sp. strain Fi:7, an aquatic psychrotolerant from Southern Ocean, shows enzyme activity upto 50% at 5°C and optimal activity at temperature 35°C [38]. Marine bacteria, *Vibrio harveyi*, utilize only chitin polymer as the only source of carbon and able to produce chitinase A [39]. Soil bacterial isolate *Bacillus thuringiensis* from Mexico produces endochitinases, chitobiosidases, and N-acetyl-β-glucosaminidases *in vitro* in presence of colloidal chitin as carbon source. Another *B. thuringiensis* HD-1 cell-free supernatant contains exochitinase, which is able to hydrolyze the disaccharide 4-Methylumbelliferyl N-acetyl-β-D-glucosaminide [4-MU (GlnAc)] pH and temperature optima at 6.5 and 65°C, respectively [40]. Five extracellular chitinases were extracted from *Bacillus cereus* 6E1 using carboxymethyl-chitinremazol brilliant violet 5R (CM-chitin-RBV) as carbon substrate through a novel in-gel chitinase assay and named as Chi36 [41]. First chitinases from *B. amyloliquefaciens* V656 was identified as FI and FII having mol. wt. of 14.4 and 16.9 kDa, respectively, in neutral pH [42].

15.3.2 Fungal Chitinases

Chitinases of fungal origin belong to GH 18 along with bacterial chitinases [19, 43] and expresses sequence similarity with class III plant chitinases [11]. It consists of five domains, namely, N-terminal signal peptide region, CBD, catalytic domain, C-terminal extension region, and serine/threonine-rich region. Former two domains are essential for chitin degradation while later three shows no/sparse chitinolytic activity. Subgroups A, B, and C are the major divisions of fungal chitinases which obviously result of alterations in amino acid sequences. The variations in the subgroups extended upto difference in substrate-binding site architectures ultimately alter catalytic activities (exo vs. endo) and different CBMs [20, 36]. CBMs are absent in Subgroup A chitinases but has a catalytic domain having deep substrate binding site and has molecular mass of 40–50 kDa. Subgroup B chitinases are nonprocessive chitinases with uneven in sizes and have a CBM or a serine/threonine rich domain on C-terminal. Their molecular mass ranges from 30 to 90 kDa. Subgroup C is the latest classification of subgroup. These are having deep substrate binding site and processive in nature. Having CBM 50 else known as lysine motifs (LysM) which is special feature of this [36] and present in *N*-terminal of catalytic domain withy molecular mass in the range of 140–170 kDa. As a special feature, it degrades both self and non-self-cell walls in *Trichoderma* sp. (mycoparasitic fungi) [44].

Fungal chitinases are significant for nutrition and morphogenesis and are synthesized during different fungal growth stages [15]. Additionally, they have important role in autolysis. Many fungal populations synthesized 20 different types of chitinases [36]. Fungal chitinases interact with their substrate/s through CBD as a mechanism of action. CTS1 and K1Cts1p as CBD are concerned with protein-protein interaction or tertiary structure formation through disulphide bonds. These two are having a six-cysteine conserved region [45, 46]. *Agaricus* sp., *Aspergillus* sp., *Beauveria* sp., *Conidiobolus* sp., *Fusarium* sp., *Lecanicillium lecanii*, *Lycoperdon* sp., *Metharhizium* sp., *Mucor* sp., *Mortierella* sp., *Myrothecium* sp., *Neurospora* sp., *Penicillium* sp., *Stachybotrys* sp., and *Trichoderma* sp. are the major chitinases producers [47]. *Aspergillus fumigatus* is having 11 conserved active site domains for chitinases. Further they are divided into "fungal/bacterial" chitinases having similarity to bacteria and "fungal/plant" chitinases and maximum toward later. Reports are available on significant large size of "fungal/plant" chitinases to "fungal/bacterial" chitinases. *Trichoderma harzianum* synthesizes all three chitinases in seven types as one chitobiosidase, two *N*-acetylglucosaminidases, and four endochitinases [48]. In the sediments of Lake Chełmzynskie, 32%–40% of molds were able to decompose chitin [21].

Fungal chitinases are naturally inducible (adaptable) in nature. They are expressed only under certain conditions induced by a certain factor(s) and are regulated by a repressor/inducer system chitooligosaccharides (as NAG) and chitin acts as an inducers whereas glucose and easily assimilable other carbon sources acts as a repressor for the activity. It is observed that in absence of these substrates, no production of chitinase occur [18]. Decrease of chitinase production up to 86% was observed in presence of alanine as a substrate [49]. Chitinases, specifically GH-18, are strongly inhibited by allosamidin which are specific and competitive in nature. During hydrolysis, the carbonyl oxygen of the *N*-acetyl group and the C-1 of GlcNAc form an intermediate oxazoline ring and are structurally similar to allosamidin [19, 50]. Many of the multiple chitinase-encoding genes of moulds may encode secreted enzymes having nutritional roles for metabolism and growth. During budding stage of *Saccharomyces cerevisiae*, the degraded chitin monomers get deposited at the newly formed bud due to endochitinases of 130-kDa mass. Both *cst1* and *cst2* genes encoding for this enzyme were identified in *S. cerevisiae*. Several genes from *Trichoderma* sp. such as *chit33*, *ech42*, *chit42*, and *nag1* have been identified as chitinase encoding genes when cultured in medium containing chitin as a main carbon source. These genes are involved in mycoparasitism in *T. harzianum*, *T. atroviride*, and *T. virens*. Another gene *sechi44* encoding for endochitinases having 44 kDa from *Stachybotrys elegans* also exhibited mycoparasitic activity and similar to *T. atroviride* of *ech42*. *Hebeloma syrjense* secretes chitinases for production N and P from organic residues. *Neotyphodium* sp., an endophytic fungus, produces endochitinase evidencing in nutrition, growth, and defense against nematodes. Under varied physiological conditions, endochitinases produced by *T. harzianum* NCIM 1185 plays different specialized roles [13]. When cultivated with insects, cuticle *Beauveria brassiana* releases carbon and nitrogen due to presence GlcNAc for fungal growth [51].

15.3.3 Actinobacteria

Soil and sediments heavily colonized by actinobacteria rather other resources. Particularly, rhizospheric soil shows abundance in number. Also, sediments of Lake Chełmzynskie have 45%–69% of actinobacteria involved in chitinolytic activity [21]. *Streptomyces* sp. comprises of approximately 90% of the soil actinobacterial population and degrading chitins and similar polymers [52]. GH 19 chitinases are produced by actinobacteria. *Streptomyces cavourensis* SY224 and two other species produce chitinases having high antifungal potential and used as biocontrol agents in agriculture [22, 53, 54]. The optima pH, temperature, and pI range vary from 8.0 to 10.0, 28°C to 80°C, and 4.5 to 8.5, respectively [55]. *Streptomyces violaceusniger* 66 secretes endochitinase and *Streptomyces thermoviolaceus* OPC-52075 releases the same having activity in the foresaid range [56]. The first report on hydrolysis of β-glucosaminidic linkages in partially *N*-acetylated chitosan is due to two novel chitinases, C-1 and C-2 of *S. griseus* HUT 6037. These are having mol. wt. of ~27 kDa and optimum temperature 55°C [57]. A soil bacterium *S. griseus* MG3 is able to synthesize chitinase IS having significant activity in wide range of pH and applied in biocontrol activity [58]. Two chitinases, A and B having mol. wt. 43 and 45 kDa, isolated from cell-free supernatant of *S. albovinaceus* S-22 having optimum temperature 40°C and pH 5.6 expresses potential antimycotic activity [59]. An endochitinase isolated from *S. violaceusniger* XL-2 of 28.25 kDa able to biocontrol *Phanerochaete chrysosporium*, a wood-rotting phytopathogen [60]. Several genes encoding chitinases have been cloned from *Streptomyces* sp. Colloidal chitin, GlcNAc, and chito-bios/triose/tetrose induces chitinase production in *S. lydicus* WYEC108 while glucose and carboxy methyl cellulose (CMC) repress the gene activity [61]. Protein engineering studies revealed the presence of a two-component signal transduction system which regulates the synthesis of chitinases in *Streptomyces thermoviolaceus* OPC-520 and other strains. The systems contain a histidine kinase and a response regulator. While initiating a signal from the environment through the presence of chitin or its derivatives as a main carbon source, an actinobacterial kinase undergoes autophosphorylation at a histidine residue and subsequently catalyzes the transfer of a phosphate group to an asparagine residue in the sequence of a regulator (response). Finally, the phosphorylated response regulator in combination with a promoter activates the transcription of chitinase genes [12, 62].

15.3.4 Viruses/Others

Very few researchers have been interested in viral chitinases. Thus, obscurity in information is on chitinases of viral origin. Few viruses like *Baculoviruses* are able to synthesize naturally chitinases which are subsequently used for pathogenesis [63].

15.4 Genetics of Microbial Chitinase

For microbial degradation of chitin various strains, detailed processes have been characterized approximately 3 decades ago, and picture is clearer with the advent of culture independent approaches and applied to various resources. Due to availability of many other organic carbon sources, chitinases are treated as nonessential in individual cases while chitin and similar high-molecular-weight biopolymer hydrolysis is a primary step in organic matter degradation which requires these enzymes. This also leads to occurrences of large pool of uncultured chitin-degrading bacteria in aquatic systems [64]. Though there are various studies involved and techniques applied to get the details on microbial chitinases, a few which are more predominantly used are construction of metagenomics library, analysis of direct DNA, and microbes with chitinase activities. Different investigators opted for different methods for chitinase gene detection and expression, considering specific biases that need to be taken into account for each method. Different screening methods were adopted to screen chitinase genes from soil sample using a combination of molecular approaches [65]. To screen the previously isolated bacterial DNA, T-RFLP was carried out. A fosmid library was prepared for screening the resulting soil metagenome having chitinase genes. A similar earlier study with bulk and maize rhizospheric soil carried out using T-RFLP and clone library analysis and resulting in variation in chitinase gene diversity in both the samples [66]. These methodologies were also implemented in marine environment analysis. In both the environments chitinase genes of the isolates were finely distributed among metagenomics and directly extracted soil DNA clusters [67]. Highest numbers of dominant chitinase gene variant resided in *Streptomyces* sp., *Stenotrophomonas* sp., *Pseudomonas* sp., and *Bacillus* sp. [22, 55, 68, 69].

Chitinase genes are obtained from many types of γ-proteobacteria, including members of Enterobacteriaceae, Alteromonadaceae, *Aeromonas* sp., *Vibrio* sp., *Shewanella* sp., and *Pseudoalteromonas* sp. Also, PCR products were obtained from *Roseobacter* group concluding culture-dependent marine α-proteobacteria that possess group I chitinases. *Roseobacter* group clone family A was identical to *Sagittula stellata* strain E37. PCR results confirmed the presence of chitinase genes in α- and γ-proteobacteria instead of *Cytophaga-Flavobacter* from marine environment [64]. With supplement of chitin, high levels of *chi* gene expression were reported in *Metarhizium anisopilae* isolates. *T. harzianum* CECT 2413 synthesizes a 33-kDa chitinase which is repressed by monosaccharide and reverted back to normal in the presence of mycelia and the gene responsible is *CHIT33*. *CHIT42*, another gene, produces 42 kDa from the same organism is weakly depressed under starvation conditions and shows independent regulation by induction [70]. The *ech42* chitinase gene was induced by growing *T. harzianum* IMI 206040 in a minimal medium containing chitin as main source of carbon. A gene for *S. cerevisiae* endochitinase denoted as *CTS1* was isolated and cloned into a *Schizosaccharomyces pombe* shuttle vector. *Enterobacter agglomerans* synthesizes endochitinase due to presence of *chiA* genes having 562–amino acid sequence open reading frame and formed 61-kDa precursor protein having 86.8% homology with the *chiA* of *S. marcescens* chitinase. Chitinase from *Janthinobacterium lividans* observed due to the presence of *chi69* genes. A 1,424-nucleotide sequence in *Trichoderma hamatum* synthesizes the chitinase enzyme of 42 kDa due to gene sequence *Th-ch* [6]. A detail analysis revealed that *S. maltophilia* 34S1 chitinase due to *chiA* has been transcribed monocistronically [71].

15.5 Biotechnological Advances in Microbial Chitinase Production

Microbial chitinase is basically produced through fermentation, a microbial bioconversion of complex substrates to simple compounds. It is the key process of enzyme production from microbial sources. Physiological parameters like pH, temperature, water availability, and aeration will surely enhance the yield along with parameters like media component and substrates, and much ongoing researches are predicting so. In addition to that, media optimization will increase production to several folds. Presently, computer-based statistical methodologies were implimented to optimize the media components to maximize chitinase production. A soil isolate *Chitinolyticbacter meiyuanensis* SYBC-H1 increases 15.5-fold chitinase in optimized media than unoptimized media [72]. Significant increase, i.e., 56.1-fold in chitinase production was observed through response surface methodology (RSM) applied for *Lysinibacillus fusiformis* B-CM18, a chick pea rhizospheric isolate [73].

15.5.1 Media Components

Extracellular chitinase production is affected with concentration of carbon and nitrogen substrates and salts in synthetic medium. Solely, chitin plays a role of inducer for chitinase production, while along with colloidal chitin, crab and shrimp shells and chitin flakes, i.e., from cheap agricultural sources like wheat bran and rice bran, intensify the production. Researchers have also used the fungal cell walls as carbon supplements. Chitoligomers are necessary in the medium for direct hydrolysis of colloidal chitin. Also, catabolite repression was observed for chitinase production in presence of glucose and chitin [20]. Compared to colloidal chitin, lactic acid processed chitin induces better yield [74]. When switching from raw chitin of different other bioresources to colloidal chitin, 14% increase in chitinase production was observed. Inoculation of mycelia as a replacement for spores in *M. anisopliae*, *B. brassiana*, and *A. flavus* also increased production and reduction in inoculating time [49]. Firmicutes efficiently utilize shrimp shells for chitinase production than colloidal chitin. *Bacillus* sp. and *Pseudomonas* sp., the common soil inhabitants, effectively use shrimp wastes to produce chitinases. In contrary with shrimp wastes, *Aspergillus* sp. releases more enzymes in growth media. *Streptomyces aureofaciens* CMUAc130, an actinobacterial plant endophyte, induces chitinase production with GlcNAc while increases the production along with colloidal chitin. Also, surplus of CMC, starch, and divalent cations (Mg^{2+}) increases chitinase activity [75]. Highest enzyme activity obtained with 1% colloidal chitin along with strong repression was observed in presence of polysaccharides in *Streptomyces lydicus* WYEClO8 [61]. *Bacillus* sp. BG-11 responded to tryptophan, tyrosine, glutamine, arginine, and their analogs at a concentration of 0.1 mM in culture media [76]. It is pertinent to mention here that starch and yeast extract are considered as good carbon and nitrogen sources.

Both organic and inorganic compounds treated as enzyme inhibitors and are also oxidizing/reducing agents. *Streptomyces* sp. produces a competitive inhibitor, allosamidin which plays specific inhibitors for yeast, mould, and insect chitinases. It acts as a non-hydrolyzable analog of the oxazolinium ion intermediate leading to inhibitory effect [50]. Similarly, Psammaplin A, a brominated tryrosine-derived compound, has also recognized in the form of a non-competitive inhibitor of chitinase B of GH 18 produced by *S. marcescens*. A detail crystallographic study suggested that a draggle-tailed Psammaplin A binds in the range of the active site of the enzyme [77]. Another chitinase inhibitor from *Clonostachys* sp. FO-731 is Argadin also work in a similar mode [78].

15.5.2 Physical Parameters

Chitinase production is heavily reliant on pH, incubation temperature, time, water activity inoculum size, and aeration. The incubation time, in general, for chitinolytic activity of filamentous fungi ranges from 20 to 40 hours in the course of exponential growth phase. Several fungi including *A. fumigatus*, *Lecanicillium lecanii*, *L. ungicola*, and *Metarhizium* sp. synthesize maximum enzyme in 72 hours. Addition of sugar source delayed the production upto 24 hours but improved the yields. Reports are available on production of endochitinases in the early growth stages followed by exochitinases. Enzyme production and activity are not free from variation in pH and temperature. The optimum pH is the range in which the enzyme production is maximum and at par with enzyme activity. Chitinases are produced in 4.0 to 8.5 of broad pH range. *Stachybotrys elegans* significantly declined the enzyme activity at pH 8.0–9.0, whereas maximum enzyme activity was noticed at pH 5.0. Highest activity at pH 5.0 and production in a range of 4.0 to 7.0 was obtained in *L. lecanii*. In this condition, pH affects the chitinase gene expression in *M. anisopliae*, transcription at pH 5.0, and also detected at moderate alkaline pH 8.0, whereas it stops at pH 3.0. Variation in enzymatic activity of *L. fungicola* was marked in changes due to pH and resulting in production of endochitnase at acidic and proteolytic enzymes at higher pH [63]. *Streptomyces* sp. ANU 6277 produces chitinase of 45 kDa at temperature of 35°C and pH 6, which was a laterite soil isolate [79]. Marine isolates *Streptomyces canus*, *Streptomyces pseudogriseolus*, and *Micromonospora brevicatiana* were showed optimum activity from 40°C to 60°C and 8.0 of temperature and pH, respectively [80]. Another soil isolate from Riyadh, UAE, expresses activity at temperature of 35°C and pH 8.5 with incubation period of 3 days [81]. Production of chitinase in *Streptomyces lydicus* WYEClO8 was highest at 25°C to 30°C [61]. Likewise, optimum temperature for *Serratia marcescens* QMB 1466 was 30°C and pH 4.0–7.0. *Vibrio alginolyticus*, a proteobacteria, has a pH range of 4.0 to 9.0 and temperature 40°C for chitinase production [82]. On account of structural crystalinity in chitin, it often required thermostable enzyme to free the monomers in shrimp and prawns waste shells. Thermophiles *Streptomyces thermoviolaceus*, *Bacillus* sp. BG 11, *Bacillus licheniformis* X-7u, and *Bacillus stearothermophilus* CH-4 are key sources of thermostable chitinase [83].

15.5.3 Modes and Methods of Fermentation

Production of enzymes from microbial sources followed two types of cultivation methods, *viz.*, submerged fermentation (SubF) and solid-state fermentation (SolF). SubF is being carried out in the vicinity of high oxygen concentrated [O_2] in liquid nutrient media. Cons of the method lie in viscosity of broth with the fungal SubF as the cell mass and mycelium hinder impeller action and also limit the O_2 and mass transfer. Whereas SolF is carried out in low liquid, using agricultural residues, ease in downstream processing, and with feasible economy. It creates at par natural environment for efficient native and GMO fungal growth. The pros also include high volumetric productivity, significantly higher yield concentration, and simpler fermentation apparatus. It is documented that 20%–30% substrate can be amended leads to increase in end product [84]. Chitinases from *B. brassiana* are produced through SolF koji cultures, wheat bagasse, and colloidal chitin within pH 6.0–9.6. Use of mycelia increased the production from 117.2 U/g initial dried substrate (IDS) to 109.2 U/g IDS with reduced time of 20 hours [85]. Based on mode of operation, chitinases are produced in liquid batch, fed batch, and continuous fermentations. Types of bioreactors used for the processes are air lift bioreactors (ALBs), air lift with net draft tube bioreactors (ALndtBs), bubble column reactors (BCRs), and stirred tank bioreactors (STBs). Significant increase in chitinase production and activity was recorded with STBs in continuous cultivation mode. Optimized batch cultivation using *T. harzianum* increased 0.384 U production of exochitinase [6]. In contrast, *Paenibacillus* sp. CHE-N1 produces chitinase 78% higher than batch mode of operation using crab shell chitin in continuous mode [86]. *Verticillium lecanii* chitinase activity in traditional method (shaker flask) was 9.95 mU/ml which was further increased to 18.2 mU/ml with 5-L STBs and 19.9 mU/ml in 30-L ALBs [87]. *Beauveria bassiana*, a chitin-degrading fungus, which was isolated from the sediments of sea water, was optimized for the optimum process parameters using wheat bran as a substrate in SolF. The maximum yield of chitinase from marine fungus is 246.6 U/g of dry substrate. Soil containing waste of shrimp shells has higher chitinase than normal soil. Two strains of *Penicillium chrysogenum* able to produce chitinase of 3,809 U/g and 2,516 U/g IDS *in vitro*, respectively [88].

15.5.4 Advances Biotechnological Methods

Along with the methods and/or processes mentioned above few more advanced biotechnological methods have been adopted for improved chitinase production. Through the method of co-culture chitinase production can be increased. More often than not, microbes enhance self-efficiency when grown in inter or intra communities [89]. Using cheaper substrates (low capita investments), optimum quality control, and maximum product generation can be achieved with co-culture. This technique is practically successful for several hydrolytic enzymes and is at the infant stage for chitinase. Microbial consortia are another option in chitinase enhancement. Few studies highlighted the synergistic effect of chitinases and Bacillus thuringiensis endotoxins effectual than single strain against larvae of Spodoptera exigua and Helicoverpa armigera. Complete inhibition is also marked for Rhizoctonia solani and Botrytis cinerea spores following similar treatment [90]. Whole-cell immobilization was also done for chitinase production and increase in yield was also recorded. *Micromonospora chalcae* immobilized with calcium alginate and chitin. Production of immobilized cells is 30% higher than free cells and self-life is also higher. Taiwanese soil isolate *Pseudomonas aeruginosa* K-187 immobilized on a polymeric support and retains its activity even after 10 batches of fermentation in STBs [91] and making it economical.

There was an uprise in enzyme production through fermentation by employing recombinant strains at the present time. In order to carryout cloning and expression of diverse genes from *S. plymuthica*, *B. circulans* WL-12 and *Aeromonas hydrophilia* into *E. coli*, several attempts have been made [92, 93]. Chitinase from thermophilic *Rhodothermus marinus* is expressed in *E. coli* and found to be utmost thermostable bacterial chitinase [94]. Recombinant *E. coli* have *A. hydrophilia* genes for enhancing chitinase production [95].

Processivity of chitinase can be upgraded through placing hydrophobic amino acid at the edge of active site to make a narrow groove above active site. Due to this narrow groove, polymeric chain after every cleavage will not be released by chitinase while breaking of glycosidic bonds carried out uninterrupted. Next, another technique is domain swapping for improving chitinase activity. It is observed that chitinases express low activity due to absence of CBD. In this process, CBD of an active chitinase is swapped by advanced protein engineering and leads to better substrate binding of newly engineered chitinase. This technique itself carries huge potential for the improvement of chitinolytic enzymes for future endeavors [96].

15.6 Applications of Microbial Chitinases

The hydrolyzing property of chitinase is the unique property by which the insoluble form of chitin is transferred into simpler monomer and oligomer forms. The hydrolysis of chitin by the enzymatic process plays a vital role in healthcare and industrial realm because of antibacterial, fungicidal, antihypertensive, and the food quality enhancement property, and these are described in detail below.

15.6.1 Agricultural

15.6.1.1 Biopesticides

Pests are major insects which destroy the crops and vegetables and reduce the farming products. Their exoskeleton and the cuticular lining of the gut are made up of chitin so the chitinase enzymes play a vital role for the control of the pests and insects. The chitinase activity helps in the degradation of chitin present in the wall of insects via forming pores in the gut and exoskeleton, resulting in less survivability *in vivo*. Also, the growth of mites, pests and the housefly larvae are ceased due the same mechanism [97]. Entomopathogenic fungi outwardly produce chitinases to overcome the physical barrier, penetrate the exoskeleton for infection, and play a greater part in plant defenses. These are maximally comes under Deuteromycotina and Zygomycotina subdivisions. Wax moth *Galleria mellonella* Linn. is biologically controlled through *B. brassiana* that produces chitinolytic enzymes. Similarly, *S. marcescens* and *B. thuringiensis* subsp. *aizawai* produce chitinase toxic to both *G. mellonella* and gypsy moth larvae (*Lymantria dispar* Linn.). *B. thuringiensis* has been practically applied globally as microbial control agent and bioinsecticides [98, 99]. Bacterial chitinases have also been shown to be potential bioinsecticides when coupled with other suitable Cry proteins [100].

15.6.1.2 Biocontrol

Bacterial and fungal chitinases have been in single or with synergism antagonist to phytofungal pathogens. In biocontrol of *R. solani* and *Fusarium oxysporum* in cotton and *Sclerotium rolfsii* in bean seeds, chitinase from *Aeromonas caviae* is effectively involved [101]. Though the chitinases affects various parts of complex fungal structure, it is also counted as cell wall degrading enzymes (CWDEs). It has been reported that the chitinase from *S. marcescens* helps in the suppression of disease caused by *S. rolfsii*. The development in the genetic engineering helped to successfully introduce the chitinase producing gene into the *E. coli* which helped in the control of the disease triggered by *S. rolfsii* and *R. solani* in the cotton [102, 103]. Cell-free supernatant of *Alphanocladium album* inhibited strongly the growth of *Nectteri haematcocca* fungus, a pathogen of pea [13]. *S. lydicus* WYEC108 was capable of deforming the germinating oospores of *Pythium ultimum* and damaging the hyphal cell walls [61]. Chitinases from *Trichoderma* sp. have been characterized and it is considered as one of the potential biocontrol agents on soil borne fungal pathogens. A specific group of bacteria named as *Streptomyces violaceusniger* produces the chitinase and some antifungal compounds and shows the anti-fusarium activity (AFA) against plant pathogenic fungal groups [104]. Transgenic rice plants with the *chiC* gene of *Streptomyces griseus* HUT6037 showed high resistance toward *Magnaporthe grisea* causative agent of leaf blast disease [27]. It is also stated that chitinous wastes are utilized as biofertilizers.

15.6.2 Biomedical

The most important role of the chitinase enzyme is in the field of human healthcare, healthy lifestyle, and cosmetic product manufacture along with in the development of nano-medicines [105, 106]. The antifungal activity and highly biocompatible quality make chitinase and its derivatives a part of biomedical applications, including wound healings, drug delivery, cartilage tissue engineering, and nerve generation. The chitinase can also be used as additives during the therapy for enhancing the activity and effect of antifungal drugs. In humans, chitinase is also being suggested to be used for detection of invasive mycotic infection. Similarly, it can be directly used the supplement during the manufacture of antifungal lotions and creams for topical administrations [107]. An enormous biomedical potential has also been detected in chitooligosaccharides (chitohexaose and chitoheptaose) for antitumor activity, wound healing property, and antihypertensive activity and used in human medicines. GlcNAc is also reported to be useful as anti-inflammatory agents. The need for highly purified enzymes for maximal production of chitin and chitosans, which is used as membrane for drug delivery and in tissue engineering, has increased. Chitinases are also used in anti-cancer therapy. Reports stated that making ophthalmic preparations from chitinases with other microbicides can be prepared.

15.6.3 Pharmaceutical

The chitinous waste products can be used for the production of single-cell protein (SCP) by the bioconversion of chitin to its simpler forms by chitinase. Chitinases produced by *S. marcescens* are used for the hydrolysis of chitin present in the wastes and the yield of SCP with the help of yeast *Pichia kudriavzevii* [6]. The protein and nucleic acid content in these SCP are found to be 45% and 8%–11%, respectively [108]. Maximally used fungal source for SCP are *Candida tropicalis*, *Hansenula polymorpha*,

Myrothecium verrucaria, and *S. cerevisiae*. Chitinolytic enzyme complexes from *M. verrucaria* and *S. cerevisiae* were used for the production of SCP from chitinous waste and the protein content was approximately 61% [109]. More than 60% SCP and 1% to 3% of nucleic acid were produced from *S. cerevisiae* [20]. The prospective applications of these unique chitinase enzymes are therapeutic drugs for diseases like asthma and chronic rhinosinusitis (CRS), an antineoplastic drug [110], and as a general ingredient to be used in protein engineering.

15.6.4 Industrial

Chitinase is highly essential for the enzymatic degradation of fungal cell wall and leads to protoplast separation for further study without causing any damage to its components. Chitinase from members of *Enterobacter* sp. is most effective for cell wall degradation and removal of protoplast from the fungal hypha of *Aspergillus niger*, *Trichoderma reesei*, *Agaricus bisporus*, *Pleurotus florida*, etc. [20]. Sometimes, Chi-Ag (chitinase-gold) complex can be used for similar purpose. Enzymatic hydrolysis of the cell walls using chitinase preparation was found to be effective in the recovery of tannase enzyme also. *B. circulans* WL-12 having elevated chitinase activity was potentially employed in protoplasts generation from *Phaffia rhodozyma* [93]. Chitinases from *Streptomyces* sp. were found to be effective in generation of protoplast from *Aspergillus oryzae* and *Fusarium solani*. Additionally, chitinase from *T. harzianum* showed most efficiency in generating protoplast from different fungi [13]. Likewise the other prospective applications are it can be used as a flavor enhancer in food and energy drinks [111].

15.6.5 Environmental

Morphogenesis is the process of origin and development of morphological characteristics in an organism; it has been reported that the chitinase enzyme plays a key role during the morphogenesis process in case of insects and *S. cerevisiae* [112]. The functional expression of chitinase and chitosanase and their effects on morphogenesis in *S. pombe* have reported in details as expression of *chiA* gene in *S. pombe*; cells grew slowly and became elongated while the expression of *choA* gene leads to swollen cells. In conclusion, both *chiA* and *choA* genes expression resulted in elongated and fat cells. Nowadays, mosquitoes are the main source of disease transmission from one place to other and hence can be used as a suitable vector for carrying different agents to control various pests. It has been observed that the first and fourth instar larvae of *Aedes aegypti* mosquito, a vector of dengue and yellow fever, can be killed in 24 and 48 hours, respectively, when delivered with pure chitinase enzyme from *Myrothecium verrucaria* owing to lipolytic activity [113].

15.6.5.1 Waste Management

Recombinant chitinases in comparison to pure chitinases have been utilized in chitinous biomass conversion which resulting in production of depolymerized components from chitinous waste of aquatic organism and reduces environmental pollution. They have been effectively used in processing of shellfish waste to obtain value added products. Chitin monomers such as NAG have been used in food industry for sweeteners, and chitinases were successfully used for NAG production from shell fish waste. Moreover, shell fish wastes have been converted to SCPs.

15.6.6 Others

There is a relation between the fungal strain thrive in soil and its surroundings which gives a standard measure for the actively populating fungal biomass in that areas. The association of chitinase activity with the content of fungus-specific indicator molecules using specific methylumbelliferyl substrates aids fungal biomass estimation [114]. Likewise, some proteins with chitin binding property and the enzyme showing chitinase activity synthesized by soil bacteria are extensively used in detection of diseases in human caused by fungus [115]. Apart from biopesticides, they have also been used for terrestrial and marine animal feed, preparation of bioactive chito-oligosaccharides (COS), and improvising plant/host defense responses for developing transgenic plants (at few instances). GlcNAc, COS, and glucosamine oligomers have broad range of applications in every essential sector. Chitinase has been recognized as one of the pathogenesis related proteins (PR proteins) produced by higher eukaryotes.

At present, the chitinase research focuses on improving its catalytic activity. Directed evolution and site directed mutagenesis are the two major approaches of protein engineering. Placing hydrophobic amino acid at the edge of active site to make a narrow groove above active site enhances chitinase efficiency. This results in continuous activity of chitinase and not escape from the polymeric chain in a processive fashion. Another approach is domain swapping. Many chitinases showed low catalytic activity due to lack of CBD. The CBD from an active chitinase can be swapped through protein engineering. Enhancing substrate binding of the engineered protein results in increase in catalytic activity. These approaches hold huge potential for the improvement of chitinolytic enzyme applications practically [96].

15.7 Conclusion

Chitin is the second most polysaccharide in nature and exists in marine invertebrates to higher plants including algae, bacteria, fungi, insects, and viruses. Being chitin-degrading enzyme, chitinases have wider application in agriculture, medicine, biochemical processing, engineering, waste management, pesticide control, food and feed, sweetness, and cell wall degradation. Practical feasibility in applications of chitinases is possible through higher yield and improved catalytic activity. Microbes are act as chief source of chitinases, while present day research are focusing on its catalytic activity improvement. Advances in gene cloning and protein expression through modern biotechnological approaches facilitate the upgrading and increase the industrial produce which ultimately leads to uprise in the socioeconomical standards.

References

1. Liu, X. and Kokare, C., Microbial Enzymes of Use in Industry, in: *Biotechnology of Microbial Enzymes*, G. Brahmachari (Ed.), pp. 267–298, Academic Press, Cambridge, United States, 2017.
2. Thakur, N., Nath, A.K., Chauhan, A., Parmar, S.C., Pandey, H., Thakur, K., Chitinases from microbial sources, their role as biocontrol agents and other potential applications. *J. Entomol. Zool. Stud.*, 7, 837, 2019.
3. Nagpure, A., Choudhary, B., Kumar, S., Gupta, R.K., Isolation and characterization of chitinolytic *Streptomyces* sp. MT7 and its antagonism towards wood-rotting fungi. *Ann. Microbio.*, 64, 531, 2014.
4. Nayak, S.K. and Mishra, B.B. (Eds.), *Frontiers in soil and Environmental Microbiology*, CRC Press, Boca Raton, Unites States, 2020.
5. Nayak, S.K., Dash, B. *et al.*, Chitinase producing soil bacteria: Prospects and Applications, in: *Frontiers in soil and Environmental Microbiology*, S.K. Nayak and B.B. Mishra (Eds.), pp. 289–298, CRC Press, Boca Raton, 2020.
6. Felse, P.A. and Panda, T., Production of microbial chitinases: a revisit. *Bioprocess. Eng.*, 23, 127, 2000.
7. Roopavathi, A.S., Vigneshwari, R., Jayapradha, R., Chitinase: Production and applications. *J. Chemi. Pharma. Res.*, 7, 924, 2015.

8. Gohel, V., Singh, A., Vimal, M., Ashwini, P., Chhatpar, H.S., Bioprospecting and antifungal potential of chitinolytic microorganisms. *Afr. J. Biotechnol.*, 5, 54, 2006.
9. Patil, R.S., Ghormade, V., Deshpande, M.V., Chitinolytic enzymes: an exploration. *Enzyme Microb. Techol.*, 26, 473, 2000.
10. Jones, J.D.G., Grady, K.L., Suslow, T.V., Bedrook, J.R., Isolation and characterization of genes encoding two chitinases enzymes from *S. marcescens*. *EMBO J.*, 5, 467, 1986.
11. Sahai, A.S. and Manocha, M.S., Chitinases of fungi and plants: their involvement in morphogenesis and host-parasite interaction. *FEMS Microbiol. Rev.*, 11, 317, 1993.
12. Saks, E. and Jankiewicz, U., Chitinolytic activity of bacteria. *Adv. Biochem.*, 56, 1, 2010.
13. Gokul, B., Lee, J., Song, K., Rhee, S.K., Kim, C.H., Panda, T., Characterization and applications of chitinases from *Trichoderma harzianum*– A review. *Bioprocess. Eng.*, 23, 691, 2000.
14. Henrissat, B. and Davies, G., Structural and sequence-based classification of glycoside hydrolases. *Curr. Opin. Struct. Biol.*, 7, 637, 1997.
15. Gooday, G.W., Aggressive and defensive roles for chitinases. *EXS.*, 87, 157, 1999.
16. van Scheltinga, A.C., Kalk, K.H., Beintema, J.J., Dijkstra, B.W., Crystal structures of hevamine, a plant defence protein with chitinase and lysozyme activity, and its complex with an inhibitor. *Structure*, 2, 1181, 1994.
17. Hart, P.J., Pfluger, H.D., Monzingo, A.F., Hollis, T., Robertus, J.D., The refined crystal structure of an endochitinases from *Hordeum vulgare* L. seeds at 1.8 A resolution. *J. Mol. Biol.*, 248, 402, 1995.
18. Duo-Chuan, L., Review of fungal chitinases. *Mycopathologia.*, 161, 345, 2006.
19. Matsumoto, K.S., Fungal chitinases. *Adv. Agric. Food. Biotechnol.*, 289, 2006.
20. Dahiya, N., Tewari, R., Hoondal, G.S., Biotechnological aspects of chitinolytic enzymes: a review. *Appl. Microbiol. Biotech.*, 71, 773, 2006.
21. Swiontek Brzezinska, M., Walczak, M., Lalke-Porczyk, E., Donderski, W., Microbial degradation of shrimp waste in soil of the Chełmżyńskie Lake watershed. *Pol. J. Environm. Stud.*, 19, 627, 2010.
22. Joo, G.J., Purification and characterization of an extracellular chitinase from the antifungal biocontrol agent *Streptomyces halstedii*. *Biotechnol. Lett.*, 27, 1483, 2005.
23. Poole, D.M., Hazlewood, G.P., Huskisson, N.S., Virden, R., Gilbert, H.J., The role of conserved tryptophan residues in the interaction of a bacterial cellulose binding domain with its ligand. *FEMS Microbiol. Lett.*, 106, 77, 1993.
24. Watanabe, T., Kimura, K., Sumiya, T., Nikaidou, N., Suzuki, K., Suzuki, M., Taiyoji, M., Ferrer, S., Regue, M., Genetic analysis of the chitinase system of *Serratia marcescens* 2170. *J. Bacteriol.*, 179, 7111, 1997.
25. Ueda, M., Fujiwara, A., Kawaguchi, T., Arai, M., Purification and some properties of six chitinases from *Aeromonas* sp. No. 10S–24. *Biosci. Biotechnol. Biochem.*, 59, 2162, 1995.

26. Donderski, W. and Trzebiatowska, M., Influence of physical and chemical factors on the activity of chitinases produced by planktonic bacteria isolated from Jeziorak Lake. *Pol. J. Environ. Stud.*, 9, 77, 2000.
27. Itoh, Y., Takahashi, K., Takizawa, H., Nikaidou, N., Tanaka, H., Nishihashi, H., Watanabe, T., Nishizawa, Y., Family 19 chitinase of *Streptomyces griseus* HUT6037 increases plant resistance to the fungal disease. *Biosci. Biotechnol. Biochem.*, 67, 847, 2003.
28. Baliyarsingh, B., Nayak, S.K., Mishra, B.B., Soil Microbial Diversity: An Ecophysiological Study and Role in Plant Productivity, in: *Advances in Soil Microbiology: Recent Trends and Future Prospects, Microorganisms for Sustainability, vol 4,* T.K. Adhya, B.B. Mishra, K. Annapurna, D.K. Verma, U. Kumar (Eds.), pp. 1–17, Springer Nature, Singapore, 2017.
29. Sturz, H. and Robinson, J., Anaerobic decomposition of chitin in freshwater sediments, in: *Chitin in nature and biotechnology*, R. Muzzarelli, C. Jeuniaux, G.W. Gooday (Eds.), pp. 531–538, Plenum Press, New York, United States, 1985.
30. Van Aalten, D.M.F., Synstad, B., Brurberg, M.B., Hough, E., Riise, B.W., Eijsink, V.G.H., Wierenga, R.K., Structure of a two-domain chitotriosidase from *Serratia marcescens* at 1.9-A° resolution. *PNAS*, 97, 5842, 2000.
31. Vaaje-Kolstad, G., Horn, S.J., Sørlie, M., Eijsink, V.G.H., The chitinolytic machinery of *Serratia marcescens*—a model system for enzymatic degradation of recalcitrant polysaccharides. *FEBS J.*, 280, 3028, 2013.
32. Ruiz-Sanchez, A., Cruz-Camarillo, R., Salcedo-Hernandez, R., Barboza-Corona, J.E., Chitinases from *Serratia marcescens* Nima. *Biotechnol. Lett.*, 27, 649, 2005.
33. Gal, S.W., Choi, J.Y., Kim, C.Y., Cheong, Y.H., Choi, Y.J., Bahk, J.D., Lee, S.Y., Cho, M.J., Cloning of the 52 kDa chitinase gene from *S. marcescens* KCTC2172 and its proteolytic cleavage into an active 35 kDa enzyme. *FEMS Microbiol. Lett.*, 160, 151, 1998.
34. Brurberg, M.B., Eijsink, V.G.H., Nes, I.F., Characterization of a chitinase gene (chiA) from *S. marcescens* BJL200 and one-step purification of the gene product. *FEMS Microbiol. Lett.*, 124, 399, 1994.
35. Yan, Q. and Fong, S.S., Bacterial chitinase: nature and perspectives for sustainable bioproduction. *Bioresour. Bioprocess.*, 2, 31, 2015.
36. Hartl, L., Zach, S., Seidl-Seiboth, V., Fungal chitinases: diversity, mechanistic properties and biotechnological potential. *Appl. Microbiol. Biotechnol.*, 93, 533, 2012.
37. Tsujibo, H., Orikoshi, H., Baba, N., Miyahara, M., Miyamoto, K., Yasuda, M., Inamori, Y., Identification and characterization of the gene cluster involved in chitin degradation in a marine bacterium, *Alteromonas* sp. strain O-7. *Appl. Environ. Microbiol.*, 68, 263, 2002.
38. Bendt, A., Hueller, H., Kammel, U., Helmke, E., Schweder, T., Cloning, expression, and characterization of a chitinase gene from the Antarctic psychrotolerant bacterium *Vibrio* sp. strain Fi:7. *Extremophiles*, 5, 119, 2001.

39. Songsiriritthigul, C., Pantoom, S., Aguda, A.H., Robinson, R.C., Suginta, W., Crystal structures of *Vibrio harveyi* chitinase A complexed with chitooligosaccharides: implications for the catalytic mechanism. *J. Str. Biol.*, 162, 491, 2008.
40. Arora, N., Ahmad, T., Rajagopal, R., Bhatnagar, R.K., A constitutively expressed 36 kDa exochitinase from *Bacillus thuringiensis* HD-1. *Biochem. Biophys. Res. Commun.*, 307, 620, 2003.
41. Wang, S.-Y., Moyne, A.-L., Thottappilly, G., Wu, S.-J., Locy, R.D., Singh, N.K., Purification and characterization of *Bacillus cereus* exochitinase. *Enz. Microbiol. Technol.*, 28, 492, 2001.
42. Wang, S.-L., Shih, I.-L., Liang, T.-W., Wang, C.-H., Purification and characterization of two antifungal chitinases extracellularly produced by *Bacillus amyloliquefaciens* V656 in a SCSP medium. *J. Agricul. Food Chem.*, 50, 2241, 2002.
43. Seidl, V., Chitinases of filamentous fungi: a large group of diverse proteins with multiple physiological functions. *Fungal Biol. Rev.*, 22, 36, 2008.
44. Gruber, S.G. and Seidl-Seiboth, V., Self versus non-self: fungal cell wall degradation in *Trichoderma. Microbiol.*, 158, 26, 2012.
45. Kuranda, M.J. and Robbins, P.W., Chitinase is required for cell separation during growth of *Saccharomyces cerevisiae. J. Biol. Chem.*, 266, 19758, 1991.
46. Colussi, P.A. and Taron, C.H., *Kluyveromyces lactis* LAC4 promoter variants that lack function in bacteria but retain full function in *K. Lactis. Appl. Environ. Microbiol.*, 71, 7092, 2005.
47. Schlegel, H.G., *General microbiology*, Polish Scientific Publishers, Warsaw, Poland (in polish), 2006.
48. Haran, S., Schickler, H., Oppenheim, A., Chet, I., New components of the chitinolytic system of *Trichoderma harzianum. Nucleic Acid. Res.*, 99, 447, 1995.
49. Leger, S., Joshi, L., Bidochka, M.J., Rizzo, N.W., Roberts, D.W., Characterization and Ultrastructural Localization of Chitinases from *Metarhizium anisopliae, M. flavoviride*, and *Beauveria bassiana* during Fungal Invasion of Host (*Manduca sexta*) Cuticle. *Appl. Environ. Microbiol.*, 62, 907, 1996.
50. Sakuda, S., Studies on the chitinase inhibitors, allosamidins, in: *Chitin Enzymology*, R.A. Muzzarelli (Ed.), pp. 203–212, *Eur. Chitin Soc.*, Grottammare, Italy, 1996.
51. Bidochka, M.J. and Khachatourians, G.G., N-acetyl, Dglucosamine mediated regulation of extracellular proteases in entomopathogenic fungus *Beauveria brassiana. Appl. Environ. Microbiol.*, 54, 2699, 1988.
52. Paul, E.A. and Clark, F.E., *Soil Microbiology and Biochemistry*, University of Maria Curie-Sklodowska University in Lublin (in polish), Poland, 2000.
53. Gherbawy, Y., Elhariry, H., Altalhi, A., El-Deeb, B., Khiralla, G., Molecular screening of *Streptomyces* isolates for antifungal activity and family 19 chitinase enzymes. *J. Microbiol.*, 50, 459, 2012.

54. Lee, S.Y., Tindwa, H., Lee, Y.S., Naing, K.W., Hong, S.H., Nam, Y., Kim, K.Y., Biocontrol of anthracnose in pepper using chitinase, beta-1,3 glucanase, and 2-furancarboxaldehyde produced by *Streptomyces cavourensis* SY224. *J. Microbiol. Biotechnol.*, 2, 1359, 2012.
55. Zhang, Z., Yuen, G.Y., Sarath, G., Penheiter, A.R., Chitinases from plant disease biocontrol agent *Stenotrophomonas maltophilia* C3. *Phytopathology*, 91, 204, 2001.
56. Tsujibo, H., Minoura, K., Miyamoto, K., Endo, H., Moriwaki, M., Inamori, Y., Purification and properties of a thermostable chitinase from *Streptomyces thermoviolaceus* OPC-520. *Appl. Environ. Microbiol.*, 59, 620, 1993.
57. Mitsutomi, M., Hata, T., Kuwahara, T., Purification and characterization of novel chitinases from *Streptomyces griseus* HUT 6037. *J. Ferment. Bioeng.*, 80, 153, 1995.
58. Hoster, F., Schmitz, J.E., Daniel, R., Enrichment of chitinolytic microorganisms: Isolation and characterization of a chitinase exhibiting antifungal activity against phytopathogenic fungi from a novel *Streptomyces* strain. *Appl. Microbiol. Biotechnol.*, 66, 434, 2005.
59. El-Sayed, E.S.A., Ezzat, S.M., Ghaly, M.F., Mansour, M., El-Bohey, M.A., Purification and characterization of two chitinases from *Streptomyces albovinaceus* S-22. *World J. Microbiol. Biotechnol.*, 16, 87, 2000.
60. Shekhar, N., Bhattacharya, D., Kumar, D., Gupta, R.K., Biocontrol of wood-rotting fungi using *Streptomyces violaceusniger* XL-2. *Can. J. Microbiol.*, 52, 805, 2006.
61. Mahadevan, B. and Crawford, D.L., Properties of the chitinase of the antifungal biocontrol agent *Streptomyces lydicus* WYEC108. *Enz. Microb. Technol.*, 20, 489, 1997.
62. Saito, A., Fujii, T., Yoneyama, T., Miyashita, K., glkA is involved in glucose repression of chitinase production in *Streptomyces lividans*. *J. Bacter.*, 180, 2911, 1998.
63. Islam, R. and Datta, B., Diversity of chitinases and their industrial potential. *Int. J. Appl. Res.*, 1, 55, 2015.
64. Cottrell, M.T., Wood, D.N., Yu, L.Y., Kirchman, D.L., Selected chitinase genes in cultured and uncultured marine bacteria in the alpha- and gamma-subclasses of the proteobacteria. *Appl. Environ. Microbiol.*, 66, 1195, 2000.
65. Hjort, K., Bergstrom, M., Adesina, M.F., Jansson, J.K., Smalla, K., Sjoling, S., Chitinase genes revealed and compared in bacterial isolates, DNA extracts and a metagenomic library from a phytopathogen-suppressive soil. *FEMS Microbiol. Ecol.*, 71, 197, 2010.
66. Ikeda, S., Ytow, N., Ezura, H., Minamisawa, K., Miyashita, K., Fujimura, T., Analysis of molecular diversity of bacterial chitinase genes in the maize rhizosphere using culture independent methods. *Microbes Environ.*, 22, 71, 2007.

67. LeCleir, G.R., Buchan, A., Maurer, J., Moran, M.A., Hollibaugh, J.T., Comparison of chitinolytic enzymes from an alkaline, hypersaline lake and an estuary. *Environ. Microbiol.*, 9, 197, 2007.
68. Kitamura, E. and Kamei, Y., Molecular cloning, sequencing and expression of the gene encoding a novel chitinase A from a marine bacterium, *Pseudomonas* sp. PE2, and its domain structure. *Appl. Microbiol. Biot.*, 61, 140, 2003.
69. Nayak, S.K., Nayak, S. *et al.*, Antimycotic Role of Soil Bacillus sp. against Rice Pathogens: A Biocontrol Prospective, in: *Microbial Biotechnology*, vol. Vol-I, J.K. Patra *et al.* (Ed.), pp. 29–60, Springer, Singapore, 2017, https://doi.org/10.1007/978-981-10-6847-8_2.
70. Limon, M.C., Lora, J.M., Irene, G., Cruz, J., Llobell, A., Benitez, T., Pintor, J.A., Primary structure and expression pattern of the 33-kDa chitinase gene from the mycoparasitic fungus *Trichoderma harzianum*. *Curr. Genet.*, 28, 478, 1995.
71. Kobayashi, D.Y., Reedy, R.M., Bick, J., Oudemans, P.V., Characterization of a chitinase gene from *Stenotrophomonas maltophilia* strain 34S1 and its involvement in biological control. *Appl. Environ. Microbiol.*, 68, 1047, 2002.
72. Hao, Z., Cai, Y., Liao, X., Zhang, X., Fang, Z., Zhang, D., Optimization of nutrition factors on chitinase production from a newly isolated *Chitiolyticbacter meiyuanensis* SYBC-H1. *Braz. J. Microbiol.*, 43, 177, 2012.
73. Singh, R.K., Kumar, D.P., Solanki, M.K., Singh, P., Srivastva, A.K., Kumar, S., Kashyap, P.L., Saxena, A.K., Singhal, P.K., Arora, D.K., Optimization of media components for chitinase production by chickpea rhizosphere associated *Lysinibacillus fusiformis* B-CM18. *J. Basic Microbiol.*, 53, 451, 2013.
74. Stoykov, Y.M., Pavlov, A.I., Krastanov, A.I., Chitinase biotechnology: Production, purification, and application. *Eng. Life Sci.*, 15, 30, 2015.
75. Taechowisan, T., Peberdy, J.F., Lumyong, S., Chitinase production by endophytic *Streptomyces aureofaciens* CMU Ac 130 and its antagonism against phytopathogenic fungi. *Annal. Microbiol.*, 53, 447, 2003.
76. Hamid, R., Khan, M.A., Ahmad, M., Ahmad, M.M., Abdin, M.Z., Musarrat, J., Javed, S., Chitinases: An update. *J. Pharm. Bioall. Sci.*, 5, 21, 2013.
77. Tabudravu, J.N., Eijsink, V.A., Gooday, G.W., Jaspar, M., Komander, D., Legg, M., Synstad, B., van Aalten, D.M.F., Psammaplin, A., a chitinase inhibitor isolated from the Fijian marine sponge *Aplysinella rhax*. *Bioorg. Med. Chem.*, 10, 1123, 2002.
78. Arai, N., Shiomi, K., Yamaguchi, Y., Masuma, R., Iwai, Y., Turberg, A., Kolbl, H., Omura, S., Argadin, a new chitinase inhibitor, produced by *Clonostachy* sp. FO. 7314. *Chem. Pharm. Bull.*, 48, 1442, 2000.
79. Narayana, K.J. and Vijayalakshmi, M., Chitinase Production by *Streptomyces* sp. ANU 6277. *Braz. J. Microbiol.*, 40, 725, 2009.
80. Mane, U. and Deshmukh, A., Chitin degrading potential of three aquatic actinomycetes and its optimization. *Afr. J. Biotechnol.*, 8, 6617, 2009.

81. Meena, S., Gothwal, R.K., Saxena, J., Mohan, M.K., Ghosh, P., Chitinase production by a newly isolated thermotolerant *Paenibacillus* sp. BISR-047. *Ann. Microbiol.*, 64, 787, 2013.
82. Murao, S., Kawada, T., Itoh, H., Oyama, H., Shin, T., Purification and characterization of a novel type of chitinase from *Vibrio alginolyticus* TK-22. *Biosci. Biotechnol. Biochem.*, 56, 368, 1992.
83. Haki, G.D. and Rakshit, S.K., Developments in industrially important thermostable enzymes: a review. *Bioresour. Technol.*, 89, 17, 2003.
84. Singh, S., Sczakas, G., Soccol, C., Pandey, A., Production of enzymes by solid-state fermentation, in: *Current Developments in Solid-State Fermentation*, A. Pandey, C. Soccol, C. Larroche (Eds.), pp. 183–204, Springer, New York, 2008.
85. Suresh, P.V. and Chandrasekaran, M., Impact of process parameters on chitinase production by an alkalophilic marine *Beauveria bassiana* in solid state fermentation. *Process Biochem.*, 34, 257, 1999.
86. Kao, P.M., Huang, S.C., Chang, Y.C., Liu, Y.C., Development of continuous chitinase production process in a membrane bioreactor by *Paenibacillus* sp. CHE-N1. *Process Biochem.*, 42, 606, 2007.
87. Liu, B.L., Kao, P.M., Tzeng, Y.M., Feng, K.C., Production of chitinase from *Verticillium lecanii* F091 using submerged fermentation. *Enzym. Microb. Technol.*, 33, 410, 2003.
88. Patidar, P., Agrawal, D., Banerjee, T., Patil, S., Optimisation of process parameters for chitinase production by soil isolates of *Penicillium chrysogenum* under solid substrate fermentation. *Process Biochem.*, 40, 2962, 2005.
89. Tesfaw, A. and Assefa, F., Co-culture: A great promising method in single cell protein production. *Biotechnol. Mol. Biol.*, 9, 12, 2014.
90. Veliz, E.A., Martínez-Hidalgo, P., Hirsch, A.M., Chitinase-producing bacteria and their role in biocontrol. *AIMS Microbiol.*, 3, 689, 2017.
91. Wang, S.L. and Chang, W.T., Purification and characterization of two bifunctional chitinases/lysozymes extracellularly produced by *Pseudomonas aeruginosa* K-187 in shrimp and crab shell powder medium. *Appl. Environ. Microbiol.*, 63, 380, 1997.
92. Ovadis, M., Liu, X.G., Gavriel, S., Ismailov, Z., Chet, I., Chernin, L., The global regulator genes from biocontrol strain *Serratia plymuthica* IC1270: Cloning, sequencing, and functional studies. *J. Bacteriol.*, 186, 4986, 2004.
93. Mitsutomi, O., Isono, M., Uchiyama, A., Nikaidov, N., Ikegami, T., Watanabe, T., Chitosanase activity of the enzyme previously reported as β-1,3- 1,4-glucanase from *Bacillus circulans* WL 12. *Biosci. Biotechnol. Biochem.*, 62, 2107, 1998.
94. Hobel, C.F., Hreggvidsson, G.O., Marteinsson, V.T., Bahrani-Mougeot, F., Einarsson, J.M., Kristjansson, J.K., Cloning, expression and characterization of a highly thermostable family 18 chitinase from *Rhodothermus marinus*. *Extremophiles*, 9, 53, 2005.

95. Chen, S.J., Chang, M.C., Cheng, C.Y., Effect of induction conditions on production and excretion of *Aeromonas hydrophilia* chitinase by *Eschericia coli. J. Ferment. Bioeng.*, 84, 610, 1997.
96. Rathore, A.S. and Gupta, R.D., Chitinases from Bacteria to Human: Properties, Applications, and Future Perspectives. *Enzyme Res.*, 2015, 791907, 2015.
97. Sindhu, S.S., Sehrawat, A., Sharma, R., Khandelwal, A., Biological Control of Insect Pests for Sustainable Agriculture, in: *Advances in Soil Microbiology: Recent Trends and Future Prospects, Microorganisms for Sustainability, vol 4*, T.K. Adhya, B.B. Mishra, K. Annapurna, D.K. Verma, U. Kumar (Eds.), pp. 189–218, Springer Nature, Singapore, 2017.
98. Barboza-Corona, J.E., Contreras, J.C., Velazquez-Robledo, R., Bautista-Justo, M., Gomez-Ramirez, M., Cruz-Camarillo, R., Ibarra, J.E., Selection of chitinolytic strains of *Bacillus thuringiensis. Biotechnol. Lett.*, 21, 1125, 1999.
99. Sampson, M.N. and Gooday, G.W., Involvement of chitinases of *Bacillus thuringiensis* during pathogenesis in insects. *Microbiol.*, 144, 2189, 1998.
100. Ragev, A., Keller, M., Strizhov, N., Sneh, B., Prudovsky, E., Chet, I., Ginzberg, I., Koncz-Kalman, Z., Koncz, C., Schell, J., Zilberstein, A., Synergistic activity of a *Bacillus thuringiensis* δ-endotoxin and a bacterial endochitinase against *Spodoptera littoralis. Appl. Environ. Microbiol.*, 62, 3581, 1996.
101. Inbar, J. and Chet, I., Evidence that chitinase produced by *Aeromonas caviae* is involved in biological control of soil borne plant pathogen by this bacterium. *Soil Biol. Biochem.*, 23, 973, 1991.
102. Singh, S.P. and Gaur, R., Evaluation of antagonistic and plant growth promoting activities of chitinolytic endophytic actinomycetes associated with medicinal plants against *Sclerotium rolfsii* in chickpea. *J. Appl. Microbiol.*, 121, 506, 2016.
103. Selim, H.M., Gomaa, N.M., Essa, A.M.M., Application of endophytic bacteria for the biocontrol of *Rhizoctonia solani* (Cantharellales: ceratobasidiaceae) damping-off disease in cotton seedlings. *Biocontrol Sci. Technol.*, 27, 81, 2017.
104. Swarnalakshmi, K., Senthilkumar, M., Ramakrishnan, B., Endophytic actinobacteria: nitrogen fixation, phytohormone production, and antibiosis, in: *Plant Growth Promoting Actinobacteria*, G. Subramaniam, S. Arumugam, V. Rajendran (Eds.), pp. 123–145, Springer, Singapore, 2016.
105. Elieh-Ali-Komi, D. and Hamblin, M.R., Chitin and Chitosan: Production and Application of Versatile Biomedical Nanomaterials. *Int. J. Adv. Res. (Indore)*, 4, 411, 2016.
106. Azad, A.K., Sermsintham, N., Chandrkrachang, S., Stevens, W.F., Chitosan membrane as a wound-healing dressing: characterization and clinical application. *J. Biomed. Mater. Res. B. Appl. Biomater.*, 69, 216, 2004.
107. Chavan, S.B. and Deshpande, M.V., Chitinolytic enzymes: an appraisal as a product of commercial potential. *Biotechnol. Prog.*, 29, 833, 2013.

108. Mekasha, S., Byman, I.R., Lynch, C., Toupalová, H., Anděra, L., Næs, T., Vaaje-Kolstad, G., Vincent, G.H.E., Development of enzyme cocktails for complete saccharification of chitin using mono-component enzymes from *Serratia marcescens*. *Proc. Biochem.*, 56, 132, 2017.
109. Vyas, P.R. and Deshpande, M.V., Enzymatic hydrolysis of chitin by *Myrothecium verrucaria* chitinase complex and its utilization to produce SCP. *J. Gen. Appl. Microbiol.*, 37, 267, 1991.
110. Rinaudo, M., Chitin and chitosan: properties and applications. *Prog. Poly. Sci.*, 31, 603, 2006.
111. Razak, M.A., Pinjari, A.B., Begum, P.S., Viswanath, B., Biotechnological Production of Fungal Biopolymers Chitin and Chitosan: Their Potential Biomedical and Industrial Applications. *Curr. Biotechnol.*, 7, 214, 2018.
112. Indiragandhi, P., Anandham, R., Madhaiyan, M., Poonguzhali, S., Kim, G., Saravanan, V., Tongmin, S., Cultivable bacteria associated with larval gut of prothiofos-resistant, prothiofos-susceptible and field-caught populations of diamondback moth, *Plutella xylostella* and their potential for, antagonism towards entomopathogenic fungi and host insect nutrition. *J. Appl. Microbiol.*, 103, 2664, 2007.
113. Souza-Neto, J.A., Gusmão, D.S., Lemos, F.J., Chitinolytic activities in the gut of *Aedes aegypti* (Diptera:Culicidae) larvae and their role in digestion of chitin-rich structures. *Comp. Biochem. Physiol. A Mol. Integr. Physiol.*, 136, 717, 2003.
114. Miller, M., Palojärvi, A., Rangger, A., Reeslev, M., Kjøller, A., The use of fluorogenic substrates to measure fungal presence and activity in soil. *Appl. Environ. Microbiol.*, 64, 613, 1998.
115. Romaní, A.M., Fischer, H., Mille-Lindblom, C., Tranvik, L.J., Interactions of bacteria and fungi on decomposing litter: differential extracellular enzyme activities. *Ecology*, 87, 2559, 2006.

16

Lithobiontic Ecology: Stone Encrusting Microbes and their Environment

Abhik Mojumdar[1†], Himadri Tanaya Behera[1†] and Lopamudra Ray[1,2*†]

[1]School of Biotechnology, Kalinga Institute of Industrial Technology, Bhubaneswar, Odisha, India
[2]School of Law, Kalinga Institute of Industrial Technology, Bhubaneswar, Odisha, India

Abstract
In an ecological niche, communities of microorganisms associated with rocks or lithic-associated microhabitats, termed as lithobionts, are divided into different niches based on their area of colonization. Epiliths are the most conspicuous microbes that colonize the surfaces of rock, whereas endoliths predominantly penetrate and colonize the rock substratum. Establishment of lithobiontic coatings results in a deteriorative effect on the substrate; however, long-term endolithic growth also preserves the rock surface morphology, since the development of a complex cellular network of firmly woven organisms may fortify the colonized substratum. Lithobiontic coatings always grow faster than rock coatings and it is only when they are unable to colonize, rock coatings get a chance to grow. Thus, this chapter mainly focuses on microbial diversity of lithobiontic coatings and their complex colonization patterns, factors controlling their distribution, and their role in harsh environments.

***Keywords*:** Endoliths, epiliths, lithobiontic coating, diversity, colonization process

16.1 Introduction

Lithobiontic ecological niche is microhabitats of microbial communities that colonize rock or lithic environments. Microbial communities

**Corresponding author*: lray@kiitbiotech.ac.in; lopamudra.ray@kls.ac.in
†*Equal contribution as first author*

Bibhuti Bhusan Mishra, Suraja Kumar Nayak, Swati Mohapatra, and Deviprasad Samantaray (eds.) Environmental and Agricultural Microbiology: Applications for Sustainability, (341–360) © 2021 Scrivener Publishing LLC

establish on the surface by organizing themselves as thin films known as biofilms. Natural stones and rock surfaces including building materials are subjected to lithobiont colonization and, as a result, lead to biogenic weathering [1]. Some lithobionts, known as endoliths, penetrate the stone substratum and colonize the interior of the stones or rock cavities. Porous rocks consist of mineral grains that promote the growth of Chasmoendoliths and/or Cryptoendoliths. In desert ecosystems, cyanobacteria are the significant drivers of nitrogen cycling and photosynthetic carbon fixation that dominate the lithic-associated communities; however, in certain endolithic communities, lichens dominate over cyanobacteria. Endoliths and hypoliths are more extensively studied as compared to epiliths.

Lithobionts are widely spread in hot as well as cold deserts ecology and can also be termed as soil rock surface communities [2]. Harsh terrestrial landscapes, mostly comprising of rocks and mineral soils, are infuriated with conditions like water deficiency and extreme heat, moisture, and UV stress. Biotic components of the microbial ecosystems and higher forms of life are almost absent in these areas. However, lithobionts have developed strategies to cope up with this extreme environment and maintain a stable population and this threshold is called a dry limit of life. Lithobiontic ecology studies in such harsh environments including Antarctic regions have been conducted [3]. In this chapter, the microbial diversity of these lithobiontic organisms and the factors controlling their distribution and complex colonization patterns have been outlined. Also, the role of these stone encrusting organisms in extreme environments has been discussed.

16.2 Diversity of Lithobionts and Its Ecological Niche

Based on the position of colonization, lithobiontic niches are divided into different types, i.e., epiliths, endoliths, and hypoliths, as shown in Figure 16.1. The diversity of lithobionts were initially analyzed by direct microscopic observations of the isolated cultures [4]. However, many culture-independent approaches are also used for the compositional analysis of the lithobiontic communities [5–9].

16.2.1 Epiliths

Epilithic lithobionts establish their community on the surfaces of the rocks and use the rocks as their substrate for accessing sufficient quantities of

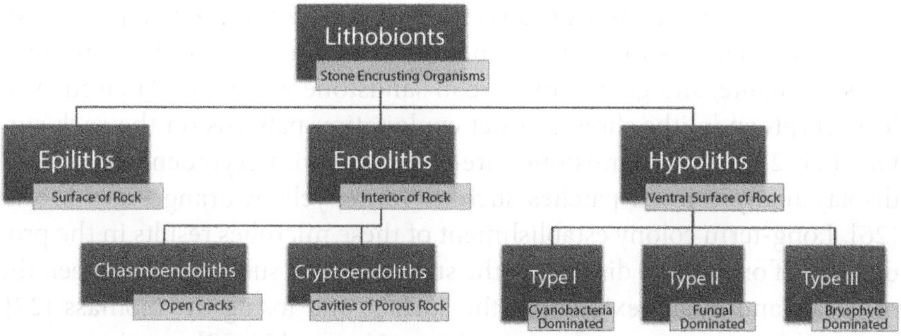

Figure 16.1 Classification of lithobiontic niches.

moisture [7, 10, 11]. Moss and lichens along with some heterotrophic fungi and cyanobacteria are the key constituents of this community [2, 12]. Among the three lithic niches, epilithic colonization is the most susceptible type, as they are constantly exposed to many environmental factors like strong winds, ultraviolet radiation, desiccation, and many other perturbations [13, 14]. Therefore, these niches are restricted only to the mild lower latitude of Antarctic regions like Mac Robertson Land and Princess Elizabeth land [15, 14].

In Princess Elizabeth Land and Mawson Rock, some species like *Chroococcidiopsis*, *Myxosarcina*, *Gloeothece*, *Plectonema*, *Lyngbya*, and *Calothrix* are identified by culture-dependent approach including 13 other cyanobacterial species in certain areas [15, 16]. Generally, the epilithic diversities are dominated by cyanobacterial groups, maybe due to their ability to resist UV irradiation [17, 18]. However, metagenomic approaches need to be carried out for the actual diversity of these epilithic communities.

16.2.2 Endoliths

These lithobiontic microbes establish their communities by colonizing in the interior of the rocks and within these, they also establish their ecological sub-niches. They are classified into two types, i.e., Chasmoendoliths and Cryptoendoliths.

Chasmoendoliths are found in the fissures and cracks of the rocks. The main habitat for this group is siliceous rocks, but in marble, granite, gypsum crusts, silicified sandstone, and anorthosite, their communities also can be established [19, 20]. Chasmoliths are not only found in the Antarctic landscape [21] but also widespread in the deserts of central Asia and central Australia, Southwestern USA, and Mexico [22, 23].

Cryptoendoliths: These groups establish their community in the porous rocks, but these also may be founded in the granite, marble, limestone, gypsum, halite, and gneiss. In Beacon sandstone in the dry Antarctic valleys, cryptoendoliths show distinct exploitation patterns on the rock surface [24, 25]. When sandstones are colonized with cryptoendoliths, they display many colored patches such as white, yellow, orange, and brown [26]. Long-term colony establishment of these microbes results in the production of oxalic acid dissolving the strengthening substances between the minerals and finally exfoliating the surface and losing the biomass [27]; hence, this process may take around 1×10^4 years [26, 28].

Endolithic communities are subjugated by cyanobacteria with the lichenized structure [29]. In Alexander Island, *Choroglea* sp., a cyanobacterial species, was discovered in translucent gypsum crusts formed on the sandstone [30]. *Chrocooccidiopsis* sp. with *Nostoc* sp. and *Cyanothece* cf. *aeruginosa* are found in the Taylor Valley [31]. In the Ross Desert, two types of cryptoendolithic communities, i.e., cyanobacterial and lichen-dominated communities, are distinguished. In McMurdo dry valleys, a total of 17 cyanobacteria were identified. Based on the dominance, three communities were surveyed which are *Gloeocapsa*, *Hormathonema-Gloeocapsa*, and *Chroococcidiopsis*.

A single green algal species *Trebouxia jamessi* [32] is a dominant species of lichenized endolithic communities and is known as a constituent of lichen associations. Actinobacteria, Planctomycete, and α- and ϒ-proteobacterial species are some of the prokaryotic members known to be associated with this community [32]. Cyanobacteria of *Chroococcidiopsis* lineage and *Gloeocapsa*, *Plectonema*, and *Hormathonema* morphologically similar to cyanobacteria are the dominant species of endolithic communities [7, 32, 33].

16.2.3 Hypoliths

These microbial communities inhabit beneath the rock surfaces at the rock-soil interface [34]. Generally, these communities colonize on the translucent rocks like quartz, gypsum, and granite and also in some cases on the opaque rocks like dolomite and gneisses [15, 20]. Photoautotrophs like cyanobacteria are the key constituents of this community and for photosynthesis, they need a minimum level of photosynthetically active radiation (PAR) [28]. In the Miers Valley, three types of hypolithic communities are found. Type I hypoliths are cyanobacterial dominated and they adhere to the rock directly to form biofilm by filamentous oscillatorian cyanobacterial monophytes [7]. Type II are fungal dominated and form a filamentous network of fungi cemented on the rock surface. However, type III are macroscopic and bryophytes dominated that entrench in the translucent soil rocks.

All these three types of hypolithic types are morphologically diverse and different in their composition. The structure of the populations can be analyzed by amplified RNA intervening sequence analysis (ARISA), terminal restriction fragment length polymorphism (T-RFLP), clone library construction, and pyrosequencing analysis of the 16S rRNA gene phylogeny [7, 9, 35]. Hypoliths are the well-studied group, of all the diversity of lithobionts [5, 7, 8, 36, 37]. Cyanobacteria are the predominant phylum comprising the hypolithic colonization globally [36, 38, 39] including Antarctica's McMurdo Dry Valleys and in other ice-free areas [7, 36, 40]. Cyanobacteria in the Antarctic and cold deserts are morphologically filamentous oscillation morphotypes [40, 41]. Other bacterial phyla such as Actinobacteria, α- and β-Proteobacteria, Planctomycetes, Firmicutes, Acidobacteria, and Verrumicrobia are also harbored in the hypolithic communities [9, 36]. Rather than the prokaryotic communities, eukaryotes like some free-living ascomycetous fungi, chlorophytes, and mosses were also present in hypoliths [5]. Many novel uncharacterized Bryophyta, fungi, and protists are reported in hypolithic communities [42]. The above study suggests that hypoliths may harbor complex food webs and plays an important role in nutrient cycling in the cold desert oligotrophic environment.

A sequential stage of succession was confirmed by the presence of different Antarctic hypolithic community morphotypes. Matrix analysis using probability dissimilarity shows that the beta diversity varies between different sites and Type I hypolithons are higher in this regard as compared to Types II and III [43]. So, it may be suggested that the cyanobacteria-dominated Type I hypolithic community is the founder community during the succession process.

16.3 Colonization Strategies of Lithobionts

Lithobiontics establish their communities on various substrates like beacon sandstone, limestone, flint, and gypsum [25, 44, 45]. Many abiotic and biotic factors such as mineral structure, ultrastructure, and pH play a significant role in contributing to the ability of microorganisms to colonize on the rock surfaces [23]. Some abiotic factors including macro- and microclimate contributing to the colonization process are further discussed in this chapter.

16.3.1 Temperature

The ambient temperature of the air and the solar radiation level are closely related to the temperature of the lithobionts [25]. Thermal buffering arises due to the black body absorption effect of the opaque rocks and the greenhouse effect of the translucent quartz rocks [4, 10, 46, 47]. The Antarctic Ross desert with temperature 15°C and the Sonoran desert of North America with 10°C temperature indicate almost similar trends of temperature difference for hot and cold deserts [4, 48]. But some reports are indicating that, in day time, the hypoliths from the hot deserts have a lower temperature at the rock interface and warmer in the night time [9, 49, 50] when compared to open soil [47].

In the McMurdo dry valleys, the annual temperatures extend from −14.8°C to 30°C having the difference of approximately 50°C between the maximum and minimum temperature [51]. This huge temperature difference imposes severe freeze-thaw stresses on the microbial communities [52]. It has been hypothesized that due to thermal buffering of the rocks, lithobionts are protected from these freeze-thaw stresses in the polar desert regions [10, 37, 53]. During polar summers, the surface of the rock temperature is almost 20°C, slightly higher than the ambient air temperatures which may lengthen the period of lithophytic microbial growth in the deserts [10, 54].

16.3.2 Water Availability

As compared to the surrounding soil, water availability is more in the rock microhabitat [37, 41, 55]. In the Antarctic Peninsula (the rainfall zones), the impact of rainfall precipitation is protected by the overlying rock. Near the sheltered margin of the stones, a narrow void of unsealed soil is formed which serves as a site for the entry of water as well as air escape for the lithic communities [56, 57]. Water during precipitation, fog, and dew events is collected in the rocks [57]. In the coastal areas and at the locations with intermittent inundation by melting of snows, or with higher atmospheric humidity, the colonization by the epiliths is highest [15, 20]. This confirms that liquid water is an important and essential need for the epilithic colonization. When the moisture gradient drops, the Antarctic epilithic diversity is also lessened. Example: the 18 genera of lichen species diversity in Dronning Maud Land was decreased to 5 genera only [58, 59].

Hypolithons can gain their water requirement from the collected runoff rainwater and by the condensed water accumulated from fog and dew events on the rock surface [57, 60]. The rock present above helps in impeding the water evaporation. In Dry Valley soils, the water content below the rocks has been found in a range of 6%–14% (w/w), while the open soil has the moisture content of 0.5%–2% w/w [40]. The porosity of the rocks also helps in retaining the moisture content [4, 54].

Harmful UV-A and UV-B radiations are filtered effectively by the quartz rocks in the Antarctic hypoliths [61]. The scattering of the light is reduced, when the air spaces are filled with moisture, so light transmittance is improved [21].

The role of extracellular polymeric substance (EPS) on lithobiontic communities has been described [2]. They are very important in retaining the water in lithobionts. In the case of endolithic cyanobacteria, EPS was speculated to preserve the microbial populations and also facilitates the photosynthetic process. The microbial EPS creates a patchwork of polyfunctional binding sites that retains excess of heavy metals and establishes growth-promoting nutrients in the sheath, promoting the fertility of the soil [62].

16.3.3 Light Availability

The translucency of the rocks is very critical for the hideaway of the lithobiontic colonization. Quartz pebbles with a thickness of 13–80 mm can filter 0.9%–2.7% of the incident sunlight [20], and much lower light levels are sufficient for the photoautotrophic process. The presence of cyanobacterial-dominated microbial communities is the confirmation of the low-level light utilization process. For the colonization of endoliths, the translucence and porosity of the rock play a significant role [63]. There is a steep gradient of lights in the interior surfaces of the rock along with depth and this is due to the debilitation of light by rock substrates [4]. Rock type, grain particle size and mineral content determine the porosity of the rock ranging between 0% and 50% [64].

The hypolithic colonization in the polar regions beneath the opaque rocks was supported by freezing and thawing of groundwater (local periglacial activities) [65]. Rocks and stones are arranged in a polygonal spatial, so the penetration of light to the bottom is facilitated through the openings around of the margins of quartz created obstruction and depletion of light by rocks minimizes the UV stress in the refuge lithic niches leading to increased microbial colonization [20].

The abundance of hypolithic colonization is dependent on the penetration of light through different rocks. So, the abundance of the colonization varies in the case of opaque and translucent rocks. Translucent rocks having the thickness from 25- to 40-mm colonization is high, while opaque rocks with the thickness between 15 and 30 mm, lower abundances of the colonization are observed [66]. However, the colonization under the thicker rocks is limited to the outer edge for the sufficient amount of light to support the photosynthetic process [20, 47].

16.4 Geography of Lithobbiontic Coatings

16.4.1 Bacteria

Bacteria are single-cell prokaryotes and two evolutionary groups, eubacteria and archaebacteria form biofilms, which may remain as isolated, cells, in aggregates, or association with other organisms [67, 68]. The decay of minerals by bacteria has been studied earlier. In the deep basal level of the water of the Columbia Plateau, bacteria establish their communities and obtain their energy from the weathering reactions [69]. In the saline environments, organic acids are secreted by the bacteria and the rocks are thought to be broken by only the salt weathering process [70]. Over the earth surface, mineral matters deposited by the bacteria for their role in the formation of largely inorganic coatings, and this is supported by six pieces of evidence, *viz.*, (i) bacteria from the rock surfaces were obtained from desert rock surfaces in controlled laboratory settings and mostly manganese oxidizers were detected [68, 71]; (ii) metabolic activity of the bacteria can be studied *in situ* and some reports provide the evidence of the importance of the bacteria in the manganese precipitation on the rock surfaces [72]; (iii) evolutionary biology and functioning of the manganese-concentrating epilithic lithobionts were studied by molecular analysis of the enzymes and DNA [68, 71–73]; (iv) after 1 or 2 weeks of thorough soaking of precipitation, budding of manganese concentrating bacteria with filamentous hyphae on the desert rock surfaces can be observed through electron microscopy; (v) although fossil bacterial structures in the rock cross-sections are rare to be found, however, by etching with hydrofluoric acid, bacterial cells can be brought out covered with iron and manganese and observed through high-resolution transmission electron microscopy; (vi) a larger structure appearing as dots on the surface of the rocks can be observed when bacterially mediated manganese grows together. The cross-sections of these dots appear like miniature stromatolites which are assumed as an indicator of a biological origin [74]. In brief, it can be concluded that bacteria play an important role in the weathering process and also help at the beginning of any new minerals in the terrestrial environment.

16.4.2 Cyanobacteria

Cyanobacteria are a group of microorganisms and important weathering agents, which are quite common on the surface of the rocks and soils. Clay minerals like vermiculite are formed by the epilithic microorganism's organic acid weathering process by removing aluminium and iron from granodiorite. On limestone, through the secretion of acids, bores a tunnel in the rock and this case cyanobacteria often have a coccoid form behaving as euendoliths [75]. New mineral deposits can be formed by cyanobacteria; hence, it is involved in the oxidation and precipitation of manganese [72]. In Aldabra Atoll, cyanobacteria were reported to deposit calcium carbonate [76, 77].

16.4.3 Fungi

Like algae and other higher plants and animals, fungi belong to the eukaryotic group with a membrane around the nucleus. Though fungi are not dominant in the cycling of nutrients in the ecosystem [78], it can form epilithic biofilms and, sometimes, chalcolithic biofilms [79].

Fungi secrets citric, gluconic, and citric acids by coating and invading the rocks and weather minerals [79]. Fungi also secrets chelating agents aiding in the removal of iron [80]. Fungi displaying hyphae-like filamentous growth [81] participate in the precipitation of manganese [68, 72]. In ancient marble structure, calcium is replaced by divalent manganese by some fungi species [82].

Most of the reports are there on the biofilm coatings of the fungi found on the stone monuments [82, 83]. In Mediterranean climates, dark red, dark brown, and black colorations are found on many rock surfaces by fungal biofilms [79]. Biofilm formation by fungi on a fresh block of marble is rapid, by black fungi approximately within 6 weeks; a dense coating of black biofilm is produced [82].

Rather than growing in a filamentous form, it can grow in a cluster form like "black globular units" or "microcolonial fungi" [84]. These microcolonies may serve as host for the development of botryoidal shapes.

16.4.4 Algae

Phaeophyta (brown algae) dominate in the shores of the rocky coast, whereas Chlorophyta (Green algae) dominate in freshwater settings, capable of interacting with anthropogenic inputs. Freshwater algal lithobionts absorb and remove the heavy metals in aquatic settings [85]. Rather than this, in the terrestrial ecosystem, non-lichenized and free-living algae also have an important role [86]. For example, on the Iceland of Thasos, Greece [87], limestone surfaces are encrusted by the algal mats. Several niches like epilithic, hypolithic, chasmolithic, and endolithic positions can be engrossed by lithobiontic algae [88]. Flaking in the rocks is observed due to the increase in the contraction of algae in the chasmolithic positions. The connection between the outer shell of the weathering coat and the host rock is loosened by the algae. The resulting flaking can intensely change the appearance of the rock by the shear stresses generated due to water and wind in cold-wet [89], cold-dry, and warm-wet climates. Hypolithic algae are most common on the semi-translucent rocks like quartz, limestone, and gypsum by taking the advantage of low intensity of light, increase in the availability of water, and reduced heat stress [88].

In the fracture of the rocks, along the separating lines of the granite boulders, chasmolithic algae can grow [34]. These chasmolithic algae live in such a position so that they can get enough penetration of light for photosynthesis along with minimization in temperature and moisture stress [88].

Within only upper few millimeters of the rock, endolithic algae live and occupy the pore spaces within the weathering coat. The most common example of this is sandstone. Wetter environments are mostly preferred in the warm deserts [90], but harsh environments like Antarctica, Nunataks in Alaska, are preferred by cryptoendolithic algae [89].

16.4.5 Lichens

Lichens are the pervasive lithobiontic coatings, and it is present even in the harshest temperature and moisture conditions on the planet [88]. Over 8% of the Earth's surfaces are dominantly covered by this type of coatings. Crustose form (limiting the exposure to the atmosphere) of the lichens are mostly found in the dry regions and firmly attached on the rock surface [88]. Foliose and fruticose lichens are attached to the rock surface by bundles of fungal hyphae [91].

A wide range of rocks like basalt, limestone [76, 88], quartz [92], gabbro [93], and syenite are covered by endolithic lichens. In extremely hot and cold deserts, endolithic lichens are common as reduction of the moisture level increases the humidity level within pore spaces of the rock [29]. Endolithic lichens induce erosion on the surface of the rock and create a characteristic colored mosaic pattern, which is a significant sign of cryptoendolithic lichen colonization [94].

Various reports are there indicating the importance of lichens as weathering agent on the natural as well as on anthropogenic rock surfaces

[93, 95–98]. Underlying rock surfaces are protected from the erosion by the lichens, stabilizing the rock surfaces [67]. The weathering process or the preservation of the host rocks by the lichens are controlled by climate up to some extent [76]. For example, in wetter climates, the erosion effect of the lichens appears to be reduced. Generally, the effect of lichens is difficult to predict as it has different roles in a different climate [99]. Sometimes, lichens help in holding the loose material in place and removal of these increases the rate of erosion. Lichens have an important role in the formation of certain textures of calcretes and oxalate precipitation [91].

16.5 Impacts of Lithobiontic Coatings

16.5.1 On Organic Remains

Lithobionts are not responsible for the organic matter deposition on the rock surface. Rocks interact with the organics and store the organic matters in pore spaces [88, 100]. The deposited organic matters gradually altered in structure. In Portugal, from a rock crevice with few centimeter openings, the sample was collected, from which dense particles (vitrinite-like) and the less dense particle has a radiocarbon age of 23,550 ± 190 (Beta 82457) and 29,990 ± 240 (Beta 86633) [101]. More fibrous material (carbonized woody remains) are found with an age of 17,460 ± 70 (Beta 86632). Concisely lithobionts can leave fossil remains which may be misinterpreted as the overlying rock coatings.

16.5.2 On Rock Weathering

The weathering effect of the lithobionts has two main effects on rock coatings.

(i) Uncoated minerals are exposed by lithobiontic weathering. A flanking pattern may be created on the rock surfaces by chasmolithic algae, mosaic patches may be left by lichens [89].

(ii) Chemolithic lithobionts are exposed due to the weathering effect along with the organisms growing within the rocks leading to a slow progression of lithobiontic organisms on the rock surface. Epilithic organisms replace the chasmolithic lithobionts as they can tolerate the surface conditions.

16.5.3 On Rock Coatings

Rather than the weathering process, lithobionts may also help in the precipitation of new mineral in the inorganic forms as a rock coating [100]. Sinks for manganese [102, 103], iron [104], phosphorous [105], and calcium [106] in different forms are created by lithobionts. Hence, it can be interpreted that the mobilization and fixation of the rock coatings are largely affected by lithobionts. Archaean cyanobacteria produce calcium carbonate stromatolites. Generally, these forms are associated with arsenic, manganese, phosphate, carbonate, and oxalate and this type of lamination present on the rocks are often considered as a proof of biological activity. Through the secretion of acids, lithobionts dominate slowly over the previously occupied rock coatings. Gradually with sun exposure, inorganic coatings are eroded physically or biochemically by epilithic lichens.

In inorganic rock coatings, lithobionts grows much faster. So, the character of the rock coatings at a given place and their biogeographic conditions of the rock encouraged the growth of the lithobionts. The lithobiontic growths are inhibited when other rock coatings are presently preventing the colonization of lithobionts.

16.6 Role of Lithobionts in Harsh Environments

In the depauperate environments, lithobionts are widely dispersed and are primary producers even in the ice-free terrestrial Antarctic environments [15, 20, 41]. About 66 $cm^2 m^{-2}$ regions were covered by Antarctic hypoliths with a contribution of up to 0.85 mg chlorophyll m^{-2} of total biomass [20]. Approximately, 0.8 ± 0.3 gm^{-2} productivity was estimated for arctic hypolithons, while approximately, 1.0 ± 0.4 gm^{-2} productivity was estimated in the same area from plants, lichens, and bryophytes [107]. So, from the above results, it can be concluded that hypolithic communities contribute a major fraction of the total photosynthetic activity in the Antarctic terrestrial ecosystems.

Acetylene reduction assay was used for determination of nitrogenase activity and it was estimated that approximately 14,200 nmol N $year^{-1}$ are contributed by hypolithic communities in the Miers Valley region. Thus, in the cold desert ecosystems, the lithic communities directly contribute to the total carbon and nitrogen turnovers [41, 61].

Using a metagenomic approach in McKelvey Valley study of functional traits driving the community assembly and microbial survival was done. From this study, a significant level of the genetic plasticity was observed in autotrophs, diazotrophs, and heterotrophs [108], and there are important differences between the endolithic, hypolithic, and open soil communities. The above study confirms that, in the Antarctic desert ecosystem, lithobionts play a critical role.

16.7 Conclusion

Lithobiontic organisms developed strategies which allowed them to survive in extreme environmental conditions. This ecosystem provides some important pieces of information about the history of the earth and as an ideal system for studies evaluating the interspecies relationships, trophic functionings and community origins and evolution. Lithobionts may be considered as "ecosystem engineers" as they can be a good marker for the aridity and change in the climate. The more extensive studies by the modern omics-based approach should be carried out for the adaptive potentials, functional capacities, and the key metabolic functions of the microbial communities involved in response to micro and macroenvironmental changes.

Acknowledgement

The authors are thankful to Kalinga Institute of Industrial Technology (KIIT), Bhubaneswar, Odisha, India. The authors are also thankful to Dr. Swati for giving us this opportunity to contribute to their book.

References

1. Wilson, M., Interactions between lichens and rocks; a review. *Cryptogamic Bot.*, 5, 299, 1995.
2. Pointing, S.B. and Belnap, J., Microbial colonization and controls in dryland systems. *Nat. Rev. Microbiol.*, 10, 551, 2012.
3. Cary, S.C., McDonald, I.R., Barrett, J.E., Cowan, D.A., On the rocks: the microbiology of Antarctic Dry Valley soils. *Nat. Rev. Microbiol.*, 8, 129, 2010.
4. Friedmann, E.I., Endolithic Microbial Life in Hot and Cold Deserts, in: *Limits of Life*, vol. 4, C. Ponnamperuma and L. Margulis (Eds.), pp. 33–45, Springer, Dordrecht, The Netherlands, 1980.
5. Khan, N., Tuffin, M., Stafford, W., Cary, C., Lacap, D.C., Pointing, S.B., Cowan, D., Hypolithic microbial communities of quartz rocks from Miers Valley, McMurdo Dry Valleys, Antarctica. *Polar Biol.*, 34, 1657, 2011.

6. Smith, J.J., Tow, L.A., Stafford, W., Cary, C., Cowan, D.A., Bacterial diversity in three different Antarctic cold desert mineral soils. *Microb. Ecol.*, 51, 413, 2006.
7. Pointing, S.B., Chan, Y., Lacap, D.C., Lau, M.C., Jurgens, J.A., Farrell, R.L., Highly specialized microbial diversity in hyper-arid polar desert. *PNAS*, 106, 19964, 2009.
8. Chan, Y., Van Nostrand, J.D., Zhou, J., Pointing, S.B., Farrell, R.L., Functional ecology of an Antarctic dry valley. *PNAS*, 110, 8990, 2013.
9. Makhalanyane, T.P., Valverde, A., Birkeland, N.-K., Cary, S.C., Tuffin, I.M., Cowan, D.A., Evidence for successional development in Antarctic hypolithic bacterial communities. *ISME J.*, 7, 2080, 2013.
10. Wynn-Williams, D., Cyanobacteria in deserts—life at the limit?, in: *The ecology of cyanobacteria*, B.A. Whitton and M. Potts (Eds.), pp. 341–366, Springer, Dordrecht, the Netherlands, 2000.
11. Smith, A., Epiphytes and epiliths, in: *Bryophyte ecology*, A.J.E. Smith (Ed.), pp. 191–227, Springer, Dordrecht, the Netherlands, 1982.
12. Staley, J.T., Palmer, F., Adams, J.B., Microcolonial fungi: common inhabitants on desert rocks? *Science*, 215, 1093, 1982.
13. Cockell, C.S., McKay, C.P., Warren-Rhodes, K., Horneck, G., Ultraviolet radiation-induced limitation to epilithic microbial growth in arid deserts–Dosimetric experiments in the hyperarid core of the Atacama Desert. *J. Photoch. Photobio. B.*, 90, 79, 2008.
14. Wynn-Williams, D. and Edwards, H., Antarctic ecosystems as models for extraterrestrial surface habitats. *Planet. Space Sci.*, 48, 1065, 2000.
15. Broady, P.A., Ecological and taxonomic observations on subaerial epilithic algae from Princess Elizabeth Land and Mac. Robertson Land, Antarctica. *Brit. Phycol. J.*, 16, 257, 1981.
16. Broady, P.A., Survey of algae and other terrestrial biota at Edward VII Peninsula, Marie Byrd Land. *Antarct. Sci.*, 1, 215, 1989.
17. Campbell, D., Eriksson, M.-J., Öquist, G., Gustafsson, P., Clarke, A.K., The cyanobacterium Synechococcus resists UV-B by exchanging photosystem II reaction-center D1 proteins. *PNAS*, 95, 364, 1998.
18. Castenholz, R.W. and Garcia-Pichel, F., Cyanobacterial responses to UV radiation, in: *Ecology of Cyanobacteria II*, B. Whitton (Ed.), pp. 481–499, Springer, Dordrecht, the Netherlands, 2012.
19. Friedmann, E.I., Endolithic microorganisms in the Antarctic cold desert. *Science*, 215, 1045, 1982.
20. Broady, P.A., The ecology of sublithic terrestrial algae at the Vestfold Hills, Antarctica. *Brit. Phycol. J.*, 16, 231, 1981.
21. Vincent, W.F., Hobbie, J.E., Laybourn-Parry, J., Introduction to the limnology of high-latitude lake and river ecosystems, in: *Polar lakes and rivers—limnology of Arctic and Antarctic aquatic ecosystems*, W.F. Vincent and J. Laybourn-Parry (Eds.), pp. 1–23, Oxford University Press, UK, 2008.

22. Friedmann, I., Lipkin, Y., Ocampo-Paus, R., Desert algae of the Negev (Israel). *Phycologia*, 6, 185, 1967.
23. Friedmann, E.I. and Ocampo-Friedmann, R., Endolithic microorganisms in extreme dry environments: analysis of lithobiontic microbial habitat, in: *Current Perspectivies in Microbial Ecology*, M.J. Klug and L.A. Reddy (Eds.), pp. 177–185, American Society for Microbiology, Washington, DC, 1984.
24. Friedmann, E.I. and Weed, R., Microbial trace-fossil formation, biogenous, and abiotic weathering in the Antarctic cold desert. *Science*, 236, 703, 1987.
25. Nienow, J.A. and Friedmann, E.I., Terrestrial lithophytic (rock) communities, in: *Antarctic microbiology*, E.I. Friedmann (Ed.), pp. 343–412, Wiley-Liss, Inc, New York, 1993.
26. Sun, H., Nienow, J., McKay, C., The antarctic cryptoendolithic microbial ecosystem, in: *Life in Antarctic Deserts and other Cold Dry Environments: Astrobiological Analogs*, P. Doran, W. Lyons, D. McKnight (Eds.), pp. 110–138, Cambridge University Press, Cambridge, 2010.
27. Sun, H.J. and Friedmann, E.I., Growth on geological time scales in the Antarctic cryptoendolithic microbial community. *Geomicrobiol. J.*, 16, 193, 1999.
28. Friedmann, E.I., Kappen, L., Meyer, M., Nienow, J.A., Long-term productivity in the cryptoendolithic microbial community of the Ross Desert, Antarctica. *Microb. Ecol.*, 25, 51, 1993.
29. Friedmann, E. and Ocampo-Friedmann, R., The Antarctic cryptoendolithic ecosystem: relevance to exobiology. *Origins of Life*, 14, 771, 1984.
30. Hughes, K.A. and Lawley, B., A novel Antarctic microbial endolithic community within gypsum crusts. *Environ. Microbiol.*, 5, 555, 2003.
31. Büdel, B., Bendix, J., Bicker, F.R., Allan Green T Dewfall as a water source frequently activates the endolithic cyanobacterial communities in the granites of taylor valley, Antarctica (1). *J. Phycol.*, 44, 1415, 2008.
32. José, R., Goebel, B.M., Friedmann, E.I., Pace, N.R., Microbial diversity of cryptoendolithic communities from the McMurdo Dry Valleys, Antarctica. *Appl. Environ. Microbiol.*, 69, 3858, 2003.
33. Wong, F.K., Lau, M.C., Lacap, D.C., Aitchison, J.C., Cowan, D.A., Pointing, S.B., Endolithic microbial colonization of limestone in a high-altitude arid environment. *Microb. Ecol.*, 59, 689, 2010.
34. Golubic, S., Friedmann, E.I., Schneider, J., The lithobiontic ecological niche, with special reference to microorganisms. *J. Sediment. Res.*, 51, 475, 1981.
35. Wood, S.A., Rueckert, A., Cowan, D.A., Cary, S.C., Sources of edaphic cyanobacterial diversity in the Dry Valleys of Eastern Antarctica. *ISME J.*, 2, 308, 2008.
36. Bahl, J., Lau, M.C., Smith, G.J., Vijaykrishna, D., Cary, S.C., Lacap, D.C., Lee, C.K., Papke, R.T., Warren-Rhodes, K.A., Wong, F.K., Ancient origins determine global biogeography of hot and cold desert cyanobacteria. *Nat. Commun.*, 2, 1, 2011.

37. Cowan, D.A., Khan, N., Heath, C., Mutondo, M., Microbiology of Antarctic terrestrial soils and rocks, in: *Polar Microbiology–The Ecology, Biodiversity and Bioremediation Potential of Microorganisms in Extremely Cold Environments*, A.K. Bej, J. Aislabie, R.M. Atlas (Eds.), pp. 1–29, CRC Press, Boca Raton, USA, 2010.
38. Lacap, D.C., Warren-Rhodes, K.A., McKay, C.P., Pointing, S.B., Cyanobacteria and chloroflexi-dominated hypolithic colonization of quartz at the hyper-arid core of the Atacama Desert, Chile. *Extremophiles*, 15, 31, 2011.
39. Pointing, S.B., Warren-Rhodes, K.A., Lacap, D.C., Rhodes, K.L., McKay, C.P., Hypolithic community shifts occur as a result of liquid water availability along environmental gradients in China's hot and cold hyperarid deserts. *Environ. Microbiol.*, 9, 414, 2007.
40. Smith, M.C., Bowman, J.P., Scott, F.J., Line, M.A., Sublithic bacteria associated with Antarctic quartz stones. *Antarct. Sci.*, 12, 177, 2000.
41. Chan, Y., Lacap, D.C., Lau, M.C., Ha, K.Y., Warren-Rhodes, K.A., Cockell, C.S., Cowan, D.A., McKay, C.P., Pointing, S.B., Hypolithic microbial communities: between a rock and a hard place. *Environ. Microbiol.*, 14, 2272, 2012.
42. Gokul, J.K., Valverde, A., Tuffin, M., Cary, S.C., Cowan, D.A., Micro-eukaryotic diversity in hypolithons from Miers Valley, Antarctica. *Biology*, 2, 331, 2013.
43. Fierer, N., Nemergut, D., Knight, R., Craine, J.M., Changes through time: integrating microorganisms into the study of succession. *Res. Microbiol.*, 161, 635, 2010.
44. Nienow, J.A., Friedmann, E.I., Ocampo-Friedmann, R., Endolithic micro-organisms in arid regions, in: *Encyclopedia of Environmental Microbiology*, G. Bitton (Ed.), Wiley-Liss, Inc, New York, 2003.
45. Cockell, C.S. and Stokes, M.D., Widespread colonization by polar hypoliths. *Nature*, 431, 414, 2004.
46. McKay, C.P., Friedmann, E.I., Gómez-Silva, B., Cáceres-Villanueva, L., Andersen, D.T., Landheim, R., Temperature and moisture conditions for life in the extreme arid region of the Atacama Desert: four years of observations including the El Nino of 1997–1998. *Astrobiology*, 3, 393, 2003.
47. Schlesinger, W.H., Pippen, J.S., Wallenstein, M.D., Hofmockel, K.S., Klepeis, D.M., Mahall, B.E., Community composition and photosynthesis by photo-autotrophs under quartz pebbles, southern Mojave Desert. *Ecology*, 84, 3222, 2003.
48. McKay, C.P. and Friedmann, E.I., The cryptoendolithic microbial environment in the Antarctic cold desert: temperature variations in nature. *Polar Biol.*, 4, 19, 1985.
49. Azúa-Bustos, A., González-Silva, C., Mancilla, R.A., Salas, L., Gómez-Silva, B., McKay, C.P., Vicuña, R., Hypolithic cyanobacteria supported mainly by fog in the coastal range of the Atacama Desert. *Microb. Ecol.*, 61, 568, 2011.

50. Makhalanyane, T.P., Valverde, A., Lacap, D.C., Pointing, S.B., Tuffin, M.I., Cowan, D.A., Evidence of species recruitment and development of hot desert hypolithic communities. *Environ. Microbiol. Rep.*, 5, 219, 2013.
51. Doran, P.T., McKay, C.P., Clow, G.D., Dana, G.L., Fountain, A.G., Nylen, T., Lyons, W.B., Valley floor climate observations from the McMurdo Dry Valleys, Antarctica, 1986–2000. *J. Geophys. Res.*, 107, 4772, 2002.
52. Nienow, J.A., McKay, C.P., Friedmann, E.I., The cryptoendolithic microbial environment in the Ross Desert of Antarctica: light in the photosynthetically active region. *Microb. Ecol.*, 16, 271, 1988.
53. Cowan, D.A., Khan, N., Pointing, S.B., Cary, S.C., Diverse hypolithic refuge communities in the McMurdo Dry Valleys. *Antarct. Sci.*, 22, 714, 2010.
54. Kappen, L., Friedmann, E., Garty, J., Ecophysiology of lichens in the dry valleys of Southern Victoria Land, Antarctica I. Microclimate of the cryptoendolithic lichen habitat. *Flora*, 171, 216, 1981.
55. Tracy, C.R., Streten-Joyce, C., Dalton, R., Nussear, K.E., Gibb, K.S., Christian, K.A., Microclimate and limits to photosynthesis in a diverse community of hypolithic cyanobacteria in northern Australia. *Environ. Microbiol.*, 12, 592, 2010.
56. Thomas, D.S.G., Arid environments: their nature and extent, in: *Arid zone geomorphology: process, form and change in drylands*, D.S.G. Thomas (Ed.), pp. 1–16, Wiley, West Sussex, UK, 2011.
57. Warren-Rhodes, K.A., Rhodes, K.L., Pointing, S.B., Ewing, S.A., Lacap, D.C., Gomez-Silva, B., Amundson, R., Friedmann, E.I., McKay, C.P., Hypolithic cyanobacteria, dry limit of photosynthesis, and microbial ecology in the hyperarid Atacama Desert. *Microb. Ecol.*, 52, 389, 2006.
58. Cameron, R.E., Farthest south algae and associated bacteria. *Phycologia*, 11, 133, 1972.
59. Engelskjøn, T., Botany of two Antarctic mountain ranges: Gjelsvikfjella and Muhlig-Hofmannfjella, Dronning Maud Land. *Polar Res.*, 4, 205, 1986.
60. Bullard, J.E., Vegetation and dryland geomorphology, in: *Arid zone geomorphology; process, form and change in drylands*, D.S.G. Thomas (Ed.), pp. 109–131, Wiley, Chichester, 1997.
61. Cowan, D.A., Pointing, S.B., Stevens, M.I., Cary, S.C., Stomeo, F., Tuffin, I.M., Distribution and abiotic influences on hypolithic microbial communities in an Antarctic Dry Valley. *Polar Biol.*, 34, 307, 2011.
62. Gadd, G.M., Heavy metal accumulation by bacteria and other microorganisms. *Experientia*, 46, 834, 1990.
63. Cowan, D.A. and Tow, L.A., Endangered antarctic environments. *Annu. Rev. Microbiol.*, 58, 649, 2004.
64. Šperl, J. and Trčková, J., Permeability and porosity of rocks and their relationship based on laboratory testing. *Acta Geodyn. Geomater.*, 5, 41, 2008.
65. Cockell, C.S., McKay, C.P., Omelon, C., Polar endoliths–an anti-correlation of climatic extremes and microbial biodiversity. *Int. J. Astrobiol.*, 1, 305, 2002.

66. Berner, T. and Evenari, M., The influence of temperature and light penetration on the abundance of the hypolithic algae in the Negev Desert of Israel. *Oecologia*, 33, 255, 1978.
67. Danin, A., Patterns of biogenic weathering as indicators of palaeoclimates in Israel. *Proceedings of the Royal Society of Edinburgh, Section B: Biological Sciences*, vol. 89, p. 243, 1986.
68. Krumbein, W.E. and Jens, K., Biogenic rock varnishes of the Negev Desert (Israel) an ecological study of iron and manganese transformation by cyanobacteria and fungi. *Oecologia*, 50, 25, 1981.
69. Stevens, T.O. and McKinley, J.P., Lithoautotrophic microbial ecosystems in deep basalt aquifers. *Science*, 270, 450, 1995.
70. Viles, H., Ecological perspectives on rock surface weathering: towards a conceptual model. *Geomorphology*, 13, 21, 1995.
71. Dorn, R.I. and Oberlander, T.M., Microbial origin of desert varnish. *Science*, 213, 1245, 1981.
72. Grote, G. and Krumbein, W., Microbial precipitation of manganese by bacteria and fungi from desert rock and rock varnish. *Geomicrobiol. J.*, 10, 49, 1992.
73. Dorn, R.I., DeNiro, M., Ajie, H., Isotopic evidence for climatic influence on alluvial-fan development in Death Valley, California. *Geology*, 15, 108, 1987.
74. Grotzinger, J.P. and Knoll, A.H., Stromatolites in Precambrian carbonates: evolutionary mileposts or environmental dipsticks? *Annu. Rev. Earth Planet Sci.*, 27, 313, 1999.
75. Danin, A., Gerson, R., Garty, J., Weathering patterns on hard limestone and dolomite by endolithic lichens and cyanobacteria: supporting evidence for eolian contribution to terra rossa soil. *Soil Sci.*, 136, 213, 1983.
76. Viles, H., Blue-green algae and terrestrial limestone weathering on Aldabra Atoll: An SEM and light microscope study. *Earth Surf. Process Landf.*, 12, 319, 1987.
77. Walter, H. and Box, E., Global classification of natural terrestrial ecosystems. *Vegetatio*, 32, 75, 1976.
78. Harley, J., Fungi in ecosystems. *J. Appl. Ecol.*, 8, 627, 1971.
79. Urzi, C.E., Criseo, G., Krumbein, W.E., Wollenzien, U., Gorbushina, A., Are colour changes of rocks caused by climate, pollution, biological growth, or by interactions of the three? *Conservation of stone and other materials. Vol. One: causes of disorders and diagnosis. Vol. Two: prevention and treatments. Proceedings of the international RILEM/UNESCO congress*, Paris, pp. 279–286, 1993.
80. Silverman, M.P. and Munoz, E.F., Fungal attack on rock: solubilization and altered infrared spectra. *Science*, 169, 985, 1970.
81. Dorn, R.I. and Oberlander, T.M., Rock varnish. *Prog. Phys. Geogr.*, 6, 317, 1982.

82. Gorbushina, A., Krumbein, W., Hamman, C., Panina, L., Soukharjevski, S., Wollenzien, U., Role of black fungi in color change and biodeterioration of antique marbles. *Geomicrobiol. J.*, 11, 205, 1993.
83. Wollenzien, U., De Hoog, G., Krumbein, W., Urzi, C., On the isolation of microcolonial fungi occurring on and in marble and other calcareous rocks. *Sci. Total Environ.*, 167, 287, 1995.
84. Taylor-George, S., Palmer, F., Staley, J., Borns, D., Curtiss, B., Adams, J., Fungi and bacteria involved in desert varnish formation. *Microb. Ecol.*, 9, 227, 1983.
85. Sneddon, J. and Pappas, C.P., Binding and removal of metal ions in solution by an algae biomass. *Am. Environ. Lab.*, 3, 9, 1991.
86. Hunt, C.B. and Durrell, L.W., Plant ecology of Death Valley, California, with a section on distribution of fungi and algae. pp. 1–68, *US Govt. Print. Off.*, Washington, D.C., 1966.
87. Price, C.A. and Doehne, E., *Stone conservation: an overview of current research*, Getty publications, Los Angeles, California, 2011.
88. Friedmann, E.I. and Galun, M., Desert algae, lichens, and fungi, in: *Desert biology*, G.W. Brown (Ed.), vol. II, pp. 166–212, Academic Press, New York, 1974.
89. Hall, K. and Otte, W., A note on biological weathering on nunataks of the Juneau Icefield, Alaska. *Permafr. Periglac.*, 1, 189, 1990.
90. Bell, R.A. and Sommerfeld, M.R., Algal biomass and primary production within a temperate zone sandstone. *Am. J. Bot.*, 74, 294, 1987.
91. Gehrmann, C. and Krumbein, W.E., Interaction between epilithic and endolithic lichens and carbonate rocks, in: *The Conservation of Monuments in the Mediterranean Basin, Proceedings of the 3rd International Symposium*, V. Fassina, H. Ott, F. Zedda (Eds.), pp. 311–316, Venice, Italy, 1994.
92. Cooks, J. and Otto, E., The weathering effects of the lichen Lecidea aff. Sarcogynoides (Koerb.) on magaliesberg quartzite. *Earth Surf. Process.*, 15, 491, 1990.
93. Cooks, J. and Fourie, Y., The influence of two epilithic lichens on the weathering of quartzite and gabbro. *South African geographer/Suid-Afrikaanse geograaf*, vol. 17, p. 24, 1990.
94. Ascasoa, C., Wierzchosb, J., Castelloa, R., Study of the biogenic weathering of calcareous litharenite stones caused by lichen and endolithic microorganisms. *Int. Biodeterior. Biodegr.*, 42, 29, 1998.
95. Jackson, T.A. and Keller, W.D., A comparative study of the role of lichens and "inorganic" processes in the chemical weathering of recent Hawaiian lava flows. *Am. J. Sci.*, 269, 446, 1970.
96. Price, M., Jones, E.W., Calnan, C., MacDonald, D., Lichen aureus: a localized persistent form of pigmented purpuric dermatitis. *Br. J. Dermatol.*, 112, 307, 1985.
97. Danin, A. and Caneva, G., Deterioration of limestone walls in Jerusalem and marble monuments in Rome caused by cyanobacteria and cyanophilous lichens. *Int. Biodeterior.*, 26, 397, 1990.

98. Romão, P.M. and Rattazzi, A., Biodeterioration on megalithic monuments. Study of lichens' colonization on Tapadao and Zambujeiro dolmens (southern Portugal). *Int. Biodeterior Biodegr.*, 37, 23, 1996.
99. Piervittori, R., Salvadori, O., Seaward, M.R.D., Lichens and monuments, in: *Biodeterioration of Stone Surfaces: Lichens and Biofilms as Weathering Agents of Rocks and Cultural Heritage*, L. St. Clair and M. Seaward (Eds.), pp. 241–282, Springer, Dordrecht, the Netherlands, 2004.
100. Krumbein, W.E. and Dyer, B.D., This Planet is Alive —Weathering and Biology, A Multi-Facetted Problem, in: *The Chemistry of Weathering, Nato ASI Series (C: Mathematical and Physical Sciences)*, J.I. Drever (Ed.), pp. 143–160, Springer, Dordrecht, the Netherlands, 1985.
101. Welsh, P. and Dorn, R., Critical analysis of petroglyph radiocarbon ages from Côa, Portugal and Deer Valley, Arizona. *American Indian Rock Art*, 23, 11, 1997.
102. Ghiorse, W.C. and Ehrlich, H.L., Microbial biomineralization of iron and manganese, in: *Biomineralization. Processes of Iron and Manganese*, H.C.W. Skinner and R.W. Fitzpatrick (Eds.), pp. 75–99, Catena, Cremlingen, Germany, 1992.
103. Uren, N. and Leeper, G., Microbial oxidation of divalent manganese. *Soil Biol. Biochem.*, 10, 85, 1978.
104. Nealson, K.H., Microbial Oxidation and Reduction of Manganese and Iron, in: *Biomineralization and Biological Metal Accumulation*, P. Westbroek and E.W. (Eds.), pp. 459–479, Springer, Dordrecht, the Netherlands, 1983.
105. Konhauser, K., Fyfe, W., Schultze-Lam, S., Ferris, F., Beveridge, T., Iron phosphate precipitation by epilithic microbial biofilms in Arctic Canada. *Can. J. Earth Sci.*, 31, 1320, 1994.
106. Russell, A., Cove, D., Trewavas, A., Wang, T., Blue light but not red light induces a calcium transient in the moss *Physcomitrella patens* (Hedw.) B., S. & G. *Planta*, 206, 278, 1998.
107. Bliss, L., Svoboda, J., Bliss, D., Polar, deserts, their plant cover and plant production in the Canadian High Arctic. *Ecography.*, 7, 305, 1984.
108. Stomeo, F., Valverde, A., Pointing, S.B., McKay, C.P., Warren-Rhodes, K.A., Tuffin, M.I., Seely, M., Cowan, D.A., Hypolithic and soil microbial community assembly along an aridity gradient in the Namib Desert. *Extremophiles*, 17, 329, 2013.

17

Microbial Intervention in Sustainable Production of Biofuels and Other Bioenergy Products

Himadri Tanaya Behera[1†], Abhik Mojumdar[1†], Smruti Ranjan Das[1], Chiranjib Mohapatra[1] and Lopamudra Ray[1,2*]

[1]*School of Biotechnology, KIIT Deemed to be University, Bhubaneswar, Odisha, India*
[2]*School of Law, KIIT Deemed to be University, Bhubaneswar, Odisha, India*

Abstract

Bioenergy is sustainable energy created from natural and biological sources such as plants, animals, and their byproducts. Fossil fuels as a source of energy have contributed toward global warming and climate changes. The global CO_2 emission from fossil fuels is now approximately 7 Gt of carbon per year. Biodiesel and bioethanol as an alternative energy has been known to be utilized in some parts of the world. Microbial fuel cells that are capable of growing faster in bulk with higher metabolisms and rich in its biomass are ideal sources for such energy production. The sources for these alternative energy feedstocks have varied from plants and animals to microbes. The potential uses of bioenergy greatly reduce the greenhouse gas emission, release of harmful pollutants associated with global warming and climate change. Microorganisms play a crucial role in the production of renewable biofuels as an alternative source of energy. Using microbial consortia is a feasible approach for production of clean green energy. Naturally existing microbes or genetically modified microbes can be co-cultured together. In this chapter, the recent advancements that may be applied globally for large-scale applications are discussed.

Keywords: Bioenergy, biofuels, biodiesel, biogas, bioalcohol, microbial fuel cells

**Corresponding author*: loparay@gmail.com
†*Equal contribution as first author*

Bibhuti Bhusan Mishra, Suraja Kumar Nayak, Swati Mohapatra, and Deviprasad Samantaray (eds.) Environmental and Agricultural Microbiology: Applications for Sustainability, (361–382) © 2021 Scrivener Publishing LLC

17.1 Introduction

Many years of reliance on non-renewable energy sources have acknowledged that a looming reduction of the current non-renewable assets is inescapable. Rapid increase in industrialization and increase in world population has increased the energy consumption [1]. Fossil fuel consumption as an alternate fuel source has contributed immensely toward global warming and climate changes. The global CO_2 emission from fossil fuels is now approximately 7 Gt of carbon per year and the atmospheric CO_2 concentration is 400 parts per million (ppm) [2]. CO_2 emission, increase in atmospheric CO_2 concentration, rise in surface temperature are potential risk for the environment, society, and the economy and are too high to be ignored. The available amount of non-renewable energy sources is decreasing along with increasing global demand for energy. This has gained attention toward search for alternate energy sources which eventually will provide solution to the problem of greenhouse gas emission caused by conventional petroleum products.

Sustainable energy created from renewable, biological sources such as plants, animals, and their byproducts are referred as bioenergy. To meet the requirement for energy, bioenergy is one of many available different renewable sources. Origin of most bioenergies is from forests, agricultural farm, and wastes or lignocellulosic biomass (LCB). Bioenergy is also considered as indispensable contribution to the environment as the continued use of fossil fuels can be a main reason behind significant environmental issues by generating greenhouse gases and harmful pollutants eventually contributing to global warming. However, the potential uses of bioenergy can be a better option for greatly reduction in the greenhouse gas emission, release of harmful pollutants associated with global warming and climate change.

Biofuels are obtained from biomass including fuels such as solid biomass, liquid fuel, and biogases [3]. Methods for the production of biofuels by microorganisms have accelerated advancement in microbial technologies and related fields [4]. As a renewable energy, biodiesel and bioethanol are reported to be used in some parts of the world. However, interest has also been gained in other biofuel or bioenergy products, such as biogas, biohydrogen, biobutanol, syngas, and biobutanol.

From plants and animals to bacteria, the origins of these sustainable energy feedstocks that are capable of replacing traditional fossil fuels differ. Microorganisms themselves have gained significant attraction in current history as biomass for bioenergy or biofuel production. The ideal sources for such energy production are microbial fuel cells (MFC) that are capable of growing faster in abundance with augmented metabolisms and enriched in their microbial biomass. Here, in this modern scenario, significant consideration has been given to how microorganisms play a crucial role in the production of renewable biofuels as a substitute energy source of along with various forms of biofuels and bioenergy items.

17.2 Biomass

Residual biomass is potentially the largest and earliest origin of biomass to generate beneficial energy. Municipal solid waste, waste oils and animal fat, farm waste and energy crops, fertilizer, wastewater sludge, and industrial waste waters provide ample energy to satisfy a large fraction of energy demand, provided that energy forms could be competently reformed. The modification of these biomass energies to valuable forms could compensate for about 7% of the total yearly energy expenditure of the United States (approximately, 3.3 TW) (Energy Information Administration, 2005). These wastes, however, also cause considerable environmental damage, so their storage and conversion to energy will greatly improve the quality of the environment. It typically includes a mixture of proteins, carbohydrates, lipids, and nucleic acids that are the residues of living organisms. Cellulose and lignin also produce plant derived materials, and almost all of the available biomass is found in solid-form macromolecules.

In order to transform complex organic matter into bio-energy production, there are three stages. In the initial step, the complex materials needed to be broken into simple chemical forms that can be taken up by microorganisms by combining mechanical, chemical, and enzymatic attacks and this step is called pre-treatment. Many therapies are used for the bioavailability of complex biomass for potential energy conversion, such as high temperature, high or low pH, hydrolytic enzymes, ultrasound, radiation, and microwaves. In the second stage, once the organic macromolecule is bio-available, fermentation to simpler products was employed. In the third phase of microorganism stabilization, the production of an energy form which commonly generates water. The benefit of using microorganisms for electricity generation is that the extraction of fuel from water prevents high energy costs. An electron sink that generally generates water and can be preserved for human use is produced by different groups of microorganisms found in nature. Methane gas (CH_4), hydrogen gas (H_2), and the electron themselves (i.e., electricity) are three best electron and energy outputs.

17.3 Biofuel

As shown in Table 17.1, biofuels may be classified into two categories: primary and secondary biofuels. Natural biofuels typically produced from plants, trees, firewood, animal waste, and crop residue are the main biofuels. The secondary biofuels are directly generated from microorganisms and plants and can be further classified into three groups. Production of ethanol from starch rich crops such as sugarcane, wheat, barley, maize, potato, or biodiesel from sunflower, soybean, and animal fat is the first generation of biofuels. The generation of bioethanol and biodiesel from many species of plants is the second generation of biofuels. The processing of biodiesel from microalgae and microbes is the third generation of biofuels [5, 6].

Table 17.1 Classification of biofuels (adapted from ref. [5]).

Primary	Secondary		
	First Generation	Second Generation	Third Generation
Firewood, wood chips, pellets, animal waste, forest and crop residues, landfill gas.	Bioethanol or Biobutanol by fermentation of starch (from wheat, barley, corn, potato) or sugars (from sugar cane and sugar beet); Biodiesel by transesterification of oil crops (rapeseed, soybeans, sunflower, palm, coconut, used cooking oil, and animal fats)	Bioethanol and biodiesel produced from conventional technologies but based on novel starch, oil and sugar crops such as Jatropha, Cassava, or Miscanthus; Bioethanol, Biobutanol, syndiesel produced from lignocellulosic materials (e.g., straw, wood and grass)	Biodiesel from microalgae; Bioethanol from microalgae and seaweeds; Hydrogen from green microalgae and microbes

17.3.1 Biodiesel

Biodiesel is a monoalkyl ester of fatty acids from animal fats, algal, and other microbial lipids, vegetable oils, or waste grease and is produced by catalytic transesterification with alkaline catalyst and petrochemical derived alcohol catalyst such as methanol [1, 7]. The first-generation biodiesel production heavily depended on edible vegetable oils. In the United States and Brazil, soya bean oil is the major source of biodiesel production, whereas, in Europe and other tropical countries, palm oil and rapeseed is mostly used [8]. For large-scale biodiesel production, utilization of edible oils cannot fulfill the demand. As an alternative, waste cooking oil, animal fats, and oil derived from non-edible energy crops like *Pongamia*, *Argemone mexicana*, and *Jatropha* are utilized which also reduces the cost of biodiesel production [9, 10]. However, in recent years, the attention has been shifted to biomass-based production of biodiesel using microbial resources.

Guo *et al.* (2015) documented that higher lipid or oil producing algal species like *Scenedesmus obliquus* and shrub such as *Calluna vulgaris* containing over 20% of lipid and biomass fuel of more than 20 dry t ha^{-1} per annum were considered for biodiesel production [7]. Microbial-based lipid production for biodiesel has several advantages over other sources for example reduced biomass doubling time in exponential growth, flexibility in terms of season and climate, less labor-intensive, simple scale-up, and higher yield of almost 100 times than plant oils in L/ha/year [10, 11]. Microorganisms like microalgae, yeasts, filamentous fungi, and bacteria can generate significant quantity of lipids and could be considered as a promising agent for biodiesel production.

17.3.1.1 Microalgae in Biodiesel Production

Microalgae biodiesel processing is a third generation of biofuel that has advantages over the first and second phases of renewable energy [5, 12]. Microalgae biofuel processing requires various phases, such as planting, harvesting, fermentation, drying, cell disruption, extraction of lipids, and transesterification reaction [13]. As compared to petroleum diesel, biodiesel produced by microalgae is similar mostly on grounds of [14]. Microalgae utilizes light as the energy source more effectively than higher plants for the transformation of carbon dioxide into organic compounds [15]. Many species of microalgae may generate biodiesel by transesterification by lipid conversion. It can be used for methanol and alcohol energy generation due to the abundance of proteins and carbohydrates in microalgae [16]. It is possible to classify microalgae into five classes, i.e., blue-green algae (cyanobacteria), green algae, red algae, and brown algae [17]. *Auxenochlorellaprotothecoides* is a facultative heterotrophic green alga that can contain up to 55% of lipid under nitrogen limited conditions [18]. Biodiesel, hydrocarbons, and biocrude oil can be produced under 25°C by another green microalgae, *Botryococcus braunii* 765, while *Chlorella minutissimacan* produces further lipids at the same temperature when grown in the basic medium [19, 20]. For biofuel development, however, cyanobacteria are probably the most dominant.

17.3.1.2 Oleaginous Yeasts in Biodiesel Production

Oleaginous yeasts for the lipid production offer certain advantages over microalgal feedstock as they are heterotrophic organisms, they have shorter generation time but can reach much highercell densities [21]. Moreover, yeasts are less susceptible to virus infections and the chances of bacterial contamination can be reduced by lowering the pH. Only 70 out of 1,600 known species of yeast are reported to be oleaginous [22]. Species of yeast such as *Rhodotorula* sp., *Rhodosporidiumtoruloides*, *Cryptococcus curvatus*, *Lipomycesstarkeyi*, and *Yarrowialipolytica* can acquire lipids as discrete, intracellular lipid bodies, sometimes up to 80% of their dry weight [23]. Kuan *et al.* (2018) reported that *Rhodotorulaglutinis* oleaginous yeast can be harvested for producing biodiesel as feedstock material. Specific transesterification reaction, rather than lipid recovery, was carried out by *R. glutinis* and by an acid-based catalytic system the biomass yield was of 111% fatty acid methyl ester (FAME) when compared to traditional methods. Under environmental stress conditions such as nitrogen limitation, phosphorus limitation coupled with excess of ammonia in the medium, presence of metal nanoparticles, temperature, and pH variations are reported to be responsible for lipid accumulation in oleaginous yeast [24–26].

17.3.1.3 Oleaginous Fungi in Biodiesel Production

Another class of oleaginous microorganisms that are known as a source of lipids and can be a possible feedstock for biodiesel production are filamentous fungi. Most oleaginous fungi can result up to 20%–25% of total of their dry weight basis in lipids, although some species may have more than 25% of the oil content. Huang *et al.* (2016) reported that *Mortierellaisabellina* can accumulate oils upto 86% of its biomass [27]. In the area of nutraceuticals, oleaginous fungi were used for the synthesis of polyunsaturated fatty acidssuch as docosahexaenoic acid, eicosapentaenoic acid, gamma linolenic acid, and arachidonic acid from *Mucorcircinelloides, Mucorrouxii,* and *Mortierella alpine* [28, 29].

17.3.1.4 Bacteria in Biodiesel Production

Bacteria have superiority in the production of biodiesel because of its higher growth rate and easy culture methods. *Streptomyces* sp., *Mycobacterium* sp., *Nocardia* sp., and *Rhodococcus* sp. are certain bacterial species known for storage of high concentrationintracellular lipids [23]. However, bacteria do not accumulate large quantities of fatty acids as studies demonstrate that bacterial strains *Dietziamaris* sp. S1, *Micrococcus* sp. AG10, *Nocardioides* sp. S3, *Sphingomonas* sp. AG6, *Stappia* sp. AG2, *Oceanicaulisalexandrii* sp. AG4, and *O. alexandrii*sp. AG7 separated from oceanic phytoplankton living cells, *Emilianiahuxleyi* has a fatty acid content ranging from 0.3% to 4% dry weight [30].

17.3.2 Bioalcohol

Ethanol, propanol, butanol, and isobutanol can be produced by fermentation of food products such as sugarcane, potato, corn, and cassava. It is one of the most significant liquid transportation fuels that can replace conventional petroleum fuels. Among all bioalcohols, bioethanol is widely produced followed by biobutanol which is also considered as a green fuel with greater importance. Microbial transformation of LCB into second generation liquid biofuels such as bioethanol and biobutanol is still a matter of intense research; however, the recalcitrant nature of the lignocelluloses makes it very difficult to hydrolyze it using biological enzymes without prior pretreatment [31].

17.3.2.1 Bioethanol

Bioethanol is a major petrol substitute that is produced by microbial saccharification and fermentation of carbohydrate rich biomass. The United States and Brazil are the largest producers of bioethanol and contributes 85% of the world's bioethanol production [32]. It can be used as pure ethanol or can be blended with gasoline for limitation of emission of gases [33]. Due to its higher octane number, heat of vaporization, and expanded inflammability limits, it is considered as a better fuel than gasoline.

In recent years, production of second generation biofuels, such as ethanol production from LCB has gained research interest as LCB is abundant and cheap. In the hydrolysis and fermentation of cellulosic biomass for bioethanol production, microorganisms are involved. Macroalgal cultures are abundant in carbohydrate sources and are therefore the most common raw material for larger production of bioethanol with simplistic growth conditions [34]. Filamentous fungi like *Aspergillus*, *Penicillium*, and *Humicola* have been studied for cellulolytic enzyme production. It was found that *Penicillium* cellulases has higher efficiency as it is rich in β-glucosidase during saccharification of biomass, which significantly reduced the cellulose requirements, hence reducing the cost of cellulases in the process.

Yeasts have significant role in bioethanol production by fermentation of wide variety of sugars produced after the hydrolysis of pretreated biomass [32]. *Saccharomyces cerevisiae* RL-11, *Pichiastipitis* NRRL-Y7124, and *Kluyveromycesfragilis* Kf1 are some commercially used yeast strains that are good producers of bioethanol from different types of sugars [35]. However, in recent studies, thermotolerant cellulolytic enzymes has been highlighted over mesozymes due to the fact that thermostable cellulases have greater conversion rates, lower chances of contamination at high temperature and due to enhanced stability, it has better hydrolysis performance [36, 37].

17.3.2.2 Biobutanol

In recent years, biobutanol has gained significant importance due to its production by microbial fermentation of renewable sources. It is considered as a superior fuel in contrast to bioethanol. It has higher density (36 MJ/kg) than ethanol (24.5 MJ/kg) and also has greater similarity to gasoline but does not increase Reid vapor pressure (RVP) of gasoline [38, 39]. As a result, it can be combined with gasoline and can used to power the combustion engines.

In 1912, the technology for biosynthesis of butanol from starch was first developed by Chaim Weizmann, using a strain of *Clostridium acetobutylicum* [40] and the fermentation process was known as acetone, butanol, and ethanol (ABE) fermentation. *C. acetobutylicum* can change from acidogenesis (acetate/butyrate pathway) to solventogenesis metabolism (acetone and butanol production) [41]. Some saccharolytic *Clostridium* strains can produce butanol from cellulosic substrates.

The capability of ionic liquid derived algae and hexane derived algae in transforming *Chlorella vulgaris* strain UTEX2714 starch components to direct butanol was compared and reported by Gao *et al.* (2016). In addition, 4.99 and 6.63 g/L of butanol were retrieved from ionic liquid derived and hexane derived algae, respectively, without any detoxification [42].

17.3.3 Biogas

Biogas is a promising source of alternate energy that can replace fossil fuels [43]. Biogas plants produce a mixture of energy effective gases such as methane (CH_4) along with carbon dioxide (CO_2) and other small gases by microbial degradation of organic material under oxygen depleting conditions [44]. Desulfurized and purified biogas can be utilized as a natural gas [45]. Methane has a higher octane ring as compared to gasoline and produces less CO_2 compared to conventional fossil fuels [46]. The microbial conversion of complex organic matter is carried out under anaerobic conditions. The anaerobic digestion (AD) process of biodegradation comprises of five steps: hydrolysis, fermentation, acidogenesis, dehydrogenation (or acetogenesis), and methanogenesis. The first step involves the breakdown of complex organic matters by hydrolysis involving certain bacteria belonging to class *Clostridia* and *Bacilli* [47]. Acidogens are fermentative bacteria that further degrade the hydrolyzed products and the process is known as acidogenesis. In the next step, organic acids and alcohols are converted into acetate as well as CO_2 and water by the process called acetogenesis. *Clostridium aceticum* and *Acetobacteriumwoodii* are highly acetogenic and undergo diverse fermentative pathway. The last step involves two groups of organisms, i.e., acetoclastic methanogens that gains energy by splitting acetate to CH_4 and CO_2 and another group is CO_2 reducing methanogens that uptake H_2 to produce methane [48, 49].

The benefit of the biogas process is that it is relatively easy to handle and allows small industrial units to use the polysaccharide components of plant materials to generate energy, such as electricity and heat. Alternatively, after purification and enrichment, the gas can be compressed and then fed to the gas grid or used in combustion engines or vehicles as a fuel [50].

17.3.4 Biohydrogen

Biohydrogen is a recently developed commercially available biofuel. Molecular hydrogen (H_2) has highest energy content than that of other gaseous fuels and the only biofuels that does not contribute to greenhouse emission and ozone layer depletion as it is oxidized to water. It can be mixed with methane or otherwise used independently. H_2 has lower solubility which can be an output from anaerobic microbial systems [51]. H_2 has the similar combustion rate as methane and combustion of $H_2$and CH_4 mixture reduces production of NOx which is the major source of air pollution [52]. However, H_2 has more added advantage which can be utilized as fuel for traditional fuel cell, and electricity can be generated by omitting the combustion step. This electricity produced is a pollution-free energy with productivity of 50% more than the combustion steam-turbine approach. Nearly all of hydrogen gas comes from reformation of the fossil fuels. So, biomass source of H_2 could be a huge benefit for undepletable natural energy [51].

Bacterial fermentation, a truncated version of methanogenesis, is the most accepted way for the conversion of biomass to H_2. Even though fermentation of H_2 is very simple, the main drawback is only that a small fraction of the electrons in the organic matter ends up in generation of H_2 even when glucose is the primary source. The major issue lies in fermenting bacteria channeled by most of their electrons to natural products and not to H_2. When glucose is fermented, only 4 mol H_2/mol glucose is produced.

Photosynthetic microorganisms may directly transform solar energy from organic or inorganic substrates or water into hydrogen. Economically, it is not adoptable to grow photosynthetic bacteria in photobioreactors using synthetic culture media. So, it may be more realistic to generate H_2 by anaerobic fermentation of organic substrates than by photo biological conversion and also fermentative route of conversion does not require the light source as well as it works with various types of Substrates. A combination of dark and photo fermentation in a two-stage hybrid system, however, will improve the overall hydrogen production [53, 54].

(i) Stage I—dark fermentation (facultative anaerobes)

$$C_6H_{12}O_6 + 2H_2O \rightarrow 2CH_3COOH + 2CO_2 + 4H_2$$

(ii) Stage II—Photo fermentation (photosynthetic bacteria)

$$2CH_3COOH + 4H_2O \rightarrow 8H_2 + 4CO_2$$

Increased hydrogen production can be achieved by using appropriate microbial strain, process adjustment, and efficient design of the bioreactor.

17.4 Other Bioenergy Products

17.4.1 Microbial Fuel Cells

The energy demand around the world continues to increase and, therefore, is the reason behind world energy catastrophe and environmental deterioration. But the dependence on fossil fuels is not sustainable due to its finite, depleting impact on the environment. So, for the environmental and economic sustainability an alternative, renewable natural energy sources should be focused. MFC is a bioreactor which converts the chemical energy that is present in the organic and inorganic compound to electrical energy by catalytic reactions of the microorganisms. These substrates include carbohydrates, cellulose, proteins, and waste waters [55].

There is no need for tremendous pre-processing of fuel or other catalyst in MFC and the best thing is that the atmosphere is not provided with gross carbon dioxide by oxidizing organic carbon sources. These MFCs can be a major advantage over the hydrogen fuel cells. Along with the degradation of the organic materials and waste-products, the electricity can be generated by using bacteria in the MCFs. Figure 17.1 shows the diagram of a typical MFC which is made up of anodic and cathodic chambers separated by a proton exchange membrane (PEM) [56]. In the anodic chamber of MFC, microbes oxidize the substrate and generate electrons and protons in the process and as the oxidation product carbon dioxide is produced. There is no carbon emission as the carbon dioxide in the reproducible biomass comes from the photosynthetic process. Exceptionally, the anodic electrons are assimilated by directed combustion method and those electrons are shuttled through an external electric field to the cathode. When passed through a salt bridge present in the cathodic chamber, the proton enters in to it and forms water by joining with the oxygen present there. The electrons and protons are extracted by the microorganisms in the anode chamber by the dissimilative mechanism of oxidation of the organic substrates.

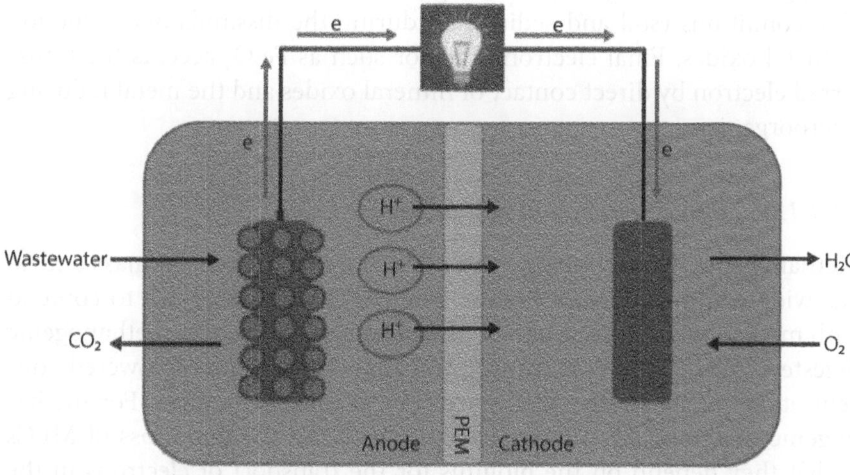

Figure 17.1 Dual MFC with a proton exchange membrane.

In anaerobic anodic chamber, generation of electric current is facilitated by holding microbes isolated from oxygen or another end marginal a receiver besides anode. The electrode reactions are shown below:

Anodic reaction: $CH_3COO^- + 2H_2O \rightarrow 2CO_2 + 7H^+ + 8e^-$
Cathodic reaction: $O_2 + 4e^- + 4H^+ \rightarrow 2H_2O$

Thus, recognizing the complete reaction is the breakdown of the substrate into water and carbon dioxide as a by-product accompanying the output of energy. Thus, electricity can be produced in the external circuit based on the electrode reaction due to electron flow from anode to cathode [57].

17.4.1.1 Microbes Used in MFCs

Abundant microorganisms are capable of transferring electrons from the metabolism of organic materials to anode. Some microorganisms used in the MFCs are listed in Table 17.2.

Rich sources of microorganisms are aquatic slit, soil, waste water, fresh water slit, and activated sludge. The key issue in accepting the theory of working mechanism of MFC is the anodic electron transfer. Through an electron transfer system, microbes transfer electrons to the electrode. The transfer mechanism either comprises of a set of constituents in the extracellular matrix of bacteria or with electron transports solubilized in the bulk solution. *Geobacter* sp. (dissimilatory metal reducing microorganisms) generates biologically advantageous as ATP. This is formed in anaerobic conditions (soil and sediments) during the dissimilatory reduction of metal oxides. Final electron acceptor such as Fe_2O_3 accepts the transferred electron by direct contact of mineral oxides and the metal reducing microorganisms.

17.4.1.2 Future Aspects of Microbial Fuel Cells

Methanogenic AD technology can make use of the same biomass and so has wide commercial applications. The MFC technology has to contend with methanogenesis technology. The benefit of MFCs over methanogenic digesters is that at temperatures lower than 20°C and with lowered concentrations of substrates, MFCs are able to convert biomass. For methanogenic digesters, all of these are uncertain. But the main loss of MFCs is that they depend on the biofilms for the transport of electrons in the absence of mediator, while the methanogenic digesters exclude this need

Table 17.2 Microbes used in MFCs (adopted from ref. [75]).

Microbes	**Substrate**	**Applications**
Actinobacillus succinogenes	Glucose	Neutral red or thionin as electron mediator [58, 59]
Aeromonas hydrophila	Acetate	Mediator-less MFC [60]
Alcaligenes faecalis, Enterococcus gallinarum, Pseudomonas aeruginosa	Glucose	Self-mediate consortia isolated from MFC with a maximal level of 4.31 W m^{-2} [61]
Clostridium butyricum	Starch, glucose, lactate, molasses	Fermentative bacterium [62]
Desulfovibrio desulfuricans	Sucrose	Sulfate/sulfide as mediator [63]
Escherichia coli	Glucose sucrose	Mediators such as methylene blue needed [63, 64]
Geobacter metallireducens	Acetate	Mediator-less MFC [65]
Gluconobacter oxydans	Glucose	Mediator (HNQ, resazurin or thionine) needed [66]
Klebsiella pneumoniae	Glucose	HNQ as mediator biomineralized manganese as electron acceptor [67]
Lactobacillus plantarum	Glucose	Ferric chelate complex as mediators [68]
Proteus mirabilis	Glucose	Thionin as mediator [69]
Pseudomonas aeruginosa	Glucose	Pyocyanin and phenazine-1-carboxamide as mediator [61]
Rhodoferax ferrireducens	Glucose, xylose sucrose, maltose	Mediator-less MFC [70]

(*Continued*)

Table 17.2 Microbes used in MFCs (adopted from ref. [75]). (*Continued*)

Microbes	Substrate	Applications
Shewanella oneidensis	Lactate	Anthraquinone-2,6-disulfonate (AQDS) as mediator [71]
Shewanella putrefaciens	Lactate, pyruvate, acetate, glucose	Mediator-less MFC [72, 73]; but incorporating an electron mediator like Mn into the anode enhanced the electricity production [74]

by competently reusing the microbial consortium without immobilization of cells. So, it is hopeful that the MFC technology will co-exist with the methanogenic digesters in the future. To obtain some "super bugs" for MFCs recombinant DNA technology and mutagenesis can possibly use in future. In a consortium one type of bacteria may contribute the mediators of electron which are used by another type of bacterium to transport their electrons to an anode more quickly. MFCs can be used for different applications like as a biosensor, wastewater treatment, biohydrogen production, and bioelectricity generation; till now, some of the fundamental information has been known in the research for MFC, and there is still more to be acquired in MFC research and its operations in large-scale conditions.

17.4.2 Microbial Nanowires in Bioenergy Application

Microorganisms share electrons with their surrounding world, and this is the key strategy for microorganisms to create bioenergy. Other than MFCs, electrically conductive filaments, called microbial nanowires (MNW), are more controversial pathways for extracellular electron exchange. In several species, MNWs have been implicated in extracellular electron transfer. Different microorganisms have been known to develop MNWs, but only two microorganisms have been analyzed in detail, i.e., *Geobactersulfurreducens* and *Shewanellaoneidensis*, and in these two species, the nanowire work of is entirely different. MNWs may be divided into three categories based on convenient details [58, 59].

17.4.2.1 Pili

Type IV pili (TFP) are the most widespread pili found in *G. sulfurreducens* and *Synechocystis* species. TFP retains unique characteristics rather than common functions like adhesion and biofilm formation. Unique characteristics include twitching like motility, DNA uptake during transformation and phage attachment, and most importantly electron carrying capacity [60, 61]. *G. sulfurreducens* MNWs are polymers of PilA subunit and *Synechocystis* MNWs are composed of PilA1 subunit [62, 63]. In *G. sulfurreducens* in association with MNWs, cytochromes are present whether *Synechocystis* MNWs are embedded with cytochromes. MNWs of different microorganisms vary in length and width because of two reasons, i.e., (1) observed width may vary as TFP have bundle forming ability; and (2) depending on age of culture and sample preparation methods, delicate pili may lead to breakage, so that length may vary [64, 65].

17.4.2.2 Outer Membranes and Extended Periplasmic Space

Three different types of extracellular proteinaceous appendages are present in *S. oneidensis*, i.e., (1) Mshpili, (2) TFP, and (3) flagella. Mshpili is involved in the transfer of extracellular electron. TFP and flagella have been shown to be expendable. However, unlike pili and flagella, MNWs in *S. oneidensis* are a mixture of different cytochromes and periplasmic and outer membrane components. The formation of MNWs in case of *S. oneidensis* is completely different and the reason behind it is completely unknown [66].

17.4.2.3 Unknown Type—MNWs Whose Identity to be Confirmed

Conductive structures like pili have been identified in *A. hydrophila, R. palustris, D. desulfuricans* whose complete identity is not known [67–69]. Two specific types of MNWs (short/thin and long/thick) were identified in *N. punctiforme. M. aeruginosa* possess MNWs similar to an unnamed protein. *P. thermopropionicum* in monoculture as well as in coculture produce electrically conductive flagellum like appendages. These appendages may be are flagella but the further study has shown that *P. thermopropinocium* in coculture with *M. thermoautotrophicus* produce flagella involved in symbiosis [70, 71].

17.4.3 Microbial Nanowires in Bioenergy Production

MNWs play an important role in bioenergy production. MNWs play a relevant role in increasing the efficiency of MFCs in which electron transfer should happen via biofilms so that micro-organisms which are distant from the anode can transfer electrons to it. In *G. sulfurreducens*, MNWs help the cells to make productive contact with electrodes by acting as a bridge between electrodes and cells and enable long range of electron transfer through biofilm. Electricity production is increased 10 times by long range of electron transfer. In improving the efficiency of photosynthetic MFCs and microbial solar cells, MNWs play an important role. MNWs have been known to be involved in methane production in anaerobic digesters [72, 73].

The capacity of microorganisms in producing MNWs improves the potential to impact their surrounding environment further enhancing their status as "tiny but still powerful organisms". Depending on the niche and physiological needs, microorganisms employ diverse functions like extracellular electron transfer to metals, tolerance to toxic metals preventing photo damage, and cell communication. These functions in microorganisms may be due to the occurrence of MNWs. But still more attempts are needed to analyze the mechanism behind the electron flow via different MNWs which would considerably help in the intonation of electro conductive properties of MNWs [74].

17.5 Conclusion

Exhausting supplies of conventional fossil fuels and the global concern over greenhouse gas emission and its effects are raising concern for a green and clean renewable source of energy. Using microbial consortia is a feasible approach for production of clean green energy. Naturally existing microbes or genetically modified microbes can be co-cultured together for improvement of certain processes and better yield of products. The recent advancements discussed in this chapter may be applied globally for large-scale application with socio-economic feasibilities.

References

1. Akinsemolu, A.A., The role of microorganisms in achieving the sustainable development goals. *J. Cleaner Prod.*, 182, 139–155, 2018.

2. Liao, J.C. *et al.*, Fuelling the future: microbial engineering for the production of sustainable biofuels. *Nat. Rev. Microbiol.*, 14, 5, 288, 2016.
3. Janda, K., Kristoufek, L., Zilberman, D., Biofuels: policies and impacts. *Agric. Econ.*, 58, 8, 372–386, 2012.
4. Al Makishah, N.H., Bioenergy: Microbial Biofuel Production Advancement. *Int. J. Pharm. Res. Allied Sci.*, 6, 3, 93–106, 2017.
5. Dragone, G. *et al.*, Third generation biofuels from microalgae, in: *Current research, technology and education topics in applied microbiology and microbial biotechnology*, A Méndez-Vilas (Eds.), pp. 1355-1366, Badajoz, Spain : Formatex Research Center, 2010.
6. Abdelaziz, A.E., Leite, G.B., Hallenbeck, P.C., *Addressing the challenges for sustainable production of algal biofuels: II. Harvesting and conversion to biofuels. Environ. Technol.*, 34, 13–14, 1807–1836, 2013.
7. Guo, M., Song, W., Buhain, J., Bioenergy and biofuels: History, status, and perspective. *Renew. Sustain. Energy Rev.*, 42, 712–725, 2015.
8. Singh, S. and Singh, D., Biodiesel production through the use of different sources and characterization of oils and their esters as the substitute of diesel: a review. *Renew. Sustain. Energy Rev.*, 14, 1, 200–216, 2010.
9. Kumar, A. and Sharma, S., Potential non-edible oil resources as biodiesel feedstock: an Indian perspective. *Renew. Sustain. Energy Rev.*, 15, 4, 1791–1800, 2011.
10. Atabani, A.E. *et al.*, A comprehensive review on biodiesel as an alternative energy resource and its characteristics. *Renew. Sustain. Energy Rev.*, 16, 4, 2070–2093, 2012.
11. Huang, C. *et al.*, Single cell oil production from low-cost substrates: the possibility and potential of its industrialization. *Biotechnol. Adv.*, 31, 2, 129–139, 2013.
12. Nigam, P.S. and Singh, A., Production of liquid biofuels from renewable resources. *Prog. Energy Combust. Sci.*, 37, 1, 52–68, 2011.
13. Alam, F. *et al.*, Biofuel from algae-Is it a viable alternative? *Procedia Eng.*, 49, 221–227, 2012.
14. Schenk, P.M. *et al.*, Second generation biofuels: high-efficiency microalgae for biodiesel production. *Bioenergy Res.*, 1, 1, 20–43, 2008.
15. Rodionova, M.V. *et al.*, Biofuel production: challenges and opportunities. *Int. J. Hydrogen. Energy*, 42, 12, 8450–8461, 2017.
16. McGinn, P.J. *et al.*, Integration of microalgae cultivation with industrial waste remediation for biofuel and bioenergy production: opportunities and limitations. *Photosynth. Res.*, 109, 1–3, 231–247, 2011.
17. Bhatt, N.C. *et al.*, Coupling of algal biofuel production with wastewater. *Sci. World J.*, 1–10, 2014.
18. Patel, A. *et al.*, Heterotrophic cultivation of Auxenochlorella protothecoides using forest biomass as a feedstock for sustainable biodiesel production. *Biotechnol. Biofuels*, 11, 1, 169, 2018.

19. Ge, Y., Liu, J., Tian, G., Growth characteristics of Botryococcus braunii 765 under high CO2 concentration in photobioreactor. *Bioresour. Technol.*, 102, 1, 130–134, 2011.
20. Li, Z. *et al.*, Optimization of the biomass production of oil algae Chlorella minutissima UTEX2341. *Bioresour. Technol.*, 102, 19, 9128–9134, 2011.
21. Li, Q., Du, W., Liu, D., Perspectives of microbial oils for biodiesel production. *Appl. Microbiol. Biotechnol.*, 80, 5, 749–756, 2008.
22. Sitepu, I.R. *et al.*, Oleaginous yeasts for biodiesel: current and future trends in biology and production. *Biotechnol. Adv.*, 32, 7, 1336–1360, 2014.
23. Meng, X. *et al.*, Biodiesel production from oleaginous microorganisms. *Renewable Energy*, 34, 1, 1–5, 2009.
24. Huang, X. *et al.*, Enhancement of lipid accumulation by oleaginous yeast through phosphorus limitation under high content of ammonia. *Bioresour. Technol.*, 262, 9–14, 2018.
25. Sibi, G. *et al.*, Metal nanoparticle triggered growth and lipid production in Chlorella vulgaris. *Int. J. Sci. Res. Environ Sci. Toxicol.*, 2, 1, 1–8, 2017.
26. Brown, C. and Rose, A., Fatty-acid composition of Candida utilis as affected by growth temperature and dissolved-oxygen tension. *J. Bacteriol.*, 99, 2, 371–378, 1969.
27. Huang, G. *et al.*, Novel fungal lipids for the production of biodiesel resources by Mucor fragilis AFT7-4. *Environ. Prog. Sustainable Energy*, 35, 6, 1784–1792, 2016.
28. Du Preez, J. *et al.*, Production of gamma-linolenic acid by Mucor circinelloides and Mucor rouxii with acetic acid as carbon substrate. *Biotechnol. Lett.*, 17, 9, 933–938, 1995.
29. Koike, Y. *et al.*, Effect of consumed carbon to nitrogen ratio of mycelial morphology and arachidonic acid production in cultures of Mortierella alpina. *J. Biosci. Bioeng.*, 91, 4, 382–389, 2001.
30. Zabeti, N. *et al.*, Fatty acid composition of bacterial strains associated with living cells of the haptophyte Emiliania huxleyi. *Org. Geochem.*, 41, 7, 627–636, 2010.
31. Adsul, M. *et al.*, Blending of cellulolytic enzyme preparations from different fungal sources for improved cellulose hydrolysis by increasing synergism. *RSC Adv.*, 4, 84, 44726–44732, 2014.
32. Azhar, S.H.M. *et al.*, Yeasts in sustainable bioethanol production: A review. *Biochem. Biophys. Rep.*, 10, 52–61, 2017.
33. Pejin, D. *et al.*, Fermentation of wheat and triticale hydrolysates: A comparative study. *Fuel*, 88, 9, 1625–1628, 2009.
34. Kang, A. and Lee, T.S., Converting sugars to biofuels: ethanol and beyond. *Bioengineering*, 2, 4, 184–203, 2015.
35. Mussatto, S.I. *et al.*, Sugars metabolism and ethanol production by different yeast strains from coffee industry wastes hydrolysates. *Appl. Energy*, 92, 763–768, 2012.

36. Akram, F. *et al.*, Insight perspectives of thermostable endoglucanases for bioethanol production: A review. *Renewable Energy*, 122, 225–238, 2018.
37. Bhalla, A. *et al.*, Improved lignocellulose conversion to biofuels with thermophilic bacteria and thermostable enzymes. *Bioresour. Technol.*, 128, 751–759, 2013.
38. Xue, C. *et al.*, Recent advances and state-of-the-art strategies in strain and process engineering for biobutanol production by Clostridium acetobutylicum. *Biotechnol. Adv.*, 35, 2, 310–322, 2017.
39. Aziz, Z. and Gozan, M., Production of gasohol from isobutanol. *IOP Conference Series: Earth and Environmental Science*, IOP Publishing, 2017.
40. Dürre, P., New insights and novel developments in clostridial acetone/butanol/isopropanol fermentation. *Appl. Microbiol. Biotechnol.*, 49, 6, 639–648, 1998.
41. Lütke-Eversloh, T. and Bahl, H., Metabolic engineering of Clostridium acetobutylicum: recent advances to improve butanol production. *Curr. Opin. Biotechnol.*, 22, 5, 634–647, 2011.
42. Gao, K., Orr, V., Rehmann, L., Butanol fermentation from microalgae-derived carbohydrates after ionic liquid extraction. *Bioresour. Technol.*, 206, 77–85, 2016.
43. Amigun, B., Sigamoney, R., von Blottnitz, H., Commercialisation of biofuel industry in Africa: a review. *Renew. Sustain. Energy Rev.*, 12, 3, 690–711, 2008.
44. Kovács, K.L. *et al.*, Biogasproducing microbes and biomolecules, in: *Biofuels: from microbes to molecules*, X. Lu (Eds.), 47, Caister Academic Press, UK, 2014.
45. Bharathiraja, B. *et al.*, Biogas production–A review on composition, fuel properties, feed stock and principles of anaerobic digestion. *Renew. Sustain. Energy Rev.*, 90, 570–582, 2018.
46. Korakianitis, T., Namasivayam, A., Crookes, R., Natural-gas fueled spark-ignition (SI) and compression-ignition (CI) engine performance and emissions. *Prog. Energy Combust. Sci.*, 37, 1, 89–112, 2011.
47. Wirth, R. *et al.*, Characterization of a biogas-producing microbial community by short-read next generation DNA sequencing. *Biotechnol. Biofuels*, 5, 1, 41, 2012.
48. Schink, B., Energetics of syntrophic cooperation in methanogenic degradation. *Microbiol. Mol. Biol. Rev.*, 61, 2, 262–280, 1997.
49. Zabranska, J. and Pokorna, D., Bioconversion of carbon dioxide to methane using hydrogen and hydrogenotrophic methanogens. *Biotechnol. Adv.*, 36, 3, 707–720, 2018.
50. Ryckebosch, E., Drouillon, M., Vervaeren, H., Techniques for transformation of biogas to biomethane. *Biomass Bioenergy*, 35, 5, 1633–1645, 2011.
51. Rittmann, B.E., Opportunities for renewable bioenergy using microorganisms. *Biotechnol. Bioeng.*, 100, 2, 203–212, 2008.

52. McCormick, R.L. *et al.*, *Fuel additive and blending approaches to reducing NOx emissions from biodiesel*, SAE Technical Paper, US, 2002.
53. Kotay, S.M. and Das, D., Biohydrogen as a renewable energy resource—prospects and potentials. *Int. J. Hydrogen. Energy*, 33, 1, 258–263, 2008.
54. Das, D. and Veziroglu, T.N., Advances in biological hydrogen production processes. *Int. J. Hydrogen. Energy*, 33, 21, 6046–6057, 2008.
55. Ditzig, J., Liu, H., Logan, B.E., Production of hydrogen from domestic wastewater using a bioelectrochemically assisted microbial reactor (BEAMR). *Int. J. Hydrogen. Energy*, 32, 13, 2296–2304, 2007.
56. Hu, Z., Electricity generation by a baffle-chamber membraneless microbial fuel cell. *J. Power Sources*, 179, 1, 27–33, 2008.
57. Parkash, A., Microbial fuel cells: a source of bioenergy. *J. Microb. Biochem. Technol.*, 8, 247–255, 2016.
58. El-Naggar, M.Y. *et al.*, Electrical transport along bacterial nanowires from Shewanella oneidensis MR-1. *Proc. Natl. Acad. Sci.*, 107, 42, 18127–18131, 2010.
59. Yuzvinsky, T. *et al.*, Electrical Transport along Bacterial Nanowires. *Biophys. J.*, 100, 3, 132a, 2011.
60. Pelicic, V., Type IV pili: e pluribus unum? *Mol. Microbiol.*, 68, 4, 827–837, 2008.
61. Proft, T. and Baker, E., Pili in Gram-negative and Gram-positive bacteria—structure, assembly and their role in disease. *Cell. Mol. Life Sci.*, 66, 4, 613, 2009.
62. Reguera, G. *et al.*, Extracellular electron transfer via microbial nanowires. *Nature*, 435, 7045, 1098, 2005.
63. Sure, S. *et al.*, Inquisition of Microcystis aeruginosa and Synechocystis nanowires: characterization and modelling. *Antonie Van Leeuwenhoek*, 108, 5, 1213–1225, 2015.
64. Feliciano, G., Steidl, R., Reguera, G., Structural and functional insights into the conductive pili of Geobacter sulfurreducens revealed in molecular dynamics simulations. *Phys. Chem. Chem. Phys.*, 17, 34, 22217–22226, 2015.
65. Li, Y. and Li, H., Type IV pili of Acidithiobacillus ferrooxidans can transfer electrons from extracellular electron donors. *J. Basic Microbiolv*, 54, 3, 226–231, 2014.
66. Bouhenni, R.A. *et al.*, The role of Shewanella oneidensis MR-1 outer surface structures in extracellular electron transfer. *Electroanal. Int. J. Devoted Fundam. Pract. Aspects Electroanal.*, 22, 7-8, 856–864, 2010.
67. Castro, L. *et al.*, Aeromonas hydrophila produces conductive nanowires. *Res. Microbiol.*, 165, 9, 794–802, 2014.
68. Eaktasang, N. *et al.*, Production of electrically-conductive nanoscale filaments by sulfate-reducing bacteria in the microbial fuel cell. *Bioresour. Technol.*, 210, 61–67, 2016.

69. Sure, S. *et al.*, Identification and topographical characterisation of microbial nanowires in Nostoc punctiforme. *Antonie Van Leeuwenhoek*, 109, 3, 475–480, 2016.
70. Shimoyama, T. *et al.*, Flagellum mediates symbiosis. *Science*, 323, 5921, 1574–1574, 2009.
71. Gorby, Y.A. *et al.*, Electrically conductive bacterial nanowires produced by Shewanella oneidensis strain MR-1 and other microorganisms. *Proc. Natl. Acad. Sci.*, 103, 30, 11358–11363, 2006.
72. Angenent, L.T. *et al.*, Production of bioenergy and biochemicals from industrial and agricultural wastewater. *Trends Biotechnol.*, 22, 9, 477–485, 2004.
73. De Mes, T. *et al.*, Methane production by anaerobic digestion of wastewater and solid wastes, in: *Bio-methane Bio-hydrogen,* J.H. Reith, R.H. Wijffels, H. Barten (Eds.), pp. 58–102, Dutch Biological Hydrogen Foundation - NOVEM, The Netherlands, 2003.
74. Sure, S. *et al.*, Microbial nanowires: an electrifying tale. *Microbiology*, 162, 12, 2017–2028, 2016.

18

Role of Microbes and Microbial Consortium in Solid Waste Management

Rachana Jain[1]*, Lopa Pattanaik[2], Susant Kumar Padhi[3] and Satya Narayan Naik[1]

[1]*Centre for Rural Development and Technology, Indian Institute of Technology, Delhi, Hauz-Khas, New Delhi, India*
[2]*Department of Chemical Engineering, Indian Institute of Technology Guwahati, Guwahati, Assam, India*
[3]*Civil Engineering Department, Shiv Nadar University, Greater Noida, Uttar Pradesh, India*

Abstract

Worldwide, solid waste disposal is a serious problem rising day by day with the increase in population. Open dumping and incineration is a common practice for municipal solid waste disposal in India and other developing countries. This practice raises the acute pollution problem as well as a health risk. There is an urgent need for sustainable techniques to address this problem to generate a minimal impact on the environment. There are many sustainable methods for solid waste management, *viz.*, sanitary landfilling, composting, vermicomposting, anaerobic digestions, and ethanol production. Among all the microbes and biological agents are playing a very important role. This chapter presents a detailed review of solid waste management and sustainable solution to this problem with the help of microbes and biological agent.

***Keywords*:** Bioethanol, composting, microbial consortium, pretreatment, waste management, anaerobic digestion

**Corresponding author*: rachana_nbs@yahoo.co.in

Bibhuti Bhusan Mishra, Suraja Kumar Nayak, Swati Mohapatra, and Deviprasad Samantaray (eds.) Environmental and Agricultural Microbiology: Applications for Sustainability, (383–422) © 2021 Scrivener Publishing LLC

18.1 Introduction

Worldwide solid waste is a serious problem and getting bigger as a result of increase in population, urbanization, industrialization, and unsystematic waste disposal. It is a root of pollution and climate change. Gradually, waste management taking place of topmost global agenda and need a holistic solution. Waste management is inferior in developing countries than the developed ones due to poor economic resources, political, technical, and operational limitations. A practical, economic, and sustainable operative system of waste management is the demand of the time.

Solid waste normally cover household, agricultural, industrial, and commercial refuse that is comprised of degradable, partially degradable, and non-degradable materials [1]. According to the report of World Bank 2018, global municipal solid waste (MSW) generation is around 2.01 billion metric tonnes if the same condition remain, it is estimated to increase 70% by 2050 [2]. The composition of waste varies from cities to cities. According to the report of Waste to Energy, Planning Commission, 2014, MSW of Indian cities contain 51% biodegradable, 10% plastic, 7% paper, and 32% other kind of non-degradable content [3]. However, by following five "R" strategies, *viz.*, reduce, reuse, recycle, recover, and restore of waste can control the waste in sustainable way [4]. Microbes play a very important role in recycling of organic proportion. Several naturally occurring microorganisms are able to change organic waste into valuable resources such as compost, biofuel, and other valuable products. There is a need to identify these microbes with special capacity and need to apply them in a needed place in a properly designed way. Microbes can be big tool to solve the waste management problem in a sustainable way. In this chapter, the mitigation strategies for the solid waste such as recycling and valorization by microbes are discussed in detailed.

18.2 Types of Solid Waste

It had been observed over the period with the increase of the anthropogenic activities leading to industrialization and commercialization, which increases a burden in urban wastes. These wastes can be classified as biodegradable and non-biodegradable wastes. Some of the common types of solid wastes are as follows:

18.2.1 Domestic Wastes

Housekeeping activities like food preparation, sweeping, gardening, wastes from repair and maintenance work, old items like clothes, and furnishing generate a massive amount of domestic solid wastes.

18.2.2 Institutional and Commercial Wastes

Solid wastes originate from the stores, shops, schools, offices comprising of papers, typewriter ribbon, punch-cards, tapes, etc. In addition to this, the wastes generated from hospitals, hotels, and restaurants can be listed in this category.

18.2.3 Wastes From Street Cleansing

Papers, vegetable waste, small containers, stones, dust, and debris are the major contributors of wastes generated during street cleansing.

18.2.4 Industrial Wastes

Construction and demolition wastes, including the solid waste materials from the building sites and the wastes from the factory like packaging materials, food wastes, ruined materials of wood, plastic, metal, and cardboard are the primary forms of waste generated from industries. Further, it can be sub-divided into solid, liquid, and sludge based on the physical property. It can also be categorized based on various other properties like inert, non-flammable, combustible, or biodegradable.

18.2.5 Nuclear Wastes

Wastes obtained from the nuclear power plants like radioactive materials, contain unstable isotopes, which decay over the period and emits harmful radiations.

18.2.6 Agricultural Wastes

Agricultural wastes include rice-husk, sugarcane bagasse, sawdust, wheat shell, wood chips, corn starch, and so on derived from plants and animal manures. The details of various sources and types of solid waste are summarized in Table 18.1.

Table 18.1 Sources and types of solid waste.

Sources	Origin of waste	Solid wastes
Residential	Single and multifamily residency	Plastics, paper, cardboard, metals, glass, food wastes, e-waste, and other household hazardous wastes
Commercial	Office buildings, markets place, restaurants, storehouses, hotels	Wood, plastic, paper, cardboard, metals, and hazardous waste
Institutional	Hospitals, prisons, schools, government centre	Wood, plastic, paper, cardboard, metals, hazardous waste, and biomedical waste
Municipal utility	Recreational areas, landscaping, parks, beaches	Used polyethylene, tissues, napkins, organic wastes, wood trash, paper, cards, and plastic
Industrial	Light and heavy manufacturing, fabrication, refineries, power and chemical plants, mineral extraction and processing	Housekeeping wastes, packaging, food wastes, hazardous wastes, ashes, special wastes, scrap material, slay, tailings
Construction and demolition	construction sites, road repair, renovation sites, demolition of buildings	Wood, metal, concrete, dirt, etc.
Agriculture	Crops, orchards, vineyards, dairies, feedlots, farms	Spoiled food wastes, agricultural wastes, hazardous wastes (e.g., pesticides)

18.3 Waste Management in India

In India, a rapid rise in the population and unplanned urbanization is the major contributor to increased solid waste. The present Indian population is 1,366 million compared with 1,028 million in 2001. Presently, solid waste management (SWM) in India is in the nascent stage; still, old ways of waste collection and disposal are underuse. According to the Press Information

Bureau, India produces 62 million tonnes (MT) of waste annually. Only 69% is collected; out of this, 19% is treated, and 50% is directly dumped into the landfill [5]. The main obstacle in waste management in urban India is nearly no segregation of solid waste, all kind of waste, *viz.*, construction and demolition debris (C&D), plastic, commercial and industrial refuses, and e-waste dumped altogether [6]. MSW generation in small towns ranges around 0.17 kg per person per day, while in big cities, it reaches approximately 0.62 kg per person per day [7]. According to the NDTV report [8], most of the Indian are unaware of waste management and leave it to authority. Therefore, the very primary challenge is to create awareness on waste management.

In India, SWM regulation and administration were majorly governed by several government bodies like the Ministry of Environment and Forests and Climate Change (MoEF), Central Pollution Control Board (CPCB), Ministry of Urban Development (MoUD), National Environmental Engineering Research Institute (NEERI), and State Pollution Control Board (SPCB). Whereas, Urban Local Bodies (ULBs) are responsible for implementation at the ground level. For the first time, under the fourth five-year plan (1969–1974), the Government of India (GoI) initiated better MSW management facilities by providing loans and grants to the state government to set up MSW composting facilities. Further, under the National Scheme of Solid Waste Disposal (1975–1980), mechanical composting facilities (Capacity 150–300 tonnes/day) were set up in several cities. At present, these facilities are non-operational, and the primary reason of failure is poor management, inappropriate technology, and inadequate planning. However, recently, some decentralized and centralized facilities have been revived by several non-governmental organizations, community groups, voluntary groups, and individuals [9].

In 1990, the Ministry of Environment and Forests (MoEF) constituted the National Waste Management Council (NWMC), which emphasized municipal solid waste management (MSWM). The objective of NWMC was to estimate the quantity of recyclable waste and its proper management. Further, in 1993, the national plastic waste management task force was constituted by NWMC to highlight the adverse environmental and health impacts caused due to plastic recycling. Later, in 1998, Recycled Plastic Usage Rules was constituted by MoEF, which bans recycled plastic bags in terms of storing, carrying, and packing of food items.

The outbreak of an epidemic in Surat in 1994 due to improper urban SWM leads to constitution of a high-powered committee Urban Solid Waste Management (USWM) under the Chairmanship of Dr. Bajaj (Bajaj committee) in the year 1995. This committee suggested the need for source segregation, community-based door-to-door collection and transportation, and charging user-fees. This committee also standardized the design of municipal vehicles used for transportation of waste and suggested appropriate technologies for treatment and disposal of MSW through composting process [9]. In 2000, the MSW (Management and handling) rules were formulated by GoI. This rule stated that, the implementation, provisions, and any infrastructure development for collection, storage, segregation, transportation, processing, and disposal of MSW were laid on the shoulder of local municipal authorities. This rule implicated the management of biodegradable waste through pelletization, vermicomposting, composting, and anaerobic digestion (AD), etc., and inert, non-biodegradable waste and other appropriately stabilized biological waste through landfilling [9]. The MSW Rules 2000 underwent revisions by MoEF in year 2013 and 2015 and the new revised rule has been named as Solid Waste Management Rules, 2016. Besides these rules, the GoI and the state governments have drafted several other acts and rules to deal with other categories of solid waste except MSW, such as (i) Hazardous Wastes (Management and Handling) Rules, 1989 and Amendment Rules, 2000 and 2003, (ii) The Bio-Medical Waste (Management and Handling) Rules, 1998 and Amendment Rules, 2003, (iii) The Batteries (Management and Handling) Rules, 2000, and Plastic Waste (Management and Handling) Rules, 2009, (iv) E-Waste (Management and Handling) Rules 2011, and (v) Construction and Demolition Waste Management Rules, 2016.

Indian Government in year 2017 made waste segregation mandatory in the country by two-bin nationwide campaign but the ground realty did not change much. The six largest metropolitan cities (Delhi, Mumbai, Kolkata, Chennai, Bengaluru, and Hyderabad) generate the maximum volume of solid waste, ranging from 4,000 TPD (tonnes per day) in Hyderabad to 9,620 TPD in Delhi [3]. MSW production and collection and segregation in different Indian cities are illustrated in Table 18.2.

Like MSW, the management of agricultural waste is one of the biggest challenges in India. A big portion of crop residue (8%–80%) underwent burning, which is otherwise known as "on-farm" crop burning [10]. Among other states of India, the major on-farm crop burning issues have been reported by Uttar Pradesh followed by Haryana and Punjab [10] during the post-harvest period, which are April–May and November–December [11]. The major reason behind on-farm crop residue burning, which is reported by National Policy for Management of Crop Residues (NPMCR) are: (i) high cost involved in the removal of the crop residue from the field either manually and in mechanized mode and (ii) the quench for higher economic returns in less time, forced farmers to cultivate more than two crops in a year and consecutively, without leaving a time gap between successive cultivations [12]. The frequent on-farm crop residues burning practices not only causes the emission of greenhouse gases (CH_4, CO, N_2O, and NOx), but also

Table 18.2 Municipal waste generation in main cities of India, collection and segregation at source.

City	Population (million)	MSW (TPD)	Door-to-door collection (%)	Segregation (%)
Delhi	19.1	9,620	39	-
Mumbai	20.0	8,600	80	-
Kolkata	14.7	6,000	-	-
Chennai	10.1	5,000	80	-
Bangalore	10.4	4,200	71	50
Hyderabad	9.1	4,000	73	-
Ahmedabad	7.5	2,500	95	-
Pune	5.8	2,300	50	52
Surat	5.8	1,680	60	12
Kanpur	3.0	1,500	-	-
Lucknow	3.3	1,200	-	-
Nagpur	2.7	1,000	-	-
Jaipur	3.5	1,000	-	-
Ludhiana	1.7	850	25	-
Indore	2.5	850	90	53
Coimbatore	2.6	850	-	-
Agra	2.0	790	-	-

impact on soil health by loss of beneficial soil microbes and essential nutrients (C, N, K, P, and S). Several initiatives have been taken by GoI to educate the agricultural community such as (i) mixing crop residue pellets with coal for power generation by The National Thermal Power Corporation (NTPC); (ii) biogas production as a renewable energy source for electricity, cooking, and lighting purposes through National Biogas and Manure Management Program; and (iii) Promotion of composting (vermicompost, biofertilizer, waste compost, and manure) through The Rashtriya Krishi Vikas Yogna (RKVY); these are some of the national programs with mission of mitigation and sustainable management of agricultural wastes [12].

18.4 Solid Waste Management

18.4.1 Municipal Solid Waste Management

MSWM is one of the most challenging problems to the environment nowadays, which need to be addressed legitimately. From the past research, it shows that more than 90% of MSW is disposed of ridiculously in an unscientific manner in open dumps or landfills, which causes various public health hazards and degrades the environment [1, 13, 14]. As per the study of [15], data related to volume and structure of waste is required to enhance the advancement of waste management structures. Poor waste management strategy at landfill sites leads to groundwater pollution and the generation of unpleasant odors, which affect the peoples staying in nearby areas [16]. Different activities such as generation, storage, collection, transfer and transport, processing, and disposal of solid wastes are the basic components of SWM. Broadly, MSWM system comprises of five activities such as:

i. Waste generation
ii. Waste collection
iii. Waste transportation
iv. Waste disposal
v. Waste recycle

However, it requires a robust system of infrastructure, maintenance, and up-gradation for all activities, which day by day becomes expensive and complicated due to the eternal and unplanned urbanization.

18.5 Solid Waste Management Techniques

MSW management methods can be broadly divided into two category, *viz.*, thermochemical and biochemical methods. Incineration, pyrolysis, and gasification are example of thermochemical ways; aerobic composting, anaerobic composting, landfills and bioethanol come in biochemical methods. Incineration is the most common thermochemical practice adopted in developed countries for the SWM as compared to pyrolysis and gasification. Whereas, aerobic composting is most used method in biochemical methods [17–20]. Details of various methods are summarized in Table 18.3.

Table 18.3 Comparison of major MSW management technology.

Technology	Advantages	Disadvantages
Incineration	Incineration is used to reduce 90% of solid waste volume. Ash generated can be used in the roads and construction industry. The heat produced during the process can be utilized for electricity or steam generation. Air emission can be controlled.	High operational and maintenance cost. High skilled worker required. Significant cost required for installing air emission control devices. Negative public perception due to release of harmful gases in air.
Landfilling	More economical. Required less-skilled workers for its implementation and maintenance. It can be used to generate energy in the form of biogas for its utility in household and small scale industry.	Day to day operation for leachate collection and treatment. Large area required for its implementation. Generates secondary pollutants that can degrade nearby groundwater, air, and soil.
Composting	Helps in the generation of high-quality organic fertilizers by utilizing biodegradable organic matter. Appreciable volumetric reduction of waste can occur.	Large space requirement for its implementation. Requires regular operational and maintenance cost with lesser environmental or economic cost. Generation of odor creating public inconvenience.

(*Continued*)

Table 18.3 Comparison of major MSW management technology. (*Continued*)

Technology	Advantages	Disadvantages
Anaerobic digestion/ Biomethanation	Reduction in waste volume. The end product is high quality fertilizer and gaseous fuel. Greenhouse gases emission to the environment is bypassed because all the gases produced are in an enclosed system.	Large space requirement for its implementation. Requires regular operational and maintenance cost with lesser
Bioethanol production	Bioethanol from MSW cannot create problem like food security. It may help in mitigation of climate change. Reduction in GHGs.	Cost of technology is higher. Production process is tedious.

18.5.1 Incineration

The combustion of MSWs at 750°C to 1,000°C is known as incineration. The end products of the process are air emission, bottom ash, fly ash residue, heat, and slag with high toxic substance content. The air emission and fly ash contain notable amounts of metals (e.g., Ar, Be, B, Ca, Cr, Co, Pb, Mg, Hg, Se, Sr, Tl, and V), minerals (e.g., SiO_2, Al_2O_3, CaO, and asbestos), dioxins, furans, and polycyclic aromatic hydrocarbons that are carcinogenic and harmful to human and animal health. This process does not eradicate waste; indeed, it only changes it in a new form. The only benefit is that it reduces the volume of waste by up to 90%. The installation of incinerators is highly expensive, with lesser economic returns. Also, plants need a regular supply of waste with high calorific value to maintain optimal combustion. These days produced ash is either disposed of in landfills or utilized in civil engineering applications [21, 22] or in agriculture as soil amendments.

18.5.2 Pyrolysis and Gasification

Pyrolysis is the thermal disintegration of solid organic waste in the absence of oxygen at a temperature between 300°C–750°C under pressure. This process gives three products: biochar, syngas, and pyrolysis liquid. Whereas, gasification is a thermal disintegration of organic waste in controlled oxygen supply at 650°C–1,000°C temperature; only two products are formed: syngas and biochar. The product's percent yield and composition depend on waste material type and reactor parameters in both processes. Biochar is a carbon-rich porous material with high calorific value with different application, *viz.*, solid fuel for boilers, as a soil amendment to increase water retention capacity, porosity, wastewater treatment, etc. Syngas and pyrolysis liquid are used to prepare valuable chemicals and fuels. But this process is costly and varies from waste to waste that make it less feasible.

18.5.3 Landfilling

The landfilling technique was first used in the United Kingdom in 1912 for SWM, which was later widely adopted in the United States for MSW disposal during the 1930s [23]. Before landfilling, open dumping is a common practice that quickly washes away heavy metals and recalcitrant pollutants in the nearby water bodies and raises the water pollution problem. Further, engineered landfilling prepared to dispose of the MSW properly, which is also called sanitary landfilling. They have engineered attenuation structures that stimulate anaerobic biodegradation and compaction of refuse materials within confining layers of soil. They did not allow direct expose of the refuge to rainfall, surface runoff, or groundwater and provide a daily cover of fresh garbage, and secure leachate and gas produced within the landfill cells [24]. The daily cover used in sanitary landfilling comprises soil, some inert materials such as construction debris, or compost residuals. It is a step by step activity that involves rigorous daily layering, compacting, soil covering of refuse into the cells, and managing surface runoff away from the waste cells [25]. Landfilling is a straightforward method of disposal and highly economical compared to other SWM techniques on at large scale, which proves to be highly effective and robust if adopted in an engineered manner [26, 27].

With time, landfills technology also evolved, and now we have many variants. They can be divided broadly into two types, *viz.*, dry tomb and bioreactor landfills. A dry tomb is an old method in which waste buried with minimum moisture that causes prolonged degradation. As a result, their aftercare periods get increased hundreds of years. Bioreactor landfills are recent advances, and they are of three types—aerobic, anaerobic, and semi-aerobic—based on their oxygen supply and microbial population. Degradation is faster in these landfills in comparison to a traditional one. That also reduces post-closure maintenance time.

Two by-products generate, *viz.*, landfill leachate (LFL), and landfill gases (LFG) during landfill disposal. LFL is a liquid derived from the landfill system either by the breakdown of waste or by the rainfall that passes through it. It contains high concentration organic and inorganic contaminants, including humic acids, ammonium, volatile fatty acids, heavy metals, xenobiotics, and inorganic salts. They need to be collected and managed to protect the environment and human health. This LFL has three stages [28] based on the age of landfill: young (less than 5 years old), medium (5–10 years old), and mature (more than 10 years old). With time, the concentration of ammonia and organics get increased in LFL that makes it a serious pollutant. There are two methods for LFL treatment: physical-chemical treatments and biological treatments [29]. Biological treatment is used more due to sustainability and cost-effectiveness. It includes four types, *viz.*, conventional nitrification-denitrification process [30], nitritation-denitritation process [31], endogenous denitritation process [32], and anammox process [33]. Table 18.4 shows the different microorganisms that are involved in different stages of landfill.

LFG consist of roughly 50%–60% methane, 40%–50% carbon dioxide (CO_2), and a small percentage of non-methane organic compounds. It is a product of organic waste decomposition by anaerobic bacteria. Its production is affected by volume, organic content, moisture content, and age of waste. The rate of emissions is affected by waste compaction, leachate recirculation [34], and aerobic landfilling [35]. These gases can be captured, converted, and used as a renewable energy source in place of release into the environment. They extracted with the help of a series of wells and a blower/flare (or vacuum) system, which include the following steps, *viz.*, flaring, adsorption, absorption permeation, and cryogenic treatment, and further used in electricity generation, an energy source for heating, and pipeline quality gas.

18.5.4 Aerobic Composting

Composting is a natural solid organic waste decomposition process by resident microbial community under the optimum environmental conditions of air in hot and moist conditions. It is pathogen-free process, which results in significant waste reduction up to 85%. The end products of the process is CO_2, water, mineral ions, and stabilized organic matter "humus" [43] also called as compost. Compost is the stabilized nutrient enrich product compatible and beneficial to plant growth [44]. Compost undergoes in the following stages:

i. An initial rapid stage of decomposition
ii. Stabilization stage
iii. Incomplete humification stage

Table 18.4 Microbes involved in landfill treatment.

Pollutant of LFL	Process	Chemical reaction	Microorganisms	References
Decomposition of biodegradable waste				
Organic matter	Aerobic	-	*Tricoderma*, *Aspergillus*, *Bacillus*, etc.	[36]
	Anaerobic	Hydrolysis, acidogenesis, acetogenesis, and methanogenesis	*Firmicutes*, *Bacteroidetes*, *Chloroflexi*, *Proteobacteria*, *Methanosaeta*, *Methanobacterium*, *Methanoculleus*, *Methanosaeta*, and *Methanosarcina*	[36]
Landfill leachate treatment				
Ammonia	Nitrification-denitification	Nitrification: $NH^+_4 + 2O_2 \rightarrow NO^-_3 + 2H^+ + H_2O + [\Delta G'_0 - 350\ kJ/mol]$ Denitrification: $NO^-_3 \rightarrow NO^-_2 \rightarrow NO \rightarrow N_2O \rightarrow N_2$	Nitrification: *Nitrosomonas* and *Nitrobacter* Denitrification: *Pseudomonas*, *Alcaligenes*, *Acinetobacter*, *Hyphomicrobium*, and *Thiobacillus*, *Lactobacillus*, and *Spirillum*	[37]

(*Continued*)

Table 18.4 Microbes involved in landfill treatment. (*Continued*)

Pollutant of LFL	Process	Chemical reaction	Microorganisms	References
	Nitritation-denitritation process	$NH^+_4 \rightarrow NO^-_2 \rightarrow NO \rightarrow N_2O \rightarrow N_2$	ammonia oxidizing bacteria (*Nitrosomonas*)	[38]
	Nitritation-endogenous denitritation process	$NH^+_4 \rightarrow NO^-_2 \rightarrow NO \rightarrow N_2O \rightarrow N_2$ phosphorus accumulating organisms (PAO) and glycogen accumulating organisms (GAOs) use the organics in raw landfill leachate to remove nitrogen	*Thauera*, *Ottowia*, *Paracoccus*, and *Comamonadaceae*	[38]
	Anammox process	$NH^+_4 \rightarrow NO^-_2$; $NH^+_4 \rightarrow N_2$; $NO^-_2 \rightarrow N_2$	*Brocadia*, *Kuenenia*, *Scalindua*, *Brasilis*, *Jettenia*, *Anammoxoglobu*, *Anammoximicrobium*	[38]
Xenobiotic organic compounds	-	-	*Pseudomonas*	[39]
Cellulose	-	-	*Fibrobacter* and *Clostridium*	[40]
Volatile fatty acid and aromatic compounds	-	-	*Thauera*, *Hydrogenophaga*, *Acidovorax*, *Comamonas*	[41, 42]

This process is a completely microbe-driven process where microbe works as an energy transducer. During this process, a massive amount of energy is generated and a portion is used by microbes. The remaining energy discharge as heat increases the pile's temperature and hastens the typical proceeding of composting. Here, an increase in temperature works as a sanitizer, while microbial activity leads to organic matter's mineralization and reduces C: N ratio. The final product was stable and safe for farming and gardening. It has many benefits, *viz.*, increases organic matter, sequesters carbon, improves plant growth, conserves water, decrease soil erosion, soil acidity, pathogen attack, and reduces dependency agrochemicals [4, 45].

Based on composting mixture temperature, microbes involved in compositing can be divided into psychrophilic (i.e., 12°C–20°C), mesophilic (i.e., 20°C–38°C), and thermophilic (i.e., 45°C–71°C) groups. Commonly occurring bacteria in composting mixture are Alcaligenes faecalis, Arthrobacter sp., Brevibacillus brevis, Bacillus circulans, Bacillus licheniformis, Bacillus megaterium, Bacillus pumilus, Bacillus sphaericus, Bacillus subtilis, Clostridium thermocellum, Flavobacterium sp., Pseudomonas sp., Thermus sp., and Vibrio sp. [46]. Whereas, fungi involved in composting are Aspergillus fumigatus, Basidiomycetes sp., Humicola grisea, H. insolens, H. lanuginosa, Malbranchea pulchella, Myriococcum thermophilum, Paecilomyces variotii, Papulaspora thermophilia, Penicillium sp., Scytalidium thermophilum, Termitomyces sp., and *Trichoderma* sp., and actinobacteria are Streptomyces sp., Frankia sp., and Micromonospora sp. One gram of compost contains approximately 10^9 bacterial, 10^8 actinobacterial, and 10^6 fungal cells [46].

These days, beneficial microbes were manually added into the compost at different steps of the process according to the requirement to fasten the process, to make more nutrient rich or to make it applicable to all kind of waste. Here is a table (Table 18.5) where such microbes and microbial consortium were listed with their utility.

18.5.5 Vermicomposting

Decompose of organic waste by aerobic microorganisms and earthworms (Oligochaete annelids), known as vermicomposting. In this process, degradation occurs in two steps, *viz.*, primary and secondary degradation process. The primary degradation of waste occurs through aerobic microorganisms. Further, secondary degradation of waste occurs in the gut of earthworm, and the final product that is called vermicompost, is nothing but the excreta of earthworm. It is a granular, odorless, and rich in essential

Table 18.5 Compost production from different solid wastes by adding microbes and microbial consortium.

Type of waste	Microbes/microbial consortium	Impact	References
Food waste	*Dysgonomonas* sp., *Pseudomonas caeni* strain, *Aeribacillus pallidus* strain, *Pseudomonas* sp., *Lactobacillus salivarius* strain, *Bacillus thuringiensis* strain, and *Bacillus cereus* strain	Anti-acidification increase organic acid degradation.	[50]
Food scraps and dry leaves	Lactic acid bacteria, photosynthetic bacteria, and yeast		[51]
Sugarcane leaves and dairy manure	*B. licheniformis* (TA65), *A. nidulans* (GXU-1), and *A. oryzae* (GXU-11)	Thermophilic and degradative enzymes were produced by microbes.	[52]
Food waste (FW) and maize straw	*Pseudomonas fragi* (KY283110), *Pseudomonas simiae* (KY283111), *Clostridium vincentii* (KY283112), *Pseudomonas jessenii* (KY283113), and *Iodobacter fluviatilis* (KY283114)	Cold adapted	[53]
Wheat straw	A cellulolytic consortium of *Trichoderma* sp., *P. chrysosporium*, and *A. oryzae*		[54]

(*Continued*)

Table 18.5 Compost production from different solid wastes by adding microbes and microbial consortium. (*Continued*)

Type of waste	Microbes/microbial consortium	Impact	References
Fruit wastes, vegetable wastes, leaves, hay, newspaper, wheat straw, and rice husks,	*Bacillus subtilis* and *Pseudomonas* sp.	Increased compost maturity	[55]
Food waste	Yeast strain *Pichia kudriavzevii*	Accelerated the composting process	[56]
OFMSW as including vegetables, food, garden, and office waste	*Phanerochaete chrysosporium* (MTCC787), *Trichoderma viride* (MTCC793), and *Pseudomonas aeruginosa* (MTCC2295)	Enzymatic activity	[57]
MSW	*Phanerochaete chrysosporium* and *Trichoderma reesei*	Cellulolytic microbial consortium increase speed of composting	[58]
MSW	Mixed culture (*Nitrobacter* and *Thiobacillus*, lignin decomposition composite, and fungi)		[59]
Organic fraction of municipal solid waste (OFMSW)	*Trichoderma viride*, *Aspergillus niger*, and *Aspergillus flavus*	High degradation of organic matter and early maturity	[60]

(*Continued*)

Table 18.5 Compost production from different solid wastes by adding microbes and microbial consortium. (*Continued*)

Type of waste	Microbes/microbial consortium	Impact	References
Kitchen-waste	*Bacillus thermoamylovorans*, Mixed *Bacillus* species (such as *B. brevis*, *B. coagulans*, and *B. licheniformis*	Composting process efficiency increased	[61]
MSW	Cellulolytic consortium of *Clostridia*		[62]
Vermicomposting			
petroleum oily sludge	*Acinetobacter radioresistens* strain KA2 and *Enterobacter hormaechei* strain KA3	Remove petroleum hydrocarbons	[63]
Paper cup waste	*Bacillus anthracis*, *B. endophyticus*, *B. funiculus*, *B. thuringiensis*, *B. cereus*, *B. toyonensis*, *Virigibacilliuschiquenigi*, *Acinetobacter baumanni*, and *Lactobacillus pantheries*	Fasten the process (19 to 12 week)	[64]
Fly ash and waste paper	*P. fluorescens*	Phosphorus enriched	[65]
Municipal green waste	*P. chrysosporium* and *A. chroococcum*	Nutrient rich	[66]

nutrients, has 1,000 fold more microbes, and low in contaminants. The microbial population in vermicompost is different from those present in the material before ingestion. Some reports show that microbes are not killed during the decomposition of organic waste in the gut while their proliferation rate increased. There are reports that gut of earthworm adds some beneficial microbes, for example, *Rhizobium japonicum*, *Pseudomonas putida*, *Azospirillum*, *Azobacter*, *Nitrobacters*, and *Nitrosomonas*, ammonifying bacteria and phosphate solubilizers [47, 48]. Vermicompost is used as a fertilizer in agriculture and gardening. Earthworms also increase the infiltration, porosity, and aeration in soil by their mobility and enhance the microbial community in field soil [49].

All kinds of organic solid waste can be vermicomposted by the addition of cattle dung in an appropriate ratio and by pre-treatment of solid waste by the suitable aerobic microbe. Vermicompost is successfully made by sewage sludge, petroleum oil sludge, leather industry waste, paper industry waste, urban residues, food industry waste, agro-industrial waste, and horticultural residues. The various author used this technology as a detoxification strategy to detox industrial waste with respect to toxins and metals [67–72].

The first stage of the vermicomposting, decomposition by aerobic microorganism is a crucial step especially in solid waste management by vermicomposting. Solid organic waste has high amount of toxic substance and by addition of suitable mesophilic microbes, it can make waste less toxic, more digestible to earthworm, and increase survivability and earthworm reproduction rate. It can also fasten the vermicomposting process and enhance nutrient enrichment. *Bacillus* and *Pseudomonas* play a crucial role in converting insoluble phosphates (tricalcium, dicalcium, and hydroxyapatite) into soluble forms by acidification, chelation, and make vermicompost phosphorus rich. Similarly, the pre-treatment of the organic mix by nitrogen-fixing bacteria enriches the final product by nitrogen and phosphorus [63, 73, 74].

Inoculation of suitable microorganisms could accelerate the vermicomposting process and improves compost quality (Table 18.5). The white-rot fungus *Phanerochaete chrysosporium* is one of the most efficient microorganisms at degrading lignin and cellulose [75]. It has been observed that inoculation with *P. chrysosporium* accelerated lignin and cellulose degradation during vermicomposting of rice straw and also improved the quality of the vermicompost product [74]. Inoculation with nitrogen-fixing bacteria could also improve compost quality. It is reported that inoculation with the nitrogen-fixing bacterium *Azotobacter chroococcum* during vermicomposting increased the nitrogen content of the final product [73].

18.5.6 Anaerobic Digestion

AD shows a promising sustainable approach to treat MSW by simultaneously producing biogas as a source of energy [76]. Due to the slow digestion rate and generation of the products that inhibit methanogenesis, it has

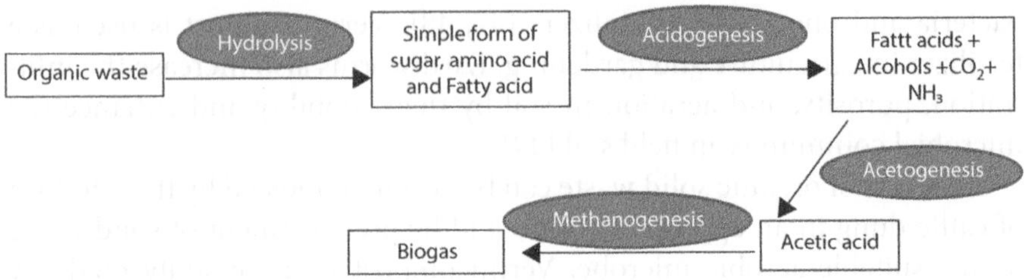

Figure 18.1 The process of anaerobic digestion in flow chart.

limited scope for scale-up [77–80]. However, in recent years, the AD process has been successfully applied to treat various agricultural wastes, food residues, and wastewater because of its capability to reduce the chemical oxygen demand (COD) and biological oxygen demand (BOD) efficiently from the waste streams and converting into biogas/methane (Figure 18.1) [81–84].

AD is a promising technique for treating MSW and producing CO_2 and CH_4, which can fulfil the energy demand along with the manure released after decomposition and can also be used as a biofertilizer. AD process can be further sub-divided into four distinct steps:

18.5.6.1 *Enzymatic Hydrolysis*

In this step, complex organic matter (protein, lipid, and carbohydrates) is broken down into simpler soluble molecules (amino acids, fatty acid, and sugars) by extracellular enzyme. Microorganisms like *Cellulomonas* sp., *Clostridium* sp., *Bacillus* sp., *Thermomonospora* sp., *Ruminococcus* sp., *Baceriodes* sp., *Erwinia* sp., *Acetovibrio* sp., *Microbispora* sp., and *Streptomyces* sp. play an important role in hydrolysis, which convert complex organic matter to soluble monomeric or dimeric substrates [85, 86].

18.5.6.2 *Fermentation*

In fermentation, reduced end products obtained after hydrolysis are converted to a mixture of the short-chain volatile fatty acids (VFAs) and other products like CO_2, hydrogen, and acetic acid by the subsistence of fermentative bacteria. Table 18.6 shows that the major fermentative bacteria played a significant role in AD, which converts the soluble end products obtained after the hydrolysis to distinct intermediates as VFAs, CO_2, alcohols, and hydrogen gas [85, 87].

Table 18.6 Important fermentative bacteria in anaerobic digestion.

Fermentation types	Genera	Important products
Acetate fermentation	*Acetobacterium, Clostridium, Sporomusa*	Acetate, CO_2
Alcohol fermentation	*Saccharomyces*	Ethanol, CO_2
Butyrate fermentation	*Butyribacterium, Clostridium*	Butyrate, butanol, isopropanol, ethanol, CO_2
Lactate fermentation	*Lactobacillus, Streptococcus*	Lactic acid, CO_2
Propionate fermentation	*Clostridium*	Propionate, acetate, CO_2

18.5.6.3 Acetogenesis

In this step, organic acids are converted to acetate, CO_2, and hydrogen using acetogenic bacteria. Wood–Ljungdahl pathway [88] is used to define the working of the acetogenic bacteria or acetogens distinct from acetate-forming fermentative bacteria as its tendency to reduce the carbon dioxide to acetate. *Acetobacterium* and *Sporomusa* are two exclusive acetogenic bacteria and, *Clostridium*, *Ruminococcus*, and *Eubacterium* genera contain both acetogenic and non-acetogenic bacteria that play an essential role in acetogenesis [85, 86].

It is well-known fact that acetogens are obligate hydrogen producers and hence are not able to survive in high partial hydrogen pressures. Thus, a mutual relationship exists between acetogens that produce hydrogen, which is consumed by methanogens.

18.5.6.4 Methanogenesis

Different types of methanogenic bacteria consume acetate, CO_2, and hydrogen to yield methane as the end product of the AD. Methanogens are obligate anaerobes, which were sensitive to environmental changes and convert the end products of acetogenic stages to methane and carbon dioxide [89]. Methane obtained from the degradation of the acetic acid weighs about 70%, while the rest are obtained from the redox reaction of the hydrogen and carbon dioxide. It is essential to maintain the dynamic equilibrium state for the effective and efficient performance of the acid formers and methane fermenters as it has been observed that methanogenic microorganisms are sensitive to acidity changes [90]. The optimum range of pH lies between 6.5–8, which is best for fermentation and production of methane gas [85, 87].

Usually, agricultural waste is recalcitrant to hydrolysis due to the presence of high lignocellulose content [91]. The theoretical yield of biogas from lignocellulosic material was measured 90% but in actuality it is only 50%. This is just because of inefficient hydrolysis of lignocellulose [92]. Lignocellulose is a complex polymer of cellulose, hemicellulose, and lignin that made up the cell wall of plant cell. To improve the hydrolysis of lignocellulose waste, various physical, chemical, and biological methods have been developed. Biological method requires less energy, more environment-friendly, and cost effective [93, 94]. In this methods microbes are used which produce the high concentration and more efficient hydrolytic enzyme and hydrolysis rate increased [93]. Table 18.7 shows the microorganism used for pretreatment and their impact on biogas production.

18.5.7 Bioethanol From Various Solid Wastes

Bioethanol, as a renewable alternative to fossil fuel, has been explored from various sources. Solid wastes generated from crop residues, agro-industrial, municipal, and livestock have been considered as one of the major sources for bioethanol production in terms of waste recycling and valorization. However, the compositional difference among various waste types and sources leads to the variation in the technologies used for bioethanol production. For example some of the major crop residues like rice straw, wheat straw, and corn stover are rich in cellulose (30%–50%), hemicellulose (20%–38%), and lignin (7%–21%), where the lignin content causes a major hindrance in hydrolysis and further biological conversion process to ethanol. Therefore, several physio-chemical pre-treatment methods have been employed on the substrate before the hydrolysis and ethanol fermentation process [109]. It has been reported that an enhancement of cellulose content by 60% using microwave-alkali-acid pre-treatment on rice straw [110]. They have obtained an ethanol yield of 18.9 g/L by simultaneous saccharification and fermentation (SiSF) method using the *B. subtilis* for hydrolysis and *S. cerevisiae* for fermentation (Table 18.8). Other than the physio-chemical method, microbial agents are also being used in the pre-treatment of lignocellulosic biomass. The predominant species used for biological pre-treatment are brown rot and white rot fungi, where the former one attacks only the cellulose, the later attack both cellulose and lignin [111]. Several fungal species, such as; *Trichoderma viride*, *Trichoderma reesei*, *Aspergillus awamori*, and *Aspergillus terreus* have been reported

Table 18.7 Effect of biological (microbial) pre-treatment of solid waste on biogas yield.

Type of waste	Microbe	Advantage	References
Wheat straw	*Polyporus brumalis* Microbial consortium TC-5 Microbial consortium	52% higher methane yield 36.6% higher methane yield 80.34% higher methane yield	[95] [96] [97]
Rice straw	*Pleurotus ostreatus* *Trichoderma reesei* *Bacillus* sp. Rumen fluid	120% higher methane yield 78.3% higher methane yield 76% higher biogas production 82.6% higher methane yield	[98] [98] [99] [100]
Corn silage	*Trametes versicolor* MC: Cellulose degrading bacteria	Increase in methane yield 38% higher methane yield	[101] [102]
Corn straw	Microbial consortium *Bacillus subtilis* MC: Yeast, cellulolytic bacteria, lactic acid bacteria	74.7% higher biogas production 17.35% higher methane yield 33.07% higher biogas production	[103] [104] [105]
Corn Stover	*Pleurotus eryngii*	19% higher biogas production	[106]
Brewery spent grain	*Pseudobutyrivibrio xylanivorans*Mz 5^T	17.8% higher biogas production	[107]
Sweet corn processing residues	MC: Predominated by genus *Clostridium*	15% higher biogas production	[108]
Cassava residues	MC: Beta proteobacterium HMD444 + *Thermoanaerobacterium thermosaccharolyticum* strain M18 + *Thermanaerovibrio acidaminovorans* DSM 6589, and *Clostridium* sp. strains LDC-8-c12, 5-8, CO6-72, etc.	96.63% higher biogas production	[105]

Table 18.8 Bioethanol production from different solid wastes.

Type of waste	Composition	Pre-treatment	Hydrolysis/Fermentation	Ethanol yield	References
Crop residues					
Rice straw	Cellulose:30.3%–52.3% Hemicellulose: 19.8%–31.6% Lignin: 7.2%–12.8%	Microwave-alkali-acid pre-treatment Microwave pre-treatment with 5 ml of 1% (w/v) NaOH (100–1,800 W, 3 min, 850 W, 150°C) Acid pre-treatment (1% v/v H_2SO_4) of dried microwave-alkali pre-treated biomass	SSF using the *B. subtilis* for hydrolysis and *S. cerevisiae* for fermentation	18.9 g/L	[109, 110]

(*Continued*)

Table 18.8 Bioethanol production from different solid wastes. (*Continued*)

Type of waste	Composition	Pre-treatment	Hydrolysis/Fermentation	Ethanol yield	References
Wheat straw	Cellulose: 32.9%–44.5% Hemicellulose: 37.8%–33.2% Lignin: 8.5%–22.3%	Alkaline peroxide pretreatment (2.15% H2O2, v/v; pH 11.5; 35°C; 24 h)	SSF using three commercial enzymes (Novozyme 188, Cellulast 1.5 L, and Viscostar), *viz.*, Cellulase, β-glucosidase, and xylanase at 45°C, pH 5.0, 120 h, Fermentation using recombinant *Escherichia coli* strain FBR5	15.1 g/L	[109, 124]
Corn stover	Cellulose: 31.3%–49.4% Hemicellulose: 21.1%–26.2% Lignin: 3.1%–8.8%	Ammonia Fiber Expansion (AFEX) as the pre-treatment technology,	SHF using *Saccharomyces cerevisiae* 424A (LNH-ST)	40.0 g/L	[109, 125]

(*Continued*)

Table 18.8 Bioethanol production from different solid wastes. *(Continued)*

Type of waste	Composition	Pre-treatment	Hydrolysis/Fermentation	Ethanol yield	References
Agro-industrial residue					
Sugar cane bagasses	Cellulose: 31.3%–49.4% Hemicellulose: 21.1%–26.2% Lignin: 3.1%–8.8%	Pre-treatment with H_2SO_4 (1.25%, w/w) and autoclaved (at 121°C, 2 h) Detoxified by electro-dialysis	Fermentation with *Pachysolen tannophilus* DW06	19 g/L	[122, 126]
Olive oil palm empty fruit bunch (OEFB)	Cellulose: 23.7%–63% Hemicellulose: 21.6%–33% Lignin: 29.2%–36.6%	Acid impregnation-steam explosion pre-treatment technique. OEFB chips impregnated with acid (0.14 M H_2SO_4) for one night and then steamed (203°C, 2 min)	SHF Hydrolysis by, Celluclast 1.5L and Novozyme 188 Fermentation with *S. cerevisiae* TISTR 5339	74.43%	[123, 127]

(Continued)

Table 18.8 Bioethanol production from different solid wastes. (*Continued*)

Type of waste	Composition	Pre-treatment	Hydrolysis/Fermentation	Ethanol yield	References
Cattle manure	Cellulose: 32.7% Hemicellulose: 24.5% Lignin: 42.8%	Dilute acid pre-treatment (2.5% H_2SO_4, 90 min, 121°C) and saccharified with 50 FPU CTec 2/g glucan.	Saccharified with CellicCTec 2 (50 FPU CTec/g glucan) Fermentation with *Saccharomyces cerevisiae*	7.3 g/L	[122, 128]
Municipal waste					
Food/kitchen waste	Cellulose: 16.9% Hemicellulose: 7.7% Lignin: 17.0%	Mechanical crushing into smaller particle size	SSF Saccharification with amylolytic enzyme complex, SAN Super 240 L (a mixture of amyloglucosidase, α-amylase, and protease) Fermentation with *Saccharomyces cerevisiae*	36 g/L	[129, 130]
Waste paper	Cellulose: 40%–55% Hemicellulose: 25%–40% Lignin: 18%–30%	Dilute acid treatment (0.50 N H_2SO_4 at 120°C for 2 h)	Acid hydrolysate detoxified by dried Cao and fermented using yeast *Pichia stipitis*	3.73 g/L	[120, 131]

to be used for pre-treatment in sugar cane trash [111, 112]. Similarly, *Ceriporiopsis subvermispora, Irpex lacteus*, and *Echinodontium taxodii* have been used for pre-treatment in corn stover [113]. However, the ethanol yield obtained by the biological pre-treatment method is comparatively lower than the physio-chemical pre-treatment method and is not preferred for commercial-scale ethanol production [114, 115]. Zhao *et al.* [116] have reported a maximum of 19%–22% of ethanol obtained from corn stover by applying chemical based pre-treatment technologies (alkaline, solvent-based, and ammonia) as compared to the biological pre-treatment (using fungi) method, where a yield of only 11% has obtained. Like crop residues, Organic Fraction Municipal Solid Waste (OFMSW), which has nearly 50% of carbohydrate content, is also a potential source for bioethanol production. Mechanical size reduction, use of acid, alkaline condition, and hydrothermal treatments are effective pre-treatment methods used for OFMSW [117]. Different biodegradable fractions of MSW such as kitchen waste (carrot and potato peels), grass (garden waste), and paper/card fractions (newspaper and scrap paper) have been utilized for bioethanol production [118]. Different pre-treatment conditions subjected on these waste are dilute acid (H_2SO_4, HNO_3 or HCl, 1%–4%, 180 min, 60°C), steam treatment (121°C–134°C, 15 min), microwave treatment (700 W, 2 min), or a combination of two of them. The pre-treatment has followed by the enzymatic hydrolysis process by *Trichoderma reesei* and *Trichoderma viride* (10 and 60 FPU/g of substrate). A highest glucose yield of 72.80% was obtained with a pre-treatment condition consist of acid treatment (1% H_2SO_4), followed by steam treatment (at 121°C), and enzymatic hydrolysis with *Trichoderma viride*.

Apart from crop residue and MSW, other potential solid wastes for bioethanol are food or kitchen waste (70% carbohydrate) [119], wastepaper (40%–55% cellulose content) [120], which are part of MSW. Similarly, agro-industrial waste like coffee residue waste (37%–42% fermentable sugar) [121], sugarcane bagasse (43.6%–45.8% cellulose and 31.3%–33.5% hemicellulose) [122], oil palm empty fruit bunch (23.7% cellulose and 21.6% hemicellulose) [123], livestock waste like cattle manure (32.7% cellulose and 24.5% hemicellulose) [122] are also potential sources for bioethanol production.

In bioethanol production, apart from the pre-treatment process, the hydrolysis and fermentation process also regulate the ethanol yield. Hydrolysis or saccharification, which involves conversion of complex carbohydrates into simple monomers or sugars, is usually catalysed by enzyme or acid. However, as compared to acid hydrolysis, enzymatic hydrolysis is preferable due to less toxicity and low energy-intensive [132]. In the enzymatic hydrolysis of lignocellulosic waste or biomass, cellulase enzyme is considered as the most prominent form of enzyme complexes, which are responsible for the conversion of cellulose to glucose or galactose monomer. Majorly, the cellulases complexes are consists of i) endoglucanases (Endo-β-(1,4)-glucanases, or 1,4-β-D-glucanohydrolases, ii) exoglucanases (1,4-β-D-glucan glucanohydrolases, β-D-glucan cellobiohydrolases), and iii) β-Glucosidases (β-glucoside glucohydrolases) [114, 115]; endoglucanase generally attacks low crystalline regions of cellulose fibers, and exoglucanase removes cellobiose units which later transformed into glucose by β-glucosidase [132]. Unlike cellulase, hemicellulase enzymes are more complex as it act upon various sugars like mannan, xylan, glucan, galactan, and arbinan. Hemicellulase comprises of a combination of various enzymes like, for xylan degradation endo-1,4-β-D-xylanases, exo-1,4-β-D xylocuronidases, α-L-arabinofuranosidases, α-glucuronidase, and acetylxylan esterase, and for glucomannan degradation 1,4-β-D mannanases, β-mannosidases [114]. Generally, the cellulase enzyme resides in the microorganism in two forms, which is either a complex form or a non-complex form. The complex forms of cellulases are organized into multi-enzyme complexes called cellulosomes, majorly found in certain anaerobic bacteria, like *Clostridium thermocellum*, *Cellulomonas* sp. Whereas, the non-complexed form of cellulases, which mostly act individually and co-operatively, are found mostly in aerobic fungi, like, *Trichoderma reesei*, *Humicola grisea*, and bacteria, like, *Streptomyces lividans* and *Cellulomonas fimi* [115]. Other bacterial species like *Thermomonospora*, *Bacteroides*, *Bacillus*, *Ruminococcus*, *Acetovibrio*, *Erwinia*, *Microbispora*, and fungi *Aspergillus*, *Penicillium*, *Humicola*, *Fusarium*, and *Schizophillum*, *Phanerochaete* sp. are reported to synthesize cellulase enzyme.

Among other fungal species, *Trichoderma* is the most studied cellulase and hemicellulase producing microorganisms, which can make five endoglucanases, two cellobiohydrolases, and three endoxylanases [133, 134]. Whereas, different species of *Aspergillus* sp. have been widely investigated for β-glucosidase production [135]. It is quite evident from the report that the enzymatic hydrolysis of pre-treated maize straw (2% sodium hydroxide at 80°C for 1 h) by cellulase from *Trichoderma reesei* ZU-02 and cellobiase from *Aspergillus niger* ZU-07 with the addition of Tween 80 (5 g/L) can increase the hydrolysis yield to 7.5% [81]. Apart from the microbial-derived enzymes, commercial enzymes like Novozymes A/S, Cellic CTec3 have been used as cost-efficient enzymes for hydrolysis of lignocellulosic biomass [114]. [124] have used commercial enzymes Novozyme 188 (β-glucosidase), Cellulast 1.5 l (Cellulase), and Viscostar 150 l (xylanase) for hydrolysis of wheat straw, resulting in ethanol production of 15.1 g/L.

In other instances, Celluclast 1.5L and Novozyme 188 have been used for hydrolysis of olive oil palm empty fruit bunch (OEFB) [127]. Similarly, CellicCTec 2 (Cellulase) has been used for hydrolysis of cattle manure [128] and SAN Super 240 l (a mixture of amyloglucosidase, α-amylase, and protease) has been used for hydrolysis of food waste [130] (Table 18.8).

Like the enzymatic hydrolysis process, the sugar released after the hydrolysis process is being subjected to fermentation for ethanol production by several microorganisms (yeasts, bacteria, and fungi). The selection of a suitable microbe for ethanol production is very crucial. For commercially viable ethanol production process, the effective microbe should have the following properties such as (i) a wide range of substrate usage efficiency, (ii) more ethanol production and throughput, (iii) capability to survive under the elevated quantity of ethanol and high temperature, and (iv) resistant to inhibitors prevailing in hydrolysate with cellulolytic activity [132].

Among various microorganisms, *Saccharomyces cerevisiae* is considered to be the most efficient microbe for the production of ethanol from sugar with a conversion efficiency of 90%. However, the limitation of *S. cerevisiae* lies in the fermentation of only hexose sugar, not pentose sugar. Whereas, *P. stipitis*, *P. tannophilus*, and *Candida shehatae* are some of the yeast species, which efficiently ferment pentose sugar, especially xylose. Certain bacterial species like *C. thermocellum*, *E. coli*, *Z. mobilis*, and fungal species such as *Fusarium oxysporum*, *Neurospora crassa*, and *Paecilomyces* sp. have been effectively used in the bioethanol fermentation process either through simultaneous saccharification and co-fermentation (SSCF) or consolidated bioprocessing (CBP). For enhanced bioethanol production, genetically modified microorganisms such as *S. cerevisiae* ATCC 26603, recombinant *E. coli* KO11, *P. stipitis* NRRLY-7124, and *P. stipitis* BCC15191 have been developed [115].

Talebnia *et al.* [136] have reported various microorganisms responsible for ethanol production (65% to 99% of theoretical value) from wheat straw. Among several bacteria, yeasts, and fungi, native and recombinant strains of *S. cerevisiae*, *P. stipitis*, and *Kluyveromyces marxianus* are the profoundly studied yeast species used for the fermentation of wheat straw hydrolysate. However, the best ethanol yield with respect to final ethanol concentration and volumetric ethanol productivity has been obtained by native and non-adapted *S. cerevisiae* [136]. Similarly, among bacteria strains, *E. coli* FBR5 (a recombinant bacterium) has been reported to provide the highest ethanol yield [124]. *S. cerevisiae* has also been proven efficient for bioethanol production (36 g/L) using kitchen waste [130] (Table 18.8). An ethanol yield of 90.72 g/L has been achieved from kitchen waste by fed-batch fermentation using the yeast *S. cerevisiae*. Besides, *S. cerevisiae*, other microbial species, such as *Z. mobilis* and *C. shehatae* are also provided better ethanol yield by using kitchen waste hydrolysate [119].

18.6 Conclusion

This book chapter shows the present situation of waste management in India, the Indian government's policies, and research efforts that are going on. There are many physical, chemical, and biological methods for the treatment of organic solid waste material. Here, with a microbiological perspective, biological processes (AD, aerobic composting, vermicomposting, landfills, and bioethanol production) with their pre-treatment methods are discussed. We conclude that by the use of microorganisms, we can sustainably solve the waste management problem.

References

1. Jha, M.K., Sondhi, O.A.K., Pansare, M., Solid waste management-a caste study. *Indian J. Enviorn. Potect.*, 23, 1153, 2003.
2. Kaza, S., Yao, L., Bhada-Tata, P., Van Woerden, F., *What a waste 2.0: a global snapshot of solid waste management to 2050*, World Bank publications, Washington, DC, 2018.
3. Ahluwalia, I.J. and Patel, U., Solid waste management in India an assessment of resource recovery and environmental impact. *Indian Council for Research on International Economic Relations*, Working Paper No. 356, 2018.
4. Sharma, S., Basu, S., Shetti, N.P., Kamali, M., Walvekar, P., Aminabhavi, T.M., Waste-to-energy nexus: A sustainable development. *Environ. Pollut.*, 267, 115501, 2020.
5. PIB, *Solid Waste Management Rules Revised After 16 Years; Rules Now Extend to Urban and Industrial Areas*, Press Information Bureau, 2016, Government of India, PIB Delhi, http://pib.nic.in/newsite/PrintRelease.aspx?relid=138591.
6. Buenrostro, O. and Bocco., G., Solid waste management in municipalities in Mexico: Goals and perspectives. *Resour. Conserv. Recycl.*, 39, 251, 2003.
7. Kumar, S., Bhattacharyya, J.K., Vaidya, A.N., Chakrabarti, T., Devotta, S., and Akolkar, A.B., Assessment of the status of municipal solid waste management in metro cities, state capitals, class I cities, and class II towns in India: An insight. *Waste. Manage.*, 29, 883, 2009. http://dx.doi.org/10.1016/j.wasman.2008.04.011.
8. NDTV report by Bhatia, A. and Bhaskar, S., Fighting India's Garbage Crisis: What Are The Challenges India Faces In Managing Its Waste, 2019. https://

swachhindia.ndtv.com/fighting-indias-garbage-crisis-what-are-the-challenges-india-faces-in-managing-its-waste-32025/.
9. Talyan, V., Dahiya, R.P., Sreekrishnan, T.R., State of municipal solid waste management in Delhi, the capital of India. *Waste Manage.*, 28, 1276, 2008.
10. Jain, S., Jain, S., Wolf, I.T., Lee, J., Tong, Y.W., A comprehensive review on operating parameters and different pretreatment methodologies for anaerobic digestion of municipal solid waste. *Renew. Sust. Energ. Rev.*, 52, 142, 2014.
11. Jitendra, Venkatesh, S., Kukreti, I., Pandey, K., Niyogi, D.G., Mukerjee, P., India's Burning Issues of Crop Burning Takes a New Turn, Down to Earth, 2017. Available online: https://www.downtoearth.org.in/coverage/river-of-fire-57924 (accessed on 25 August 2020).
12. Bhuvaneshwari, S., Hettiarachchi, H., Meegoda, J.N., Crop residue burning in India: Policy challenges and potential solutions. *Inter. J. Environ. Res. Public. Health*, 16, 5, 832, 2019.
13. Diaz, L.F., Waste management in developing countries and the circular economy. *Waste. Manag. Res.*, 35, 1, 2017.
14. Zornoza, R., Moreno-Barriga, F., Acosta, J., Muñoz, M., Faz, A., Stability, nutrient availability and hydrophobicity of biochars derived from manure, crop residues, and municipal solid waste for their use as soil amendments. *Chemosphere*, 144, 122, 2016.
15. Beigl, P., Lebersorger, S., Salhofer, S., Modelling municipal solid waste generation: a review. *Waste Manag.*, 28, 200, 2008.
16. Mor, S., Ravindra, K., Visscher, A., Dahiya, R.P., Chandra, A., Municipal solid waste characterization and its assessment for potential methane generation: a case study. *Sci. Total. Environ.*, 371, 1, 2006.
17. *American Society of Mechanical Engineers Waste-to-Energy: A Renewable Energy Source from Municipal Solid Waste*, White Paper Submitted to Congress, New York, 2008.
18. Denison, R.A., Environmental life-cycle comparison of recycling, landfilling, and incineration: a review of recent studies. *Annu. Rev. Energy Environ.*, 21, 191, 1996.
19. Themelis, N.J., An overview of the global waste-to-energy industry. *Waste. Manage. World.*, 3, 40, 2003.
20. United Nations Environment Programme, *International Source Book on Environmentally Sound Technologies for Municipal Solid Waste Management*, United Nations Environment Programme, Osaka/Shiga, 1996.
21. Kan, L., Shi, R., Zhao, Y., Duan, X., Wu, M., Feasibility study on using incineration fly ash from municipal solid waste to develop high ductile alkali activated composites. *J. Clean. Prod.*, 254, 120168, 2020.
22. Zhu, W., Teoh, P.J., Liu, Y., Chen, Z., Yang, E.H., Strategic utilization of municipal solid waste incineration bottom ash for the synthesis of lightweight aerated alkali-activated materials. *J. Clean. Prod.*, 235, 603–612, 2019.

23. Hasan, S.E., *Geology and hazardous waste management*, Upper Saddle River, N.J.: Prentice Hall, 1996.
24. The Gazette of India, Ministry of Environment and Forests Notification, 2000. http://www.envfor.nic.in/legis/hsm/mswmhr.html.
25. Tchobanoglous, G., Theisen, H., Vigil, S., *Integrated solid waste management: Engineering principles and management Issues*, McGraw-Hill, New York, 1993.
26. Bashir, M.J.K., Aziz, H.A., Yusoff, M.S., Adlanm, M.N., Application of response surface methodology (RSM) for optimization of ammoniacal nitrogen removal from semiaerobic landfill leachate using ion exchange resin. *Desalination*, 254, 154, 2010.
27. Davis, M.L. and Cornwell, D.A., Introduction to Environmental Engineering, in: *International Edition*, fourth ed, McGraw Hill, New York, 2008.
28. Foo, K.Y. and Hameed, B.H., An overview of landfill leachate treatment via activated carbon adsorption process. *J. Hazard. Mater.*, 171, 54, 2009.
29. Renou, S., Givaudan, J.G., Poulain, S., Dirassouyan, F., Moulin, P., Landfill leachate treatment: review and opportunity. *J. Hazard. Mater.*, 150, 468, 2008.
30. Ma, B., Wang, S.Y., Cao, S.B., Miao, Y.Y., Jia, F.X., Du, R., Peng, Y.Z., Biological nitrogen removal from sewage via anammox: recent advances. *Bioresour. Technol.*, 200, 981, 2016.
31. Yang, Q., Peng, Y.Z., Liu, X.H., Zeng, W., Mino, T., Satoh, H., Nitrogen removal via nitrite from municipal wastewater at low temperatures using real-time control to optimize nitrifying communities. *Environ. Sci. Technol.*, 41, 8159, 2007.
32. Wang, K., Wang, S.Y., Zhu, R.L., Miao, L., Peng, Y.Z., Advanced nitrogen removal from landfill leachate without addition of external carbon using a novel system coupling ASBR and modified SBR. *Bioresour. Technol.*, 134, 212, 2013.
33. Li, H.S., Zhou, S.Q., Ma, W.H., Huang, P.F., Huang, G.T., Qin, Y.J., Xu, B., Ouyang, H., Long-term performance and microbial ecology of a two-stage PN-ANAMMOX process treating mature landfill leachate. *Bioresour. Technol.*, 159, 404, 2014.
34. Cabaraban, M.T.I., Khire, M.V., Alocilja, E.C., Aerobic in-vessel composting versus bioreactor landfilling using life cycle inventory models. *Clean. Techn. Environ. Policy*, 10, 39, 2008.
35. Read, A., Hudgins, M., Harper, S., Phillips, P.S., Morris, J., Successful demonstration of two aerobic landfills- leading towards a more sustainable solid waste manage- ment approach. *IWM Sci. Tech. Rev.*, 1, 9, 2000.
36. Buchroithner, D., Microbiology of Landfill Sites Review of the present State of Knowledge with particular Focus on the Carbon Cycle, 2015. https://publik.tuwien.ac.at/files/PubDat_238029.pdf.
37. Wiszniowski, J., Robert, D., Surmacz-Gorska, J., Miksch, K., Weber, J.V., Landfill leachate treatment methods: a review. *Environ. Chem. Lett.*, 4, 51, 2006.

38. Miao, L., Yang, G., Tao, T., Peng, Y., Recent advances in nitrogen removal from landfill leachate using biological treatments – A review. *J. Environ Manage.*, 235, 178, 2019.
39. Ghosh, P., Das, M.T., Thakur, I.S., Mammalian cell line-based bioassays for toxicological evaluation of landfill leachate treated by *Pseudomonas* sp. ISTDF1. *Environ. Sci. Pollut. Res. Int.*, 21, 8084, 2014.
40. McDonald, J.E., Houghton, J.N.I., Rooks, D.J., Allison, H.E., McCarthy, A.J., The microbial ecology of anaerobic cellulose degradation in municipal waste landfill sites: evidence of a role for Fibrobacters. *Environ. Microbiol.*, 14, 1077, 2012.
41. Trzcinski, A.P. and Stuckey, D.C., Determination of the Hydrolysis Constant in the Biochemical Methane Potential Test of Municipal Solid Waste. *Environ. Eng. Sci.*, 29, 9, 848, 2012.
42. Gabarró, J., Hernández-del Amo, E., Gich, F., Ruscalleda, M., Balaguer, M.D., Colprim, J., Nitrous oxide reduction genetic potential from the microbial community of an intermittently aerated partial nitritation SBR treating mature landfill leachate. *Water Res.*, 47, 19, 7066, 2013.
43. Riddech, N., Klammer, S., Insam, H., Characterisation of microbial communities during composting of organic wastes, in: *Microbiology of composting*, H. Insam, N. Riddech, S. Klammer (Eds.), pp. 43–51, Springer, Berlin, Heidelberg, 2002.
44. MacGregor, S.T., Miller, F.C., Psarianos, K.M., Finstein, M.S., Composting process control based on interaction between microbial heat output and temperature. *Appl. Environ. Microbiol.*, 41, 1321, 1981.
45. Rastogi, M., Nandal, M., Khosla, B., Microbes as vital additives for solid waste composting. *Heliyon*, 6, e03343, 2020, doi: 10.1016/j.heliyon.2020.e03343.
46. Carry on Composting 2020 http://www.carryoncomposting.com/416920205.
47. Madesen, E.L. and Alexander, M., Transport of Rhizobium and Pseudomonas through soil. *Soil. Sci. Soc. Am. J.*, 46, 557, 1982.
48. Pathma, J. and Sakthivel, N., Microbial diversity of vermicompost bacteria that exhibit useful agricultural traits and waste management potential. *Springer Plus*, 1, 1, 2012.
49. Joshi, R., Singh, J., Vig, A.P., Vermicompost as an effective organic fertilizer and biocontrol agent: effect on growth, yield and quality of plants. *Rev. Environ. Sci. Biotechnol.*, 14, 137, 2015, doi: 10.1007/s11157-014-9347-1.
50. Song, C., Li, M., Qi, H., Zhang, Y., Liu, D., Xia, Pan, H.W., Xi, B., Impact of anti-acidification microbial consortium on carbohydrate metabolism of key microbes during food waste composting. *Bioresour. Technol.*, 259, 1, 2018.
51. Karnchanawong, S. and Nissaikla, S., Effects of microbial inoculation on composting of household organic waste using passive aeration bin. *Int. J. Recycl. Org. Waste. Agric.*, 3, 113, 2014.
52. Xu, J., Lu, Y., Shan, G. *et al.*, Inoculation with Compost-Born Thermophilic Complex Microbial Consortium Induced Organic Matters Degradation

While Reduced Nitrogen Loss During Co-Composting of Dairy Manure and Sugarcane Leaves. *Waste. Biomass. Valor.*, 10, 2467, 2019.
53. Xie, X., Zhao, Y., Sun, Q., Wang, X., Cui, H., Zhang, X., Li, Wei., Z., A novel method for contributing to composting start-up at low temperature by inoculating cold-adapted microbial consortium. *Bioresour. Technol.*, 238, 39, 2017.
54. Lin, H., Wang, B., Zhuang, R., Zhou, Q., Zhao, Y., Artificial construction and characterization of a fungal consortium that produces cellulolytic enzyme system with strong wheat straw saccharification. *Bioresour. Technol.*, 102, 10569, 2011.
55. Pan, I., Dam, B., Sen, S.K., Composting of common organic wastes using microbial inoculants. *3 Biotech.*, 2, 127, 2012.
56. Nakasaki, K., Araya, S., Mimoto, H., Inoculation of Pichia kudriavzevii RB1 degrades the organic acids present in raw compost material and accelerates composting. *Bioresour. Technol.*, 144, 521, 2013.
57. Awasthi, M.K., Pandey, A.K., Bundela, P.S., Khan, J., Co-composting of organic fraction of municipal solid waste mixed with different bulking waste:characterization of physicochemical characterization of physicochemical parameters and microbial enzymatic dynamic. *Bioresour. Technol.*, 182, 200, 2015.
58. Raut, M.P., Prince William, S.P.M., Bhattacharyya, J.K., Chakrabarti, T., Devotta, S., Microbial dynamics and enzyme activities during rapid composting of municipal solid waste – a compost maturity analysis perspective. *Bioresour. Technol.*, 99, 6512, 2008.
59. Xi, B.-D., He, X.-S., Wei, Z.-M., Jiang, Y.-H., Li, M.-X., Li, D. *et al.*, Effect of inoculation methods on the composting efficiency of municipal solid wastes. *Chemosphere*, 88, 744, 2012.
60. Awasthi, M.K., Pandey, A.K., Khan, J., Bundela, P.S., Wong, J.W., Selvam, A., Evaluation of thermophilic fungal consortium for organic municipal solid waste composting. *Bioresour. Technol.*, 168, 214, 2014.
61. Abdullah, N., Chin, N.L., Mokhtar, M.N., Taip, F.S., Effects of bulking agents, load size or starter cultures in kitchen-waste composting. *Int. J. Recycl. Org. Waste. Agric.*, 2, 3, 2013.
62. Kinet, R., Destain, J., Hiligsmann, S., Thonart, P., Delhalle, L., Taminiau., B. *et al.*, Thermophilic and cellulolytic consortium isolated from composting plants improves anaerobic digestion of cellulosic biomass: toward a microbial resource management approach. *Bioresour. Technol.*, 189, 138, 2015.
63. Koolivand, A., Saeedi., R., Coulon, F., Kumar, V., Villaseñor, J., Asghari, F., Hesampoor, F., Bioremediation of petroleum hydrocarbons by vermicomposting process bioaugmentated with indigenous bacterial consortium isolated from petroleum oily sludge. *Ecotoxicol. Environ. Saf.*, 198, 110645, 2020.
64. Arumugam, K., Renganathan, S., Babalola, O.O., Muthunarayanan, V., Investigation on paper cup waste degradation by bacterial consortium

and Eudrilluseugeinea through vermicomposting. *Waste Manage.*, 74, 185, 2018.
65. Lukashe, N.S., Mupambwa, H.A., Green, E., Mnkeni, P.N.S., Inoculation of fly ash amended vermicompost with phosphate solubilizing bacteria (Pseudomonas fluorescens) and its influence on vermi-degradation, nutrient release and biological activity. *Waste Manage.*, 83, 14, 2019.
66. Gong, X., Wei, L., Yu, X., Li, S., Sun, X., Wang, X., Effects of Rhamnolipid and Microbial Inoculants on the Vermicomposting of Green Waste with Eisenia fetida. *PLOS ONE.*, 12, 0170820, 2017.
67. Jain, K., Singh, J., Chauhan, L.K.S., Murthy, R.C., Gupta, S.K., Modulation of flyash induced genotoxicity in Vicia faba by vermicomposting. *Ecotoxicol. Environ. Saf.*, 59, 89, 2004.
68. Srivastava, R., Kumar, D., Gupta, S.K., Bioremediation of municipal sludge by vermin-technology and toxicity assessment by *Allium cepa*. *Bioresour. Technol.*, 96, 1867, 2005.
69. Sarojini, S., Manimegala, G., Prakash, M., Ananthakrishnasamy, S., Gunasekaran, G., Vermicomposting of lignite flyash reduces its genotoxicity and phytotoxicity and improves its agronomic values for application as soil amendments in farmland: a study of vermicomposted flyash on onion crops (*Allium cepa*). *Dyn. Soil. Dyn. Plant.*, 4, 135, 2010.
70. Bhat, S.A., Singh, J., Vig, A.P., Genotoxic assessment and optimization of pressmud with the help of exotic earthworm *Eisenia fetida*. *Environ. Sci. Pollut. Res.*, 21, 8112, 2014.
71. Bhat, S.A., Singh, J., Vig, A.P., Vermistabilization of sugar beet (*Beta vulgaris* L.) waste produced from sugar factory using earthworm *Eisenia fetida*: genotoxic assessment by Allium cepa test. *Environ. Sci. Pollut. Res.*, 22, 11236, 2015.
72. Bhat, S.A., Singh, J., Vig, A.P., Genotoxicity reduction in bagasse waste of sugar industry by earthworm technology. *SpringerPlus*, 5, 1186, 2016.
73. Das, S., Deka, P., Goswami, L., Sahariah, B., Hussain, N., Bhattacharya, S.S., Vermiremediation of toxic jute mill waste employing Metaphireposthuma. *Environ. Sci. Pollut. Res.*, 23, 15418, 2016.
74. El-Haddad, M.E., Zayed, M.S., El-Sayed, G.A.M., Hassanein, M.K., El-Satar, A.M.A., Evaluation of compost, vermicompost and their teas produced from rice straw as affected by addition of different supplements. *Ann. Agr. Sci.*, 59, 243, 2014.
75. Haddadin, M.S.Y., Haddadin, J., Arabiyat, O.I., Hattar, B., Biological conversion of olive pomace into compost by using *Trichoderma harzianum* and *Phanerochaete chrysosporium*. *Bioresour. Technol.*, 100, 4773, 2009.
76. McCarty, P., The development of anaerobic treatment and its future. *Water. Sci. Technol.*, 44, 149, 2001.
77. Stams, A.J., De Bok, F.A., Plugge, C.M., Eekert, V., Miriam, H., Dolfing, J., Schraa, G., Exocellular electron transfer in anaerobic microbial communities. *Environ. Microbiol.*, 8, 371, 2006.
78. Stams, A.J. and Plugge, C.M., Electron transfer in syntrophic communities of anaerobic bacteria and archaea. *Nat. Rev. Microbiol.*, 7, 568, 2009.

79. Barredo, M. and Evison, L., Effect of propionate toxicity on methanogen-enriched sludge, *Methanobrevibacter smithii*, and *Methanospirillum hungatii* at different pH values. *Appl. Environ. Microbiol.*, 57, 1764, 1991.
80. McCarty, P.L. and Smith, D.P., Anaerobic wastewater treatment. *Environ. Sci. Technol.*, 20, 1200, 1986.
81. Chen, M., Zhao, J., Xia, L., Enzymatic hydrolysis of maize straw polysaccharides for the production of reducing sugars. *Carbohydr. Polym.*, 71, 411, 2008.
82. De Baere, L., Anaerobic digestion of solid waste: state-of-the-art. *Water. Sci. Technol.*, 41, 283, 2000.
83. Rapport, J., Zhang, R., Jenkins, B.M., Williams, R.B., *Current anaerobic digestion technologies used for treatment of municipal organic solid waste*, UC Davis prepared for California Integrated Waste Management Board, CA, USA, 2008.
84. Kwietniewska, E. and Tys, J., Process characteristics, inhibition factors and methane yields of anaerobic digestion process, with particular focus on microalgal biomass fermentation. *Renew. Sustain. Energy. Rev.*, 34, 491, 2014.
85. Lo, Y.C., Saratale, G.D., Chen, W.M., Bai, M.D., Chang, J.S., Isolation of cellulose hydrolytic bacteria and applications of the cellulolytic enzymes for cellulosic biohydrogen production. *Enzyme Microb. Technol.*, 44, 417, 2009.
86. Schmidt, S., Biegel, E., Müller, V., The ins and outs of Na+ bioenergetics in *Acetobacterium woodii. Biochim. Biophys. Acta-Bioenergetics*, 1787, 691, 2009.
87. Chipasa, K. and Mędrzycka, K., Behavior of lipids in biological waste water treatment processes. *J. Ind. Microbiol.*, 33, 635, 2006.
88. Ragsdale, S.W. and Pierce, E., Acetogenesis and the Wood–Ljungdahl pathway of CO2 fixation. *Biochim. Biophys. Acta. Proteins. Proteom.*, 1784, 1873, 2008.
89. Nayak, S.K., Baliyarsingh, B., Samantaray, D.P., Mishra, B.B., Methanogenesis: Microbiology and Ecology, in: *Biogas Technology*, S.K. Mishra, T.K. Adhya, S.K. Ojha (Eds.), pp. 1–34, New India Publishing Agency, New India Publishing Agency New Delhi, India, 2019.
90. Davidsson, Å., Lövstedt, C., la Cour Jansen, J., Gruvberger, C., Aspegren, H., Co- digestion of grease trap sludge and sewage sludge. *Waste Manage.*, 28, 986, 2008.
91. Merlin Christy, P., Gopinath, L.R., Divya, D., A review on anaerobic decomposition and enhancement of biogas production through enzymes and microorganisms. RenewableSustainable. *Energy Rev.*, 34, 167–173, 2014.
92. Azman, S., Khadem, A.F., Van Lier, J.B., Zeeman, G., Plugge, C.M., Presence and role of anaerobic hydrolytic microbes in conversion of lignocellulosic biomass for biogas production. *Crit. Rev. Environ. Sci. Technol.*, 45, 2523, 2015. 10.1080/10643389.2015.1053727.
93. Shrestha, S., Fonoll, X., Khanal, S.K., Raskin, L., Biological strategies for enhanced hydrolysis of lignocellulosic biomass during anaerobic digestion:

current status and future perspectives. *Bioresour. Technol.*, 245, 1245, 2017. 10.1016/J.BIORTECH.2017.08.089.

94. Wagner, A.O., Lackner, N., Mutschlechner, M., Prem, E.V., Markt, R., Illmer, P., Biological pretreatment strategies for second-generation lignocellulosic resources to enhance biogas production. *Energies*, 11, 1797, 2018, doi: 10.3390/en11071797.
95. Rouches, E., Zhou, S., Sergent, M., Raouche, S., Carrere, H., Influence of white-rot fungus Polyporusbrumalis BRFM 985 culture conditions on the pretreatment efficiency for anaerobic digestion of wheat straw. *Biomass Bioener.*, 110, 75, 2018.
96. Kong, X.P., Du, J., Ye, X., Xi, Y., Jin, H., Zhang, M., Guo, D., Enhanced methane production from wheat straw with the assistance of lignocellulolytic microbial consortium TC-5. *Bioresour. Technol.*, 263, 33, 2018.
97. Zhong, C., Wang, C., Wang, F., Jia, H., Wei, P., Zhao, Y., Enhanced biogas production from wheat straw with the application of synergistic microbial consortium pretreatment. *RSC. Adv.*, 6, 60187, 2016.
98. Mustafa, A.M., Poulsen, T.G., Sheng, K., Fungal pretreatment of rice straw with *Pleurotus ostreatus* and *Trichoderma reesei* to enhance methane production under solid-state anaerobic digestion. *Appl. Energy.*, 180, 661, 2016.
99. Shah, T.A., Lee, C.C., Orts, W.J., Tabassum, R., Biological pretreatment of rice straw by ligninolytic *Bacillus* sp. strains for enhancing biogas production. *Environ. Prog. Sustain. Energy.*, 38, 3, 2019.
100. Zhang, Z., Zhang, G., Li, W., Li, C., Xu, G., Enhanced biogas production from sorghum stem by codigestion with cow manure. *Int. J. Hydrogen. Energy.*, 41, 9153, 2016.
101. Tišma, M., Planinić, M., Bucić-Kojić, A., Panjičko, M., Zupančič, G.D., Zelić, B., Corn silage fungal-based solid-state pretreatment for enhanced biogas production in anaerobic co-digestion with cow manure. *Bioresour. Technol.*, 253, 220, 2018.
102. Poszytek, K., Ciezkowska, M., Sklodowska, A., Drewniak, L., Microbial Consortium with High Cellulolytic Activity (MCHCA) for enhanced biogas production. *Front. Microbiol.*, 7, 1, 2016.
103. Hua, B., Dai, J., Liu, B., Yuan, H.Z.X., Wang, X., Cui, Z., Pretreatment of non-sterile, rotted silage maize straw by the microbial community MC1 increases biogas production. *Biores. Technol.*, 216, 699, 2016.
104. Xu, W., Fu, S., Yang, Z., Lu, J., Guo, R., Improved methane production from corn straw by microaerobic pretreatment with a pure bacteria system. *Bioresour. Technol.*, 259, 18, 2018.
105. Zhang, Q., He, J., Tian, M., Mao, Z., Tang, L., Zhang, J., Zhang, H., Enhancement of methane production from cassava residues by biological pretreatment using a constructed microbial consortium. *Bioresour. Technol.*, 102, 8899, 2011.
106. Wyman, V., Henríquez, J., Palma, C., Carvajal, A., Lignocellulosic waste valorisation strategy through enzyme and biogas production. *Bioresour. Technol.*, 247, 402, 2018.

107. Čater, M., Fanedl, L., Malovrh, Š., Logar, R.M., Biogas production from brewery spent grain enhanced by bioaugmentation with hydrolytic anaerobic bacteria. *Bioresour. Technol.*, 186, 261, 2015.
108. Martin-Ryals, A., Schideman, L., Li, P., Wilkinson, H., Wagner, R., Improving anaerobic digestion of a cellulosic waste via routine bioaugmentation with cellulolytic microorganisms. *Bioresour. Technol.*, 189, 62, 2015.
109. Pattanaik, L., Pattnaik, F., Saxena, D.K., Naik, S.N., Biofuels from agricultural wastes, in: *Second and third generation of feedstocks: the evolution of biofuels*, A. Basile and F. Dalena (Eds.), p. 103, Elsevier, Amsterdam, 2019.
110. Akhtar, N., Goyal, D., Goyal, A., Characterization of microwave-alkali-acid pretreated rice straw for optimization of ethanol production via simultaneous saccharification and fermentation (SSF). *Energy. Convers. Manage.*, 141, 133, 2017.
111. Aditiya, H.B., Mahlia, T.M.I., Chong, W.T., Nur, H., Sebayang, A.H., Second generation bioethanol production: A critical review. *Renew. Sust. Eneg. Rev.*, 66, 631, 2016.
112. Singh, P., Suman, A., Tiwari, P., Arya, N., Gaur, A., Shrivastava, A.K., Biological pretreatment of sugarcane trash for its conversion to fermentable sugars. *World. J. Microbiol. Biotechnol.*, 24, 667, 2008.
113. Yang, X., Zeng, Y., Ma, F., Zhang, X., Yu, H., Effect of biopretreatment on thermogravimetric and chemical characteristics of corn stover by different white-rot fungi. *Bioresour. Technol.*, 101, 5475, 2010.
114. Saini, J.K., Saini, R., Tewari, L., Lignocellulosic agriculture wastes as biomass feedstocks for second-generation bioethanol production: concepts and recent developments. *3 Biotech.*, 5, 337, 2015.
115. Gupta, A. and Verma, J.P., Sustainable bio-ethanol production from agro-residues: a review. *Renew. Sust. Energ. Rev.*, 41, 550, 2015.
116. Zhao, Y., Damgaard, A., Christensen, T.H., Bioethanol from corn Stover review and technical assessment of alternative biotechnologies. *Prog. Energy. Combust. Sci.*, 67, 275291, 2018.
117. Barampouti, E.M., Mai, S., Malamis, D., Moustakas, K., Loizidou, M., Liquid biofuels from the organic fraction of municipal solid waste: a review. *Renew. Sust. Eneg. Rev.*, 110, 298, 2019.
118. Li, A., Antizar-Ladislao, B., Khraisheh, M., Bioconversion of municipal solid waste to glucose for bio-ethanol production. *Bioprocess. Biosyst. Eng.*, 30, 189, 2007.
119. Hafid, H.S., Shah, U.K.M., Baharuddin, A.S., Ariff, A.B., Feasibility of using kitchen waste as future substrate for bioethanol production: a review. *Renew. Sust. Eneg. Rev.*, 74, 671, 2017.
120. Byadgi, S.A. and Kalburgi, P.B., Production of bioethanol from waste newspaper. Procedia. *Environ. Sci.*, 35, 555, 2016.
121. Choi, I.S., Wi, S.G., Kim, S.B., Bae, H.J., Conversion of coffee residue waste into bioethanol with using popping pre-treatment. *Bioresour. Technol.*, 125, 132, 2012.

122. Vassilev, S.V., Baxter, D., Andersen, L.K., Vassileva, C.G., Morgan, T.J., An overview of the organic and inorganic phase composition of biomass. *Fuel*, 94, 1, 2012.
123. Omar, R., Idris, A., Yunus, R., Khalid, K., Isma, M.A., Characterization of empty fruit bunch for microwave-assisted pyrolysis. *Fuel*, 90, 1536, 2011.
124. Saha, B.C. and Cotta, M.A., Ethanol production from alkaline peroxide pretreated enzymatically saccharified wheat straw. *Biotechnol. Prog.*, 22, 449, 2006.
125. Lau, M.W. and Dale, B.E., Cellulosic ethanol production from AFEX-treated corn stover using Saccharomyces cerevisiae 424A (LNH-ST). *PNAS*, 106, 1368, 2009.
126. Cheng, K.K., Cai, B.Y., Zhang, J.A., Ling, H.Z., Zhou, Y.J., Ge, J.P., Xu, J.M., Sugarcane bagasse hemicellulose hydrolysate for ethanol production by acid recovery process. *Biochem. Eng. J.*, 38, 105, 2008.
127. Siramon, P., Punsuvon, V., Vaithanomsat, P., Production of bioethanol from oil palm empty fruit bunch via acid impregnation-steam explosion pretreatment. *Waste Biomass Valorization*, 9, 1407, 2018.
128. Vancov, T., Schneider, R.C.S., Palmer, J., McIntosh, S., Stuetz, R., Potential use of feedlot cattle manure for bioethanol production. *Bioresour. Technol.*, 183, 120, 2015.
129. Vavouraki, A.I., Angelis, E.M., Kornaros, M., Optimization of thermo-chemical hydrolysis of kitchen wastes. *Waste Manage.*, 33, 740, 2013.
130. Hong, Y.S. and Yoon, H.H., Ethanol production from food residues. *Biomass Bioenergy*, 35, 3271, 2011.
131. Dubey, A.K., Gupta, P.K., Garg, N., Naithani, S., Bioethanol production from wastepaper acid pretreated hydrolyzate with xylose fermenting *Pichia stipitis*. *Carbohydr. Polym.*, 88, 825, 2012.
132. Panpatte, D.G. and Jhala, Y.K., Agricultural Waste: A Suitable Source for Biofuel Production, in: *Prospects of Renewable Bioprocessing in Future Energy Systems*, A.A. Rastegari, N.A. Yadav, A. Gupta (Eds.), pp. 337–355, Springer, Cham, 2019.
133. Xu, J., Takakuwa, N., Nogawa, M., Okada, H., Morikawa, Y., A third xylanase from Trichoderma reesei PC-3-7. *Appl. Microbiol. Biotechnol.*, 49, 718, 1998.
134. Sandgren, M., Shaw, A., Ropp, T.H., Wu, S., Bott, R., Cameron, A.D., Ståhlberg, J., Mitchinson, C., Jones, T.A., The X-ray crystal structure of the *Trichoderma reesei* family 12 endoglucanase 3, Cel12A, at 1.9 Å resolution. *J. Mol. Biol.*, 308, 295, 2001.
135. Jäger, S., Brumbauer, A., Fehér, E., Réczey, K., Kiss, L., Production and characterization of β-glucosidases from different *Aspergillus* strains. *Biography*, 17, 455, 2001.
136. Talebnia, F., Karakashev, D., Angelidaki, I., Production of bioethanol from wheat straw: an overview on pre-treatment, hydrolysis and fermentation. *Bioresour. Technol.*, 101, 4744, 2010.

Index

β-*N*-acetylglucosaminidases, 316, 320

2-bromoethanesulfonate, 267
3-hydroxy-3-methylglutaryl coenzyme A (HMG-CoA) reductase, 267
4-methylumbelliferyl N-acetyl-β-D-glucosaminide (4-MU (GlnAc)), 319

ACC deaminase, 233, 276
Acidogens, 369
Acidothiobacillus sp., 51–52
Actinobacteria, 51, 313, 317, 321, 324
Active packaging, 186
Acute effects, 8–9
Advantages of biopolymer, 96
Agricultural application, 42
Agro-industrial, 106, 110, 116, 121
Agro-waste, 102, 104, 112, 116, 117
Air lift bioreactors (ALBs), 326
Air lift with net draft tube bioreactors (ALndtBs), 326
Algae, 349
Allochthonous substances, 276
Allosomidin, 320, 324
Alternate energy, 362, 369
Alternative, 197, 199, 204, 206, 216, 218
Anaerobic bacteria, 54–55
Anaerobic digesters, 376
Animal breeding and selection, 270
Anthropogenic activities, 41
Antibiotic resistance genes (ARGs), 41
Anti-cancer therapy, 329
Anti-fusarium activity (AFA), 329
Anti-hypertensive activity, 329
Anti-inflammatory agents, 329
Anti-neoplastic drug, 330
Anti-tumor activity, 329
Application, 197–203, 206, 208–209, 211, 215–218
Application of PHAs, 75
Applications, 151, 152, 156, 158, 162, 167, 169, 174–175, 176, 180
Arabidopsis, 250, 252
Argadin, 324
Arsenic, 42–43
Arsenic removal, 53–54
Arsenite S-adenosylmethionine ethyltransferase gene (arsM), 54
Arthobacter sp., 54
Autoclaving, 282
Auxins, 276
Azospirilium, 278, 290t
Azotobacter, 278, 289t

B. cereus, 245
Bacillus, 156, 177
Bacillus idriensis, 54
Bacillus sp., 54, 276
Bacteria, 184, 188
Bacteria used as liquid biofertilizers, 286t
Bacteriocins, 270–271
Bacteriophages, 270
Bacterivorous protozoa, 250

Beauveria bassiana, 326
Betains, 300
Bioaccumulation, 49
Bioamplification, 7
Bioavailability and production, 66
Biocapsule, 245
Biocatalyst,
  amylases, 296–299, 301
  amylopullulanase, 302
  cellulases, 296–299, 303
  chitinase, 302
  extremozymes, 293, 299–304
  glucoamylase, 298, 302
  lipases, 296–302
  pectinases, 296
  protease, 296–303
  xylanases, 296–297, 300, 302–303
Biocatalysts, 314–315
Biocontrol agent, 276, 313–314, 317, 321, 329
Biodegradability of PHAs, 76
Biodegradability of poly(lactic acid), 73
Biodegradability of polyamides, 66
Biodegradable, 183, 188–190, 200, 202, 213
Biodegradable polymer, 85–87
  cellulose, 85
  protein, 86–87
  starch, 86
Biodetoxification mechanisms, 48–55
  bioimmobilization, 54–55
  bioleaching, 50–52
  biosorption, 48–50
  biovolatilization, 52–54
Biodetoxification of heavy metals using biofilm bacteria,
  biodetoxification mechanisms, 48–55
  biofilm bacteria, 47
  interaction of metal and biofilm bacteria, 47
  introduction, 40–41
  source and toxicity of heavy metal pollution, 41–46
Biodiesel, 362, 364–367
Bioethanol, 362, 364, 367–368
Biofertilizers, treatment types, 289t–290
Biofilm bacteria, 47
  interaction of metal and, 47
Biofilms, 372, 376
Biofuels, 362–364, 367, 369
Biogases, 362
Biohydrogen, 369, 374
Bioimmobilization, 54–55
Bioinoculants, 277
Bioinsecticides, 328
Bioleaching, 50–52
Biological, 25, 26, 197–199, 202, 212, 217
Biomass, 151–155, 160, 166, 173–175
Biomedical, 313, 329
Biopesticides, 313–314, 328, 331
Bioplastic properties, 106
Bioplastics, 154, 159, 162, 168, 171, 175, 176, 179
Biopolymer, 184, 187, 188, 314, 322
Biopolymers classification,
  biomedical polymer, 109
  ecological polymer, 109
  natural polymer, 107
  photodegradable, 107
  semi bio-degradable, 107
Bioreactors, 315
Bioremediation, 25, 26, 30
  microalgal species, 9–13
  molecular aspects, 14–16
Biosensor, 186, 187, 190
Biosorbent, 26, 30, 31, 33
Biosorption, 48–50
Biosurfactants, 197–199, 202–216
Biosynthesis, 155, 161, 162, 165–166, 168–172, 174, 177–179
Biosynthesis of PHB,
  3-ketothiolase, 114–115
  acetoacetyl-CoA reductase, 115

Alcaligenes eutrophus, 114–115
PHB synthase, 115
phbA, 115
phbB, 115
phbC, 115
Biosynthesis of polyhydroxyalkanoates, 75
Biotechnological applications,
biodetergents, 299
bioremediation, 296, 299–300, 302
brewing, 303
cosmetic, 298, 302
esterification, 296
food processing, 298–299, 302
pharmacological, 302
saccharification, 302
transesterification, 296–297
Biovolatilization, 52–54
Black pepper, 248
Blending method, 73
Blue-green algae (BGA), 278
Bubble column reactors (BCRs), 326
BY medium, 261

Cadmium, 43
Carbon dioxide, 257–266
Carbon monoxide, 257
Carboxy methyl cellulose (CMC), 321, 324
Carboxymethyl-chitinremazol brilliant violet 5R (CM-chitin-RBV), 321
Carboxypolyether compounds, 267
Carrier-based biofertilizer vs liquid biofertilizers, 282–283
Catabolite repression, 324
Catastrophe, 370
Cellulolytic enzyme, 368
Chasmoendoliths, 343
Chemical fertilizer and its harmful effect, 277–278
Chemical sensor, 184, 187
Chemolithic lithobionts, 351
Chemo-lithotrophic bacteria, 51
Chitin, 314–329, 331–332
Chitin binding module (CBM), 319
Chitin binding protein, 318
Chitinase, 313–332
Chitin-binding domain (CBD), 316, 317, 319, 327, 331
Chitobiosidases, 316, 318, 320
Chito-oligosaccharides (COS), 314, 331
Chitosans, 321, 329
Chitotetraose, 315
Chitotriose, 315
Chromium, 25–33, 43–44
Chromium toxicity, 25, 26, 27
Chronic effects, 9
Chronic rhinosinusitis (CRS), 330
Chryseomonas luteola, 55
Co-culture, 326, 327
Colloidal chitin, 319, 321, 324, 326
Complexion, 49
Concentration intracellular, 367
Copolymerization method, 73
Copper, 45–46
Cr(VI), 25–32
CRISPR, 246
Cryptoendoliths, 344
Cyanobacteria, 30, 31, 32, 349
Cytocompatible, 188, 189
Cytokines, 276
Cytophaga, 318, 323

Defaunating agents, 266
Defaunation, 266
Degradation of nylon 4 under the soil, 67
Desulfovibrio sp., 54
Desulfurized, 368
Deterioration, 183, 185
Detoxification, 25, 33, 368
Deuteromycotina, 328
Diacetylchitobiose, 315, 316
Diatoms, 30, 31
Dietary manipulation, 264
Difference between biopolymer and bioplastic, 88

Dimethylarsinic acid (DMA), 42
Directed evolution, 331
Diseases, 26, 29
Domain swapping, 327, 331

Ecosystem engineers, 353
Ecotines, 300
Electricity generation, 363
Endochitinase/s, 315, 316, 318, 320, 321, 323, 325
Endoliths, 343
Entomopathogenic fungi, 328
Environment, 26, 27, 29, 33
Environmental pollution, 331
Enzyme/s, 313–316, 318, 320, 322–325, 327, 329, 331
- cell wall degrading enzymes (CWDE), 328
- chitinolytic, 315, 327, 328, 330, 332
- gycosyl hydrolase (GH), 314
- hydrolases, 314
- hydrolytic, 315, 327
- isomerases, 314
- ligases, 314
- lyases, 314
- oxidoreductases, 314
- proteolytic, 325
- tannase, 330
- thermostable, 325
- transferases, 314
- transglycosidase, 316

Epiliths, 342
Escherichia coli, 54
Essential heavy metals, 45–46
Exochitinases, 315–316, 318, 319, 325–326
Exponential growth, 365
Extracellular polymeric substances (EPS), 40, 47, 51, 55
Extremophiles,
- acidophilic, 293, 301
- alkalophiles, 298
- halophiles, 294–295, 298, 300
- osmophiles, 294
- psychrophiles, 294, 297, 299
- radioresistance, 294
- thermophiles, 294–297

Fate of pesticides, 8
Feed additives, 266
Fermentation, 155, 160, 161–162, 165–168, 171–173, 177, 180, 197–198, 204, 206–210, 285, 287, 363–365, 367–370
Ferro-chromite, 44
Firmicutes, 51, 324
Flavobacter, 323
Forage-to-concentrate ratio, 265
Formyl-methanofuran, 263
Formyl-tetrahydromethanopterin, 263
Fosmid library, 322
Fossil fuels, 362, 369–370, 376
Fungal degradation of nylon 6 and nylon 66 (synthetic polyamide), 67
Fungi, 349
Fusarium oxysporum, 249

Gamma-irradiation, 282
Gasoline, 367–369
Genetically modified organisms,
- *E. coli*, 123–124
- genetic engineering, 123

Genotoxicity, 42
Gibberellins, 276
Ginger, 251
Global warming, 362
Glucosamine, 316, 331
Glucosamine oligomers, 331
Green algae, 30, 31

HACCP, 186
Halogenated methane compound, 267
Hazardous, 151, 162
Heavy metal, 26, 30, 32
Heavy metal pollution, 40, 41–46
Hydrogen cyanide, 228
Hydrophobicity, 300

Hydrostatic pressure, 294, 298, 301
Hypoliths, 344

ICPMS, 250
Inflammability limits, 367
Initial dried substrate (IDS), 326
Invasive mycotic infection, 329
Ion-exchange method, 49
Ionome, 244, 250–251
Ionophore, 266
Itaconic acid-based heterocyclic polyamide, 68

Lasalocid compounds, 266
Lead, 44–45
Leptosperillium sp., 51–52
Lettuce, 249
Lichens, 350
Lignocellulosic biomass, 362
Lipolytic activity, 331
Liquid biofertilizer, 279
Liquid biofertilizers and their applications,
  applications of biofertilizers, 288
  benefits of biofertilizers, 278
  carrier-based biofertilizers, 279
  chemical fertilizer and its harmful effect, 277–278
  introduction, 275–277
  liquid biofertilizer vs. solid carrier-based biofertilizer, 282–283
  production of liquid biofertilizers, 285–287
  sterilization of the carrier, 282
  types of liquid biofertilizer, 283, 284t
Lithobionts, 342
Livestock, 258
Lysine motifs (LysM), 319

Mechanism, 25, 32, 33
Mercury, 45
Mesophilic, 296, 299, 302–303
Metagenomics, 246–248, 249
Metal ions, 40, 41, 42, 47, 54, 55
Metal nanoparticles, 366
Metal resistance genes (MRGs), 41
Metallothionein, 50
Metaproteomics, 248–249
Metarhizium anisopilae, 323
Metatranscriptomics, 249–250
Methane emission, 258, 264
Methanogenesis, 260–264, 369–370, 372
Methanogens, 259–264
Methenyl-tetrahydromethanopterin, 263
Methylotrophy, 250
Methyl-S-coenzyme M, 264
Methylumbelliferyl, 331
MG-RAST, 246
Microalgae, 25, 26, 30, 32, 33, 364–366
Microbacterium sp., 54
Microbial, 151–156, 159, 160–166, 169, 172, 174–177, 180–181
Microbial chitinases, 313–315, 317, 322, 323, 332
Microbial consortia, 327
Microbial control agent, 328
Microbial feed additives, 268
Microbial fuel cells, 363, 370, 372
Microbicides, 329
Microencapsulation, 184
Micronutrients, 183
Microorganism producing PHAs, 90–95
Microorganisms, 197, 199, 209–212, 217
Monensin, 266
Monomethylarsonic acid (MMA), 42
Mycoparasitic fungi, 319
Mycoparasitism, 320

Nanocomposite formation, 74
Nanomedicines, 329
Natural biofuels, 363
Nickel, 46
Nitrogen, 277
Nitrogen fixation, 224

Nitrogenous fertilizers, 277
Nitrospirae, 51
Non-essential heavy metals, 42–45
Non-renewable, 362
NPK consortia, 289t

Oats, 250
Oleaginous fungi, 366
Optimization,
one variable at a time approach, 118
response surface methodology, 118, 120–122
Organic biogenesis, 296
Organic matter, 363, 369–370
Organopollutant, 303
Osmotic balance, 300
Oxazolinium ion, 324
Oxidative stress, 28, 33

Passive packaging, 186
Pathogenesis, 317, 322
Pathogenesis related proteins (PR proteins), 331
Pea, 250
Pesticides,
chemical composition, 5
classification, 4–5
examples, 5
general characteristics, 5
PGPR, 251, 252
PHA, 151–156, 159–180
Phaffia rhodozyma, 330
PHB, 151–156, 159, 163–168, 174–176, 178
PHB applications,
agricultural and veterinary uses, 134
biomedical uses, 132–133
chiral precursor, 133
detergent applications, 134
packaging industry, 134
PHB biodegradation, 131–132
PHB characterization,
differential scanning calorimetry, 127
elemental analysis, 130
fourier transform infrared spectroscopy, 127
gel permeation chromatography, 126, 130
microscopic methods, 129
nuclear magnetic resonance spectroscopy, 128–129
polarimetry, 130
thermogravimetric analysis, 128
x-ray powder diffraction, 128
PHB granules detection, 113–114
PHB producing microorganisms,
Ralstonia eutropha, 110–112, 116
PHB production, 116–117
Phenazine, 250
Phosphate solubilizing bacteria, 289t
Phosphorus solubilization, 227
Photobioreactors, 370
Physiological changes, 185
Phytochelatins, 50
Phytohormones, 225
Phytotoxicity, 42
Planktonic bacteria, 50–51
Plant growth–promoting rhizobacteria (PGPR), 221, 276–277
Plants, 28, 29
Plastics, 151–159, 162, 168–169, 175, 176, 178
Pollution, 6
cadmium, 43
heavy metal, 40, 41–46
Polyamide, 65
Polyhydroxyalkanoates, 74, 151–152, 160, 176–181, 184
Polyhydroxybutyrate, 151–152, 166
Polyhydroxylkanoates, 88–89
Polyhydroxylkanoates applications, 89–90
Polylactic acid, 69
Polymerization method, 71
Polymers, 151, 154–159, 162, 163–164, 166–167, 169, 173–176
Polysaccharide, 26, 32

Probiotic, 245
Production, 197–204, 205–209, 211–214, 215–218
Properties, 197–204, 208–209, 211, 213–216, 218
Proteobacteria, 51, 313, 322, 323, 325
Protoplast separation, 330
Protozoa removal, 266
Psammaplin A, 324
Pseudomonas, 155, 160, 163, 168, 169–171, 173, 177–181
Pseudomonas aeruginosa, 318, 327
Pseudomonas putida, 54
Pseudomonas sp., 276
Pseudomonas syringae, 49
Pyrrolnitrin, 250
Pythium torulosum, 246

QACCP, 186
Quality indicator, 183
Quantitative trait loci (QTL), 245

R. solani, 249
Reductase enzyme, 53
Remediation methods, 6
Renewable, 184, 188
Response surface methodology (RSM), 323
Reverse micelles, 300
RFID, 186
Rhizobium, 278, 290
Rhizosphere engineering, 245–246
Rhizosphere microbial diversity, 222
Rhodopseudomonas palustris, 54
Role of microbes and microbial consortium in solid waste management, 383–422
Roseobacter, 322–323
Rumen, 260
Rumen liquor, 261

Sagittula stellata, 323
Seed coating, 245
Serratia marcescens, 318, 323–325, 328, 330
Shewanella sp., 54
Siderophores, 229, 276
Single-cell protein (SCP), 329–331
Siphophages, 270
Site-directed mutagenesis, 331
Smart packaging, 183, 190, 191
Soil metagenome, 322
Soil microbes, 276
Soil nutrients, 277, 288
Solid carrier-based biofertilizers, 279
Solid-state fermentation (SolF), 325–326
Solventogenesis metabolism, 368
Sphingomonas desiccabilis, 54
Spoilage, 184, 185, 187
STAMP, 248
Stirred tank bioreactors (STBs), 326–327
Strategies for phycoremediation,
  enzymes, 13
  genetically engineered microalgae, 13–14
Streptomyces sp., 316, 321, 322, 324, 325, 329, 330
Submerged fermentation (SubF), 325–326
Surfactants, 197–202, 205–207, 210–212, 214, 215
Synthesis, 151–174, 177–179
Synthetic polymers, 185, 187

Thermostable, 296, 299, 302–304
Thiobacillus sp., 51–52
Tissue engineering, 329
Toluene dioxygenase, 303
Toxicity, 204, 206, 209, 211, 217
Transformation, 367, 375
Trichoderma, 248
Trichoderma harzanium, 315
Trichorhizosphere, 248
Trimethylarsine (TMA) oxide, 42

Vaccination, 268
Van der Waals' force, 49

Waste management, 314, 331, 332
Water pollution, 277
Wheat, 249

Xenobiotic, 303

Zinc, 46
Zinc solubilizing bacteria, 228, 290t
Zygomycotina, 328